THE DEVELOPMENT OF QUINE'S PHILOSOPHY

BOSTON STUDIES IN THE PHILOSOPHY OF SCIENCE

VOLUME 291

For further volumes:
http://www.springer.com/series/5710

THE DEVELOPMENT OF QUINE'S PHILOSOPHY

by

MURRAY G. MURPHEY

Emeritus Professor, University of Pennsylvania

Springer

Murray G. Murphey
Emeritus Professor
University of Pennsylvania
murraymurphey200@concast.net

ISSN 0068-0346
ISBN 978-94-007-2423-5 e-ISBN 978-94-007-2424-2
DOI 10.1007/978-94-007-2424-2
Springer Dordrecht Heidelberg London New York

Library of Congress Control Number: 2011938948

Printed on acid-free paper

Springer is part of Springer Science+Business Media (www.springer.com)

This book is dedicated to Dean Morse,
Harvey Lyon, Deborah Broadnax,
Rosemary Cowler and Lillian Schlissel

Preface

This is a book about the development of the philosophy of W. V. Quine. It is not a biography of Quine in any traditional sense. Those interested in the details of Quine's life should consult his autobiography, *The Time of my Life*, or his autobiographical sketch in the Library of Living Philosophers volume on Quine. Specifically, I have had almost nothing to say about his life before he entered college; I have omitted his endless traveling from place to place, and I have also omitted the catalogue of the honors he received. All of these are covered in his autobiography, and I have nothing to add.

Quine's published output is enormous. Having read, I believe, all of it, I have dealt with what I believe to be the most important writings. There are several difficulties in doing that. First, given his huge written output, it is hardly surprising that much of it is repetitive. But these seemingly repetitive writings often contain subtle, and not so subtle, changes that reflect shifts in his thought. The result is, I fear, an account that is also repetitive in some respects, but to follow his course and to keep the chronology straight, I have not seen how to avoid that. A further difficulty has been the technical nature of his work. Quine published three different logical systems, plus a book on set theory, and many articles dealing with logical issues; even when he is being non-technical, it is usually about technical matters. I have not attempted detailed critiques of his work in logic, as I had originally intended to do; my account of his logic is descriptive rather than analytical and is relatively brief. But he was a logician and any work dealing with him must deal with his technical writings. I have therefore presented accounts of his logical and set theoretical work sufficient I hope to show what he was doing. Nevertheless, so much of Quine's work is technical in some sense that the reader should be forewarned that I have assumed familiarity with first order logic.

I have dealt with Quine as what he was – an academic. I have therefore included material concerning his academic activities that seemed important for an understanding of his overall career, and some brief notes on his families. My chief reliance has been of course on his published writings, but I have also used materials from the Quine papers at Houghton Library at Harvard and materials from the archives at Oberlin, without which I could not have written this book. Quine became a public figure, but he was not one who laid out the inner workings of his mind for public scrutiny. He was a gregarious and genial man (when he chose to be), and developed close relations with some of his students and colleagues, but he was also a very private man who did not reveal his inner thoughts to even those closest to him. But this is, after all, what one would expect of any man who won the acclaim he did.

Finally, I should add that I knew Quine personally and was once his student. Like many others, I admired him and was elated by his "Two Dogmas of Empiricism," which I now realize I misunderstood. I undertook this project because of what I believed to be his importance in the history of American philosophy. Having previously written books on Charles Peirce and C. I. Lewis, I expected to find that his logic was central to his philosophy. I was surprised to find that it was not as central as I had thought, and even more surprised as I got further into his work. My conclusions will no doubt offend some of those who have been close to him or who idolized him, but like any other scholar I have tried to tell the truth.

I am indebted to many people for their help in writing this book. I am particularly indebted to the staff of the Houghton Library at Harvard for their assistance in using the collection of Quine's papers there. I would also like to thank the archivist at Oberlin for making available to me materials relating to Quine. Professor Robert Schwartz has generously read and criticized my draft. Bruce Kuklick has given me the benefit of his careful reading of my text. I am also indebted to Mrs. Deborah Broadnax for her assistance in accessing some of Quine's writings. I am profoundly grateful to my daughters, Jesse and Kathleen, for having guided their aged father's faltering footsteps through the mystery land of computers. Finally, thanks is due, and alas overdue, to Burt Dreben, who some sixty odd years ago first drew my attention to Quine and to the importance of his work.

Introduction

This book is a study of the development of the philosophy of Willard Quine. That development took place in a context, and an understanding of Quine's philosophy requires an understanding of that context. It was an extremely complex context; many different factors were involved and a full treatment of these factors would require, and has received elsewhere, book length studies. My treatment must therefore be brief and incomplete, but I hope adequate for the purposes of this book.

Science

Of the factors that contributed to this context, perhaps the most important was the developments in science from the middle of the nineteenth century on. In 1859, Charles Darwin published *The Origin of Species* – a work that began the transformation of the biological sciences from a preoccupation with classification to the dynamic problems of change and growth. Darwin held that offspring differ from their parents by minute variations that appear to be randomly distributed but that are inheritable – hence the principle of fortuitous variation. That principle alone should produce descendants differing from their ancestors in multiple and random ways. Since that is not what the fossil record shows, Darwin needed an explanation of why only certain lines of descendants survived. He found it in the principle of natural selection. According to this principle, which he derived from Thomas Malthus' study of population, organisms reproduce faster than their supply of food and other necessities of life can be increased. The result is a competition for the resources necessary for the maintenance of life. In this competition, some organisms find that the variations they have inherited are advantageous; they survive and reproduce. Other organisms find that the variations they have inherited are disadvantageous in the competition; they perish without living long enough to reproduce. Thus only the lucky ones with the advantageous variations will survive to transmit their good fortune to their descendants.

There were problems with this theory. Darwin had no adequate explanation for fortuitous variation; the explanation he gave for it was patently inadequate. But research in heredity was stimulated by the problem, and about nineteen hundred the work of Gregor Mendel was rediscovered. The result was the new science of genetics that filled this gap in Darwin's theory. A further problem was the incompleteness of the fossil record. Darwin had no fossil remains on the basis of which he could demonstrate man's relation to the other higher primates. This problem led him to the inspired guess that the human race had originated in Africa – a guess that subsequent research has proven to be correct. But despite its problems, Darwin's theory of evolution won rapid acceptance within the biological sciences.[1]

Physics too underwent dramatic changes in this period. By the late nineteenth century, it was clear that the two basic sets of laws of physics – Newton's laws and Maxwell's equations – were not consistent with each other. The crisis was finally resolved when Albert Einstein published his Special Theory of Relativity in 1905, to be followed in 1915 by his General Theory of Relativity. Einstein's work not only transformed physics; it had profound implications for philosophy since it required changes in our basic concepts of space, time, and matter.[2] But also in 1905, Max Planck published his study of black body radiation that showed that energy was quantized. The impact of the quantum theory of physics was less immediate than that of Relativity, but in the 1920s there was an explosive growth in Quantum Mechanics, led by Niels Bohr, Werner Heisenberg, and Edwin Schrödinger. The most startling result of this work for the philosophers of the time, and the scientists, was the Heisenberg principle of indeterminacy that showed there was a fundamental indeterminacy in the foundations of the science. Since there was no such indeterminacy in Relativity Theory, the problem of how to reconcile the two was acute. But at least by 1909, Einstein recognized that there was a further problem that amounted to an outright contradiction between Quantum Mechanics and Relativity. At the Solvey Conference in 1927, Einstein pointed out that the theory of the wave collapse involved action at a distance, something that Relativity Theory forbid. In 1935, Einstein, Podolsky and Rosen published a paper that showed that Quantum Mechanics implied action at a distance in further ways.[3] It took longer for this problem to come to the fore in Quantum Mechanics than the indeterminacy issue, but it remained a crucial embarrassment. One crisis in physics had been succeeded by another.[4]

Idealism

That the stunning achievements of science from Newton on posed problems for religion and for the whole realm of values, moral and aesthetic, was all too clear to many. Among these, the man whose work had the greatest importance was Immanuel Kant. By confining science to the realm of the phenomenal, Kant hoped to guarantee that the noumenal world would be put safely beyond the reach of science, so that God, freedom, and immortality would elude the grasp of the human understanding.

It was not to be. Kant's "followers" Fichte, Schelling, and above all Hegel were inspired by Kant's work to create all-embracing Idealistic systems that swept away the barrier between the phenomenal and the noumenal. Hegel created a vast system of the world in which history became the onward march of the World-Spirit and historical figures the footsoldiers of the Absolute in its march to perfection.

Yet despite the enormous impact of German Idealism in the United States and Britain, it failed to hold its ground in Germany. The result was a plethora of movements, some of them Idealistic and some not. Ernst Haeckel brought Darwinism to Germany, playing there a role similar to that of Thomas Huxley in Britain. One result was a new direction in German science, focusing on humans as natural creatures and on the unconscious – a movement that culminated in Freud's work. A second was the metaphysics of Gustav Theodor Fechner, Rudolf Herman Lotze, and Eduard von Hartman, which sought to combine Idealism with the new biological science. A third focused on the issue of how science was possible. Of this group, Herman Cohen, the founder of the Marberg Neo-Kantian school, was the most important. Also Neo-Kantian was the Baden school, of which Heinrich Ricker was one of the best known.[5] And a fourth was a new interest in humane culture, past and present. This included of course Karl Marx, but also writers such as Jacob Burkhardt. Here also belongs the historian Wilhelm Dilthy, who coined the term *Geistewissenschaften* to embrace the humane sciences, centering on history and the problems of understanding rather than explanation, and Max Weber, whose sociology was centered on the understanding of human culture.[6]

Also of major importance was Franz Brentano, who taught at Wurzburg and Vienna. Brentano was particularly interested in mental phenomena. The characteristic that sets mental phenomena apart from the physical is its intentionality; mental phenomena such as thought are directed toward an object. Moreover, while one may simply attend to the object before consciousness, whether real or imaginary, one can also accept or reject the particular phenomenon and one can hate it or love it. The study of such phenomena Brentano calls "descriptive psychology" and he considered it an exact science which provided a basis for all philosophy. Brentano also wrote on ethics and logic, but descriptive psychology was his main interest. He was an influential teacher; among his students were Alexius Meinong, Edmund Husserl, and Kazimierz Twardowski.

While Idealism fractured in Germany, it swept the field in England. The leading figure there was F. H. Bradley whose *Appearance and Reality* was the outstanding Idealist work in the country. Following in the well worn footsteps of Kant, Bradley tried to show that the domain of appearance led to contradictions that proved its inadequacy, and that could only be resolved by invoking the Absolute. Perhaps Bradley's strongest argument was that in the realm of appearance we find many things related to one another, but we cannot conceive how this can be so. For if a stands in the relation R to b, then how is a related to R? There must be another relation R' between a and R, so that we have $aR'Rb$. But then the problem recurs; how is a related to R' ? So another relation R'' is required, and so on without end. Thus the ideality of appearance is shown by the insolvable problems it involves, and only in the Absolute can we find answers.[7]

In the United States, it was Darwin's *Origin of Species* that opened the door for Idealism. Evolutionary theory outraged Christians: not only did it contradict the story of human origins in the Bible, but it portrayed human beings as purely natural creatures with an animal ancestry. Nowhere in the line of evolutionary development did there appear to be a place for the soul. If there was no soul, there was no immortality, no post-mortem reward or punishment, and no difference in kind between man and beast. Such a result meant to many Christians that evolutionary theory implied the falsity of Christianity – something that the religious leaders were not prepared to accept. While the war between science and religion raged, philosophy, which in the United States had always been closely associated with religion, turned to Absolute Idealism for an answer. If all that exists either is, or is due to, the mind of God, then the sting of evolution was, if not drawn, at least partly removed.[8] In the United States, the foremost Idealist was Josiah Royce at Harvard. Royce's philosophic work had a religious slant from the beginning, but his Christian commitment was most clearly expressed in his two volume work, *The Problem of Christianity*, published in 1913. Royce also had a strong interest in logic, stimulated in part by Charles Peirce, but motivated primarily by his hope that he could use logic to prove his doctrines. In the years between Peirce and C. I. Lewis, Royce was the foremost logician in the United States, and he deployed his logical tools to defend not only his Idealism but his religion as well.[9]

Realism

The charms of Idealism waned, and in the late nineteenth century and the early twentieth century, it was under attack. The first movement in the United States to challenge Absolute Idealism was pragmatism. The pragmatists of the first generation – Peirce, James, and Dewey – were idealists but idealists of a different sort than Royce. Peirce was a professional physicist and a logician, whose objective was a system that would reconcile science and religion. William James was trained in physiology and biology. James saw very early that the central question in the evolutionary controversy was the nature of the human mind. He therefore moved into psychology, and his two volume *Principles of Psychology*,[10] made him *the* leading authority on psychology in the country. James was thus a scientist, but he was also a deeply religious man, and he was strongly opposed to the determinism of materialists like Herbert Spencer and what he called the "block universe" of Royce. He finally settled on a pluralist idealism but his moderate position was an attempt at a peaceful reconciliation of science and religion. Dewey began his career as a Hegelian, became a naturalist who essentially Hegelianized nature, and who made the scientific method into the general method for all inquiry.[11] James and Royce were teachers of C. I. Lewis, who became the leading American philosopher of the next generation. It was Lewis who converted pragmatism into a realist position, and who carried on the study of logic. He was one of Quine's teachers at Harvard.[12]

The attack on Idealism generally in the United States was led by the New Realists, particularly Ralph Barton Perry, William Montague, and E. B. Holt.[13] The attack was devastating; Royce was the last of the Absolute Idealists to command great influence. Meanwhile, a similar movement arose in England, led by G. E. Moore and Bertrand Russell. Moore had been trained as a classicist; he was a realist who rejected the phenomenalism of Hume. But he was also an empiricist, and most importantly an analyst. He believed that the problems of philosophy are chiefly due to the failure of philosophers to define precisely what the problems are that they need to answer before they try to answer them. He saw the task of philosophy, not as system building or speculation, but as clarifying what the ordinary statements of common sense mean, "to make clear what it is that we know and how we know it when we assert such propositions as 'Tables are real' or 'Lions exist.' "[14] Moore's devastating critique of Idealism in "The Refutation of Idealism," "External and Internal Relations" and similar pieces helped to bury English Idealism, and later bore fruit in the Oxford Natural Language philosophy.[15]

Bertrand Russell came to philosophy from mathematics. He was one of those primarily concerned with the foundations of mathematics. His early book on the foundations of geometry was followed in 1903 by *The Principles of Mathematics*,[16] which in turn led to his collaboration with Alfred North Whitehead in writing the three volume work, *Principia Mathematica*.[17] This work established Russell as the greatest logician of his age, and possibly of all time, but it also led him to apply his logical talents to science generally. In *Our Knowledge of the External World*,[18] he defended a position that he termed "logical atomism." This book was to have an immense influence; Russell argued that all our knowledge of the world has two sources; the data of sensation and the truths of logic, together with our memories of these and our ability to determine their similarities.[19] The one is purely empirical, the other purely a priori.[20] Russell's belief in what the new logic could accomplish led him to say "The old logic put thought in fetters, while the new logic gives it wings."[21] It further led him to claim

> If we knew all the atomic facts [of sense experience] ... we should, theoretically, be able to infer all truths of whatever form. Thus logic would then supply us with the whole apparatus required.[22]

This bold claim Russell did not carry out but left as a challenge for the future.

In Germany, the most influential philosopher of the time was Edmund Husserl. He had been a student of Brentano's, and subsequently served as an assistant to Karl Weierstrass, a mathematician who played a very important role in the development of the theory of functions. Husserl was greatly impressed by these issues and published a book on the philosophy of arithmetic in 1891. But the book was severely reviewed by Frege, and Husserl turned to the project he would pursue all his life – the creation of phenomenology. Like many other philosophers of that time, Husserl felt that the stunning triumphs of science were displacing philosophy from its historic role as the determiner of the basis of knowledge. In phenomenology, he sought to create a philosophy that would be prior to, and the foundation of, all science. Phenomenology was presented as a method, but it was a method that was to lead to true knowledge. From Brentano, Husserl adopted the principle that all thought is directed toward an object. His analytic method required that one put aside – bracket – all presuppositions concerning existence that we normally take as self-evident, thus reaching a pure consciousness. But since consciousness is intentional, the pure consciousness will have an object, and that object will be the essence of the things of the world. This "transcendental reduction," as he called it, brings us into contact with things as they truly are, for we so attain an intuitive and immediate evidence of things-in-themselves. Phenomenology is thus to be the "first" philosophy, the one that should precede and supply the foundation for all other knowledge, science included.[23]

Logic

The study of logic has a long history, going back to Aristotle and his precursors; one really cannot set a date for the beginning of the study of logic. But one can set a date for the beginning of the study of *mathematical logic*; it began in 1847 with the publication of George Boole's *Mathematical Analysis of Logic*.[24] What Boole did was to apply an algebra to classes and propositions in which certain types of inferences could be carried out. The *algebraic* tradition that Boole began was amplified by Stanley Jevons and John Venn, and still more by Charles Peirce whose contributions were the most important in this tradition. This work was summarized and systematized by Ernst Schroder in 1890-1895. (For a history of the algebraic tradition, see C. I. Lewis, *Survey of Symbolic Logic* Chapter 1.)[25] But at the same time a different tradition grew up in Europe concerned principally with the foundations of mathematics. Those of the algebraic tradition saw mathematical logic as the application of mathematics to logic, whereas those of the functional tradition saw logic as the foundation of mathematics. Weierstrass was one of the most important figures here.[26] Richard Dedekind's definition of real numbers as the cut between two series of rationals and his proposals concerning the basic notions of number theory were followed by Georg Cantor's theory of sets. Cantor defined numbers as attributes of sets; he then defined equality between sets in terms of a one to one correspondence between their members. Using these notions he was able to prove that there are infinite sets that have the same number of members as some of their subsets (the natural numbers and the even positive integers, for example). He also proved the existence of a hierarchy of infinite sets. Since a set of n members has 2^n subsets (including the null set and the set itself), an infinite set, such as the natural numbers, has fewer members than its power set (the set of all its subsets). Using this fact, Canter could generate an infinite series of increasing infinite sets. He also found that whereas in finite sets, the ordinal and cardinal numbers correspond one to one, this is not true for infinite sets; there are infinite sets of ordinals corresponding to each infinite cardinal number. By this means, he proved the existence of another series of increasing infinite sets. If the numbers of the sets of the latter series are called the "alephs"(Cantor's term), then the question arises, how are the two series of infinite cardinals related? Cantor hypothesized that whereas both series begin with aleph-zero (the cardinal of the set of natural numbers), the second aleph equaled the second member of the power series, and more generally that each aleph equaled the corresponding number of the power series. These conjectures, known as the special and the general continuum hypotheses, respectively, Cantor was unable to prove despite all his efforts. But he further found that his theory led to a paradox; he could prove that there was a greatest cardinal and also that there was not a greatest cardinal. For consider the set of all cardinal numbers. Being a set, this set must have a cardinal number. But since the set includes *all* cardinal numbers, it contains its own number; there cannot be a greater cardinal number than the number of that set. However, that set has a power class, and the cardinal number of its power class is greater than its own. Thus there both is and is not a greatest cardinal number. For this problem, Cantor had no solution. Meanwhile, in Italy, Guiseppi Peano succeeded in axiomatizing arithmetic and, drawing on Dedekind's work, gave five axioms that were sufficient for the whole of mathematics.[27]

Gottlob Frege brought these notions together in an attempt to prove that mathematics – represented by the Peano axioms – could be derived from logic, and that the primitive notions of Peano's system – zero, number, and successor – could be defined in purely logical terms. This Herculean task Frege thought he had accomplished, based on his set of axioms for logic. The great work was in the process of being printed when in 1903 he received a letter from Bertrand Russell showing that his axiom of abstraction led to a contradiction that became known as Russell's paradox.[28]

Beginning in 1910, there appeared Whitehead and Russell's *Principia Mathematica* – a work that C. I. Lewis described as being to intellectual labor what the pyramids are to manual labor.[29] In the three volumes of this monumental work, its authors sought to carry out Frege's planned reduction of mathematics to logic (what became known as the *logistic program*). Fully axiomatized and rigorously formulated, the authors developed the propositional calculus, the theory of quantification, the theory of classes and dyadic relations, and showed how to derive the axioms and define the terms of Peano's axioms. Russell also showed that Cantor's paradox and the Burali-Forte paradox (the paradox of the greatest ordinal number) were special cases of Russell's paradox, and he proposed the theory of types as a way of avoiding them. What the theory of types does is to stratify the universe: the lowest type contains all individuals, the next type all classes of individuals, the next all classes of classes of individuals, and so on. No class is permitted to have a member whose type is equal to or greater than its own. This prevents self-membership and so blocks the paradoxes. The *Principia* marked a new stage in the history of logic; everything that came after it built on the *Principia* as a foundation. And a steadily increasing stream of research sprang from it.

Axiomatization raises a number of problems – those of consistency, completeness, and decision. In 1908, the year Russell published the first version of his theory of types, three other major figures in the history of logic entered the field. David Hilbert was already a famous mathematician who had published the first truly adequate axioms for Euclidean geometry. Hilbert did not believe that mathematics was reducible to logic, as Frege and Russell did, but he wanted to prove the consistency of mathematics, and, together with his collaborators, such as Paul Bernays, launched a program designed to achieve this.[30] At the other extreme,

L. E. J. Brouwer launched an attack on Cantor's set theory. Brouwer took the paradoxes of set theory as evidence that the attempt to extend number theory to the infinite was misguided. He also rejected the Frege-Russell logistic position. Instead, he wrote "the neo-intuitionist considers the falling apart of the moments of life into qualitatively different parts to be reunited only while remaining separated by time, as the fundamental phenomena of the human intellect".[31] From the "falling apart" and "reuniting" come the natural numbers. For the intuitionists, there are no transfinite numbers; although the natural numbers form an infinite series, they must not be thought of as a closed totality. No existence statement is permissible unless an actual instance of it can be constructed. This led Brouwer and his followers such as Heyting to reject the law of excluded middle and so the use of indirect proofs. The attack on set theory was not limited to the Intuitionists. Mathematicians such as Henri Poincare, Emil Borel, and H. L. Lebesgue also attacked it on the ground of the antinomies.

But also in 1908, Ernst Zermelo published a set of axioms designed to avoid the paradoxes.[32] Very roughly, one can say that Russell admitted the full range of classes and then threw out those that violated the theory of types by declaring them meaningless. Zermelo took the opposite course of admitting only those that did not lead to trouble.[33] As Fraenkel wrote

> Zermelo showed that without changing its traditional logical basis, one could, with the aid of one or two undefined (primitive) relations, deduce set theory from a small number of primary assumptions (axioms). . . .With the addition, in the 1920s, of some improvements and two new axioms to Zermelo's original system, this axiom system has proved sufficient to develop classical set theory, with none of the known antinomies being derivable from it.[34]

Set creation was more difficult than in *Principia* (hereafter PM), but Zermelo avoided the complicated apparatus of the theory of types. In 1925, John von Neumann proposed a further revision of Zermelo's system that also avoided the paradoxes. He added classes more freely than Zermelo, but then restricted those capable of being a member of a class to a select type that would not lead to the paradoxes.

And here a distinction must be made. Up to this point, I have used the terms "set" and "class" as if they were synonyms. But they are not. All sets are classes, but the converse is not true. The term "set" refers to those classes only that are members of something else.

The efforts to achieve consistency were not the only by-products of PM. In the United States, C. I. Lewis thought that *Principia Mathematica's* definition of "implication" as "material implication" was unacceptable, since it failed to show the kind of connection between antecedent and consequent that he believed implication required. Lewis created an alternative system based on "strict implication" – i.e., it is *impossible* that the antecedent be true and the consequent false. Lewis's system was thus a model system; he first presented it in his *Survey of Symbolic Logic* in 1918.[35]

The problem of the grounds of the truth of mathematical statements goes back to antiquity. Those who thought it solved by axiomatizing mathematics with axioms self-evidently true suffered a rude awakening with the discovery of the non-Euclidean geometries and the realization that the axioms they had thought self evident were in fact arbitrary. But as axiomatization became increasingly the way theories were formulated, the problems became more severe, and centered on the problem of decision: was there a way to decide, for any given statement, whether it was true or false? In 1921, Ludwig Wittgenstein published *Tractatus Logico-Philosophicus*[36] in which he proposed an answer. Wittgenstein did not invent truth tables, but he used them in a way his predecessors had not.[37] He believed that there were elementary propositions that asserted the existence of atomic facts, and that all other propositions were truth functions of those elementary propositions.[38] He defined a proposition whose truth table showed that it was true for all assignments of truth and falsity to its components as a "tautology."[39] If that were the case, there would be a complete decision procedure for all the propositions of mathematics and logic. Since Wittgenstein had been a student of Russell's, he was seen by many as further developing Russell's ideas.

All of these problems were much alive in the 1930s. Hao Wang has noted that in some sciences there are particular periods of spectacular advance.

> In logic, the period was the 1930s. Godel discovered the completeness theorem (of elementary logic), the incompletability theorem (for arithmetic and set theory), an interpretation of classical arithmetic in intuitionist arithmetic and the constructible sets. J. Herbrand brought out some fine structures of elementary logic. Zermelo's 1930 publication contains a persuasive description of the concept of set underlying standard set theory. Skolem constructed an elegant non-standard model of arithmetic. G. Grentzen perfected "natural deduction" for elementary logic and gave two consistency proofs for arithmetic. Godel tightened a suggestion of Herbrand to introduce a definitive general concept of recursive functions. A.M. Turing produced a neat model of idealized computers, and gave convincing arguments to show that it captures the intuitive concept of computability. Carnap and Tarski independently made a close study of truth and validity, as they are, for example, employed only intuitively in Hilbert and Ackermann.[40]

Nor is this all. In 1936, Tarski's "The Concept of Truth in Formalized Languages" appeared in German translation.[41] And in 1940, Godel published his monograph on *The Consistency of the Axiom of Choice and the Generalized Continuum Hypothesis with the Axioms of Set Theory*,[42] showing that neither the continuum hypotheses nor the axiom of choice can be proven or disproven from the axioms of set theory.

One notes that only one of the names mentioned here is American. But Lewis's work did not end with the *Survey*. In 1932, he and C. H. Langford published *Symbolic Logic*[43] in which they presented a revised form of Lewis's theory of strict implication, and in addition presented five modal systems that have engrossed model logicians ever since. And in 1936, Alonzo Church proved that there can be no mechanical decision procedure for general logic (quantification theory). This is the state of logic that Quine found when he entered the field in 1932.

The Vienna Circle

The catastrophe of World War I played havoc with life in Europe, particularly in Germany and Austria. But philosophers continued their work even amid the ruins. In Germany, the man regarded as the foremost philosopher was Martin Heidegger. Trained in Husserl's phenomenological method, Heidegger employs it to define man and his destiny. His most important early work was *Being and Time* which appeared in 1927.[44] For Heidegger, the vocation of man is to discover his being, what he is and what his destiny is. Human individuals, he says, find themselves in a world they did not make; at birth, they are thrown into a world from which there is no escape, whether they like it or not. Men become mired in the daily round of living and getting rather than seeking to understand who and what they are. What brings them out of this state of enchantment with the trivia of life is the fear of death, for life is finite and we all face the inevitable fate of death and nothingness. Facing the inevitable, the individual seeks on the one hand his possibilities, what he can create of and in the world, but on the other hand he is bound by his ties to the world. And science as Heidegger views it is part of the world that ensnares man, that distracts him with a world of gadgets, and hinders his quest to become an authentic being. For the man facing his own end, conscience calls upon him to choose freedom and create his own existence, but he can never escape the world or the death that awaits him. Facing this doubleness of freedom and slavery, what the authentic human being must do is, recognizing his finitude and his entrapment in the world, to play freely and creatively the role that he is allotted, finding freedom in doing what he must do. In Heidegger's view of man and his fate one can hear clear echoes of Kierkegaard as well as Husserl.[45]

Austria was not Germany, and German Idealism had little influence in the Austro-Hungarian Empire. This was largely due to the fact that education in the Empire was controlled by the Jesuits who regarded German Idealism as heretical. After the dissolution of the Jesuit order, the state took control of education, and was just as reactionary as the Jesuits had been. Enlightenment ideas were considered subversive and little genuine philosophy was done until after the revolution of 1848. Even then there was little progress in philosophy, but a general liberalization of education and thought took place.[46]

The major pre-revolutionary philosopher in the Austro-Hungarian Empire was Bernard Bolzano. Born in Prague and educated there, he became a Catholic priest and taught at Karlova University in Prague. He used his position to preach a version of utopian socialism and to proclaim all human beings equal. But in 1819 he was accused of religious and political heresy and was removed from his teaching position. He continued to write but some of his writings were suppressed. He advocated a utilitarian form of ethics and did interesting work in mathematics; his *Paradoxes of the Infinite* was published in 1851, three years after his death. But some of the Bolzano circle went on to play important roles in the liberalization of Austrian education.[47]

Franz Brentano, of whom I have spoken above, was a German philosopher but he taught at the University of Vienna, and produced a number of important students, among them Husserl.[48] But an important change came at the University of Vienna with the appointment of Ernst Mach. Mach was a physicist, but he regarded such departmental categories as purely administrative divisions and thought that they hindered the advance of science, so he had no hesitation in crossing their boundaries. He was also interested in the history and philosophy of science. Mach was an empiricist who believed that all science was based on sensory experience – what he called "sensation." His views owed much to Berkeley and Hume. The sensory properties of things he termed "elements," and he took physical objects to be composed of such elements. These he believed created a basis for certainty in science, since he thought that we could not be mistaken about our own sensations. He avoids solipsism by holding that other people are known by analogy to ourselves. Scientific laws he held to be descriptions of phenomena in terms of sensations. Their chief function was to summarize past experience and to facilitate prediction, but it was the predictions rather than the laws that chiefly interested him. He criticized Newton's laws for their assumption of absolute space and time, which he held to be unintelligible. Given this extreme empiricism, his view of scientific theories was that they were purely hypothetical and were chiefly useful for allowing scientific predictions. He was skeptical of scientific constructions that did not admit of direct observation; thus he rejected the atomic theory as senseless. Since he believed that all sciences are based on sensations, he thought that all of them should be united into a single all-inclusive unified science. It was this sort of extreme empiricism that Mach brought to Vienna when he was appointed to the chair of the history and theory of the inductive sciences in 1895.[49]

When Mach retired, he was succeeded by the physicist Ludwig Boltzmann. Then in 1922, the chair passed to Moritz Schlick. Born in 1882, Schlick was educated at the University of Berlin, and took his doctorate in physics with a dissertation on the reflection of light in non-homogeneous media, under the direction of Max Planck. Before coming to Vienna, Schlick taught at Rostoch and at Keil, and published *Allgemeine Erkentnisslehre* which established his reputation as a philosopher of science. His monograph "Space and Time in Contemporary Physics" in 1917 made him one of the first interpreters of the Theory of Relativity. At the University of Vienna, Schlick ran a seminar in which Herbert Feigl and Friedrich Waissmann were students. He then began a discussion group that met on Thursday evenings, and which came to be known as the Vienna Circle. Among the members of the circle were Feigl and Waissmann, Otto Neurath (sociologist), Edgar Zilsel (mathematics and law), Bela von Juhos, Gustav Bergmann (mathematics), Felix Kaufmann (lawyer), Victor Kraft (historian and philosopher), Philip Frank (physics) Karl Menger (mathematics), Kurt Godel (logician and mathematician), Hans Hahn (mathematics), and after 1926 Rudolf Carnap (physics and mathematics), for whom Schlick arranged a position at the University of Vienna.

Interest in the Vienna Circle is currently enjoying a revival, and a number of special studies have appeared and are appearing, dealing either with the Circle itself or with particular members; for example, there is a considerable revival of interest in Otto Neurath and his work. It is odd that this revival of interest has not brought more attention to Schlick. He was, until his murder by a deranged student in 1936, the acknowledged leader of the Vienna Circle. He was a man of great kindness and generosity, and had a gift for attracting outstanding people to the Circle and giving it such unity as it had. He was its spokesman before the world until his death and was an important contributor to the philosophy of science. But what is relevant here is the position of the Circle in 1932 when Quine visited Vienna, Prague, and Warsaw. Even so, my treatment must be partial and devoted to those aspects that are of interest for Quine.[50]

Probably the most famous doctrine of the Vienna Circle was the verification theory of meaning. In the early years of the Vienna Circle, they held that only those statements that admit of empirical verification are cognitively meaningful. The basic intuition behind this doctrine is the belief that to understand a proposition is to know the conditions under which it is true. The target was metaphysics, the statements of which they believed did not permit of specified truth conditions. Thus such statements as "The Absolute is perfect" or "All events are predestined by God" admit of no conceivable empirical test, and therefore say nothing intelligible. This doctrine is not as narrow as one might think. When C. I. Lewis asked Schlick whether questions about the other side of the moon and about immortality are cognitively meaningful, Schlick agreed that they were. "Verifiable" means possible of verification, and Schlick agreed that it was possible to specify conditions under which such statements could be tested.[51] Further, statements not cognitively meaningful may still have "emotive" meaning – they may be expressions of attitudes or desires and so are not devoid of meaning, but they are not statements about which it makes sense to ask if they are true or false.[52]

The logical empiricism of the Vienna Circle – they disliked the term "logical positivism" and called themselves "Logical Empiricists" – marked a significant break from traditional empiricism. The most important break with tradition was the adoption of the logic of *Principia Mathematics* and the Frege-Russell logistic thesis that mathematics is reducible to logic. Traditional empiricism had had great difficulty in dealing with the issue of mathematics and logic; Mill had claimed that mathematical propositions were empirical. But Wittgenstein's concept of tautology offered a way out of this bind. If the propositions of logic are taken as analytic a priori statements that are tautologies, then their certainty and that of mathematics (on the logistic view) is guaranteed, but they are also meaningless and say nothing about the world. There is therefore no inconsistency in adopting logic and mathematics while at the same time holding the empiricist view that experience is the basis of all our knowledge about reality. They were then able to reject the Kantian claim that there are synthetic a priori statements, and to hold that all statements of fact are synthetic a posteriori.

What then is the proper subject of logical empiricism? It is science – all science. Philosophy is, and should be, the analysis of science: the investigation of its logical structure, presuppositions and methods, its logical form, concepts and evidence. Epistemology becomes here the analysis of the logic of science. And by science here, they meant all science – physical, natural, and social, including even history (since history makes statements about the world that are claimed to be true).

This leads to the study of the *language* of science, since science is encoded in a language. At the time Quine met the Vienna Circle, Carnap was writing *The Logical Syntax of Language*.[53] He had not yet moved on to semantics. Carnap and the Vienna Circle generally recognized that there are multiple languages, and the hope was to develop a way of dealing with general syntax – the syntax of all languages.

One of the goals of the Vienna Circle was the unity of science. This goal had been a goal of Ernst Mach, and Schlick and the others subscribed to it. Clearly it required that the same standard of verifiability must apply in all sciences. This was not seen as a problem in dealing with the physical sciences, or the natural sciences; the problem was with the social sciences, and particularly psychology. Traditional psychology talked of consciousness, thoughts, will, emotions, and other mental phenomena. But these were not, or at least were not obviously, capable of description in the same sort of language that was used in physics. This was a problem area in 1932; the Vienna Circle was moving toward the position that the language of science must be a "physicalist" language, that is, that all of its terms must refer to publicly observable phenomena. So far as psychology was concerned, that would require the acceptance of behaviorism, since only the overt behavior of people is publicly observable.

But there were serious difficulties involved in some of the doctrines of the Vienna Circle. One problem, and an embarrassing one, was that scientific laws are not completely verifiable, since they cover a potential infinity of cases. This problem would be met by changing the requirement to partial verifiability or confirmability. A more divisive issue was the debate over protocol statements. Such statements were supposed to be records of observations or experimental outcomes that verified or falsified a scientific prediction. To take a specific example from medicine, urine is supposed to be acidic; if it is alkaline, the uric acid forms crystals that can cause bladder irritation. So suppose that to test a patient's urine, Dr. X dips a strip of litmus paper in patient Y's urine, and the litmus paper turns blue. The protocol would be a proposition to the effect "At 10 a.m., on August 5, 1929, at such and such a place, Dr. X dipped a strip of litmus paper in patient Y's urine and the litmus paper turned blue." Is this statement certain or not? Neurath held that it is not; Dr. X could have been mistaken, or his interpretation of "blue" might not be the same as someone else's. He held that propositions cannot be compared with a non-linguistic given, but only with other propositions. Schlick objected that this would lead to a coherence theory of truth. Besides, "blue" is a phenomenal quality, and famously what one observer perceives as color z another may see as color w. Thus the whole issue of intersubjectivity is involved here. Carnap, as we shall see, was in the middle on this question when Quine met him.[54]

The Vienna Circle was not the only such group of philosophers and scientists functioning at that time. In 1928, a group in Berlin organized as the "Society for Empirical Philosophy." Among its leaders were Hans Reichenbach, Alexander Herzberg, Walter Dubislaw, Kurt Grelling, Kurt Lewin, Wolfgang Kohler, and Carl Hempel. The group was in close touch with the Vienna Circle, and there was frequent visiting back and forth, so each was fully aware of what the other was doing.[55] Another such group existed in Uppsala, led by Axel Hagerstrom, and Adolf Phalen, and including a number of scientists and philosophers. There was also a group in

Warsaw that, while less interested in science than the Vienna Circle, was devoted to the study of logic. It was founded by Kazimierz Twardowski who, as noted above, had been a student of Brentano's. Also in this group were Jan Lukasiewicz, Taddeusz Kotarbinski, Stanilaw Lesiewski, Kasimierz Ajdukiewicz, Leon Chiestek, Alfreed Tasrski, and Mordechaz Wajsberg.[56]

The views of the Vienna Circle were hammered out in the Thursday night meetings. It would be a mistake to think that there were no disagreements within the circle; as noted, the protocol debate was a long running one and there were others. But there was enough agreement on fundamentals so that they could unite on key matters. In 1929, Schlick spent a year at Stanford as a visiting professor. Fearing that they might lose him, the Circle presented him on his return with a pamphlet, "Wissenschaftliche Weltauffasung. Der Wiener Kreis," written by Carnap, Hahn and Neurath and dedicated to Schlick. Meanwhile, during his absence members of the Vienna Circle and the Berlin Society attended the meeting of the German Physical Society and the German Mathematical Association in Prague where they announced the organization of a congress on the epistemology of the exact sciences.[57] Thus the Vienna Circle went public. In 1928, they had organized the Verein Ernst Mach. In 1929, they founded the journal *Erkenntnis* which became an important publication outlet for them. In 1930 the Congress for the Epistemology of the Exact Sciences was held at Koenigsburg. In September of 1935, they staged an International Congress on Scientific Philosophy at the Sorbonne. The next year, an International Congress for the Unity of Science was held at the Sorbonne, and in 1938 still another Congress was held at Cambridge, England. 1939 brought the last such Congress at Cambridge, Massachusetts. Then the war put an end to these meetings. Actually, Schlick's murder in 1936 dealt the Vienna Circle a blow from which it never recovered. But the Congresses spread the message far and wide, and brought other adherents and sympathizers from other countries: Charles Morris from the University of Chicago, Susan Stebbing and A, J. Ayer from England, and of course Quine. But Hitler had no tolerance for freedom of thought, and the members of all these groups – the Vienna Circle, the Berlin Society, and the Warsaw group, had to flee for their lives. Not all of them escaped.

Contents

List of Abbreviations

1. "Auto" for "Autobiography of W. V. Quine," in *The Philosophy of W. V. Quine* (pp. 3–46), ed. Lewis Hahn and Paul Arthur Schilpp (LaSalle: Open Court, 1986).
2. "EL" for W. V. Quine, *Elementary Logic* (Boston: Ginn and Co., 1941).
3. "F&Q/c" for W. V. Quine, in *Confessions of a Confirmed Extensionalist*, ed. Dagfin Follesdal and Douglas Quine (Cambridge: Harvard University Press, 2008a).
4. "F&Q/d" for W. V. Quine, in *Quine in Dialogue*, ed. Dagfin Follesdal and Douglas Quine (Cambridge: Harvard University Press, 2008b).
5. "FLPV" for W. V. Quine, *From a Logical Point of View* (Cambridge: Harvard University Press, 1953).
6. "JSL" for *Journal of Symbolic Logic.*
7. "LSeq" for W. V. Quine, "The Logic of Sequences" (Unpublished Harvard PhD diss., 1932).
8. "ML" for W. V. Quine, *Mathematical Logic* (New York, NY: W. W. Norton, 1940).
9. "NF" for W. V. Quine, "New Foundations for Mathematical Logic," *American Mathematical Monthly* 44:70–80 (1937); reprinted in FLPV.
10. "PM" for Alfred North Whitehead and Bertrand Russell, *Principia Mathematica* (Cambridge: Cambridge University Press, 1910) revised edition 1950, Vol. I unless otherwise indicated.
11. "RR" for W. V. Quine, *The Roots of Reference* (LaSalle: Open Court, 1973).
12. "SL" for W. V. Quine, *A System of Logistic* (Cambridge: Harvard University Press, 1934).
13. "SLP" for W. V. Quine, *Selected Logical Papers* (New York: Random House, 1966).
14. "TML" for W. V. Quine, *The Time of My Life* (Cambridge: MIT Press, 1985).
15. "TT" for W. V. Quine, *Theories and Things* (Cambridge: Harvard University Press, 1981).
16. "WO" for W. V. Quine, *Word and Object* (Cambridge: MIT Press, 1960).
17. "WP" for W. V. Quine, *The Ways of Paradox* (Cambridge: Harvard University Press, 1966).
18. References beginning MSAm2587 are to the Quine papers in Houghton Library at Harvard University.

Chapter 1
Quine: Logic and Philosophy

Willard Van Orman Quine was born in Akron, Ohio, on June 25, 1908. His father, Robert Quine, worked for at the Williams Fourdry and Machine Shop, but in 1917 he left the Williams Company to start his own business – the Akron Equipment Company. He made a success of his new business – sufficiently so that he and it survived the Great Depression. Unlike her husband, Harriet Quine had gone to college, and taught in the Akron elementary schools. Robert and Harriet Quine had two sons – Robert Jr. and then Willard.

The Quines were Congregationalists, and the Quine boys attended Sunday school as well as the public schools. But the religion did not take. Quine records that he began having doubts about heaven and immortality while still a child, and the doubts only became stronger as he grew. By the time he reached adulthood, he had abandoned religion entirely.[1]

Very early in life, Quine acquired a fascination with geography. He and his friends explored the city of Akron and its environs, and Quine made maps that recorded his discoveries. Not surprisingly, this geographic interest led to stamp collecting and an interest in foreign languages. He even started a business selling stamps and published a newsletter – the "OK Stamp News." The fascination with geography later bloomed into his almost compulsive international travelling that took him to over one hundred countries, and into a passion for mastering other languages at which he excelled. Years later, when he spent a year in Brazil, he was able to lecture in Portuguese and to turn the lectures into a book in Portuguese.[2]

While in high school, Quine became a fan of Edger Allen Poe and read all his tales. It was his reading of Poe's "Eureka" that aroused his interest in philosophy. As he wrote "my philosophical interest in the nature of things was fired by his eloquent and extravagant essay." From his brother Bob, who was already attending Oberlin, he got William James' *Pragmatism* and Max Otto's *Things and Ideals*, which he said he read "compulsively and believed and forgot all."[3] Nevertheless his curiosity about philosophy persisted.

His interest in languages led to an interest in the origins of words.

M.G. Murphey, *The Development of Quine's Philosophy*, Boston Studies in the Philosophy of Science 291, DOI 10.1007/978-94-007-2424-2_1,

> "Naturally the subject proved fascinating. An interest in foreign languages, like an interest in stamps, accorded with my taste for geography. Grammar, moreover, appeals to the same sense that is gratified by mathematics, or by the structure of boundaries and road networks. Etymology, more particularly, was a bonanza. Here one can pursue scientific method without a laboratory, and check one's hypotheses in a dictionary."[3]

Etymology and philosophy were not subjects taught in Quine's high school, but mathematics was, and he made an outstanding record at the school.

When Quine entered Oberlin in September of 1926, he had three major intellectual interests: mathematics, philosophy, and language. But he was undecided on a major until a senior named Bill Bennett told him about Bertrand Russell and his "mathematical philosophy." Quine thereupon elected mathematics as his major with honors reading in mathematical philosophy.[4]

Having heard enough about Russell to whet his appetite, Quine began reading Russell's works, including the *Introduction to Mathematical Philosophy*. Meanwhile, he excelled in his mathematics courses, and kept up his linguistic interest by taking courses in Greek, German and French. Quine had a gift for languages; he could visit a country whose language he did not know and emerge six weeks later with a sufficient grasp of the local idiom to carry on a conversation in that language.

Quine also took a psychology course with Raymond Stetson, and it was there that he read Watson's *Psychology from the Standpoint of a Behaviorist*,[5] and became a devout behaviorist.[6] Stetson was a distinguished man whose chief work was in phonetics. "The terms 'phonetics', 'sound', 'articulation' refer to articulate processes of speech without regard to meaning."[7] This does not mean that Stetson was opposed to phonemics; he considered the phonemic approach useful for some purposes. But his own work was phonetic, and based on the physiological processes involved in the articulation of syllables. He was rigorously empirical, thoroughly scientific, and a behaviorist. For Quine, already interested in languages, Stetson was an ideal teacher. But there was more involved in Quine's conversion to behaviorism than this, and here some background is called for.

As was noted in the Introduction, Darwin's publication of the *Origin of Species* in 1859 led to a bitter fight between evolutionary scientists and the defenders of religion. The battle came to focus on the nature of the human mind. Nothing shows the brilliance of William James more clearly than his recognition of this fact and his efforts to deal with it – his great *Principles of Psychology* in 1890. For all of the hard science that underlay James' work, he still saw psychology as the study of consciousness; the most famous chapter in the *Principles* was chapter IX[8] on "the stream of thought." James employed introspection as a legitimate method of studying psychological phenomena, and he was a brilliant observer; he was also a trained physiologist who brought empirical science to bear on mental life. And James was a strong antideterminist; for James the mind was characterized by "free creative intelligence." James' position was thus one that sought to mediate between the extreme determinism and materialism of people like Herbert Spencer and the demands for free will and human creativity on the part of many of the religious. This position however left James with a mind-body dualism that he finally resolved by espousing a panpsychic idealism.[9]

It is against this background that one has to see the behaviorist revolution that swept American psychology, and particularly the work of John B. Watson. Born in 1878 in Greenville, South Carolina, Watson was brought up in rigid Baptist orthodoxy. Although his father was something of a rebel and a philanderer, and in 1891 walked out on his family and went to live with an Indian woman, his mother was extremely devout, and Watson had little chance to stray from the straight and narrow. He went to Furman University, which was located in Greenville, and was largely devoted to the production of Baptist ministers. Luckily for him, Gordon Moore was teaching philosophy and psychology there, and Watson received a good introduction to the history of philosophy and psychology, especially that of Wundt. Having taken his M.A. at Furman, Watson taught briefly at the nearby Batesburg Institute, but his mother's death in July of 1900 freed him to move, and he went to the University of Chicago for graduate study.

Watson found John Dewey unintelligible and turned instead to psychology where he was a student of James Angell. Psychology, as Angell and most American psychologists saw it, was the study of consciousness. Angell had been a student of William James at Harvard, and was deeply influenced by James' approach.[10] He suggested to Watson that he should concentrate on "animal education." Watson, who seems to have left his religion behind in Greenville, had at that point no objection to studying consciousness, but he chose as his specific subject how rats learn and the limits to which they can be trained.[11] It was a good choice and capitalized on growing interest in animal learning.[12] He completed his dissertation and received his doctorate in 1903.[13]

Angell fell ill, and Watson was appointed as his replacement. Desperately short of money, Watson set to work on rats. He had excellent manual skills and was able to build most of the equipment he needed, and his research soon won him national academic recognition. In 1908, Watson was appointed to the chair of psychology at Johns Hopkins University. He was not yet thirty.[14]

At Hopkins psychology was part of the department of philosophy, psychology and education, whose chairman was Mark Baldwin, who was also the editor of the *Psychological Review*. But Baldwin managed to get himself caught in a brothel, which at Hopkins was not viewed as just a philosophic error; he was forced to resign. Hopkins had fired Charles Peirce in 1884 for living with a woman he had yet to marry, and the University's code of ethics had not changed significantly since, as Watson was to discover when he himself was fired for adultery in 1920. Meanwhile, Watson found himself left as editor of the *Psychological Review*. He went to work on his animals, and began rapidly producing publications. By 1913, he was a national authority on animal behavior, and in that year he published the article "Psychology as the Behaviorist Views It" in the *Psychological Review*.[15] This article is generally seen as the opening gun in the behaviorist revolt against the psychology of consciousness that had dominated the field. And certainly Watson made the challenge clear.

> Psychology as the behaviorist views it is a purely objective experimental branch of the natural science. Its theoretical goal is the prediction and control of behavior. Introspection forms no essential part of its methods, nor is the scientific value of its data dependent upon

> the readiness with which they lend themselves to interpretation in terms of consciousness. The behaviorist, in his efforts to get a unitary scheme of animal response, recognizes no dividing line between man and brute. The behavior of man, with all its refinement and complexity, forms only a part of the behaviorist's total scheme of investigation.[16]

Behavior, Watson claimed, is the subject of psychology, not consciousness; behavior can be studied objectively whereas consciousness cannot. Introspection, he held, is not a legitimate or useful method of study, and consciousness has no value in the explanation of behavior. "One can assume either the presence or the absence of consciousness anywhere in the phylogenetic scale without affecting the problems of behavior by one jot or one title; and without influencing in any way the mode of experimental attack upon them."[17] What is needed, Watson claimed, is a psychology that deals with stimulus and response, habit formation, habit integration, and the like.[18] There is no place in such a psychology for consciousness, mental states, mind, imagery, introspection, etc.[19]

Watson followed up this article with a textbook entitled *Psychology from the Standpoint of a Behaviorist*.[20] This was the text Stetson used in the course Quine took with him. Much of the book is devoted to describing the neurophysiology underlying behavior. Watson has no objection to instincts; a chapter is devoted to them (chapter 7). But the emphasis throughout is on habit acquisition and integration. From this standpoint, language is simply another complex system of habits. "The putting on of conventional speech habits is thus an illustration of conditioned reflex level of functioning (vocal habits) plus later associative connection of the word when learned with the bodily habits connected with the object for which the word stands (true language habits)."[21] Thought according to Watson is implicit speech.

> If our view is correct, it [thought] is a constituent part of every adjustment process. It is not different in essence from tennis-playing, swimming or any other overt activity except that it is hidden from ordinary observation and is more complex and at the same time more abbreviated so far as its parts are concerned than even the bravest of us could dream of.[22]

That this behaviorist view has not been previously accepted has been due, Watson says, to religion: "the real hesitancy in making thought highly integrated bodily activity and nothing more is due to the fact that historically 'thought' has always been connected with religion."

But in the 1920s, Watson went further. In *Behaviorism*[23] Watson sought to show that language is behavior learned according to the stimulus-response design. "Soon," Watson argued, "the human has a verbal substitute within himself theoretically for every object in the world. *Thereafter he carries the world around with him by means of this organization*."[24] Connected speech is thus explained by chaining, where a first word, called out by some stimulus, then serves as the stimulus for the next response, and so on. Thought, he then claims, is "nothing but talking to ourselves"; even when we are thinking in silence, the chain of responses continues to work but "subvocally." Thus "thought" comes to be located in the musculature. "All natural thought goes on in terms of sensory-motor processes in the larynx."[25] And by 1925 Watson could openly attack the religious identification of the "soul" with consciousness. Religion, Watson holds, is a superstitious effort to control people by fear.

> No one has ever touched a soul, or has seen one in a testtube, or has in any way come into relationship with it as he has with the other objects of his daily experience ... It was the boast of Wundt's students, in 1879, when the first psychological laboratory was established, that psychology had at last become a science without a soul. For fifty years we have kept this pseudo-science, exactly as Wundt laid it down. All that Wundt and his students really accomplished was to substitute for the word "soul" the word "consciousness."[26]

One should note here that for Watson behaviorism was not just a theory of psychology; it was the only scientific approach to the study of psychology. Other approaches, such as those of "consciousness," he considered disguised forms of the study of the "soul." To be an empiricist in psychology was for him to be a behaviorist.

It is not surprising that Quine, having rejected not only the Congregationalism of his parents, but religion in general, should have found Watson appealing. From Oberlin on, Quine never wavered in his adherence to this doctrine. And its attraction was strengthened by the fact that Bertrand Russell said Watson had contributed more to psychology than any man since Aristotle[27] and that the logical positivists that Quine was soon to meet, including Carnap, adhered to it. One cannot understand Quine's post-World War II work without seeing it against this Watsonian background.[28]

But Quine's primary interest at Oberlin was in "mathematical philosophy." "It was early in my junior year that I began my honors reading, and I was fired by it." He progressed from Venn to Peano to Russell's *Principles of Mathematics* to Courturat to Whitehead, and then to *Principia Mathematica* (hereafter PM). "This great three volume work, predominantly in symbols, quickly commanded my unflagging interest and unbounded admiration."[29] At the end of his junior year, his mother gave him all three volumes of PM.[30]

Quine's department chairman, William Cairns, was sufficiently impressed with him to invite him to review Nicod's "Foundations of Geometry and Induction" for the *American Mathematical Monthly*. Quine wrote the review and was proud to see it published; it was his first publication; the first of many.[31]

Couturat had shown that "if you start with three classes, take their intersections two by two, and then form the union of these three intersections, the result is the same as if you had proceeded oppositely, taking the unions and forming their intersections."[32] Quine "educed the more general law, that the union of all intersections of m classes out of n is the intersection of all unions of n – m + 1 classes".[33] But it was not so much the proof of this theorem but proving it within the system of PM with the full rigor of that work that entranced Quine. The eighteen page proof became his senior thesis. Quine graduated summa cum laude in 1930.[34]

Even before leaving Obelin, Quine had adopted Watson's behaviorism and the logic of PM. He saw himself as a hardheaded empiricist who believed in "experimental realism and no nonsense."[35] But was this empiricism consistent with PM? Could his empiricism accept abstract classes and relations? The ontology that appealed to Quine was one of concrete objects known by the senses – in other words, nominalism. But such an ontology was inconsistent with the work of Whitehead and Russell. This was a problem over which Quine would labor for two decades.

Having been inspired by PM and knowing that Whitehead was at Harvard, Quine applied to do graduate work in philosophy there, and was not only admitted but was given a $400 dollar scholarship – the amount of Harvard tuition in 1930. The money was very welcome. Quine's father's business had survived the crash of 1929 "unscathed," so the Quine family was by no means poor, but having just married his Oberlin fiancée, Naomi Clayton, Quine's expenses were heavier than those of the ordinary graduate student, despite the fact that Naomi got a job in a Boston department store. Furthermore, the country was sinking into the worst depression in its history, and neither future money nor employment was at all certain. These factors made Quine decide to try for a two year Ph.D. It was a rather daring decision, but C. I. Lewis, having come from an impoverished background himself, had a far more realistic view of the economic situation than many of his colleagues, and he encouraged Quine to do it.[36]

At Harvard, he tells us that he took Woods on Plato, Prall on Leibniz, and Lewis on Kant.[37] What he does not say either in the published autobiographical sketch he wrote for the Schilpp volume, or in his published autobiography, is that he took Lewis's course on the theory of knowledge, in which *Mind and the World Order*[38] was the main text. Among Quine's manuscripts there are three papers that he wrote for that class. In the first of these he examined Lewis's concept of the given. "My thesis is that no analysis of given experience can yield any other experience which is, in any full sense, the 'bare datum' of the former experience; any such analysis is, rather, merely a further interpretation." He also examines the problem of what happens when a hypothesis is disconfirmed by experience. If the hypotheses is *aRb*, and a consequence of this hypothesis is false, then one may reject either the hypothesis itself or alternatively change the definitions of *a* or *b*. But note the assumption of the arbitrary character of definition. This idea of alternative responses to falsification would later receive considerable expansion in Quine's work There is no indication here that Quine was troubled by the analytic/synthetic distinction or by the a priori/a posteriori distinction.[39]

In a second paper for Lewis, Quine focused on the validity of singular judgments. This paper is addressed specifically to *Mind and the World Order*. Empirical judgments are extensional, Quine says, but abstract theoretical ones are intensional since they are based on definitions. We can make any empirical *statement* analytic by revising definitions. Thus in the singular statement "*B is an A*," we can redefine "*B*" to include the property "*A*", thereby making the statement analytic. However, assuming that the statement is taken as synthetic (as e.g. "(∃x) (*Bx* then *Ax*)") then if "*P*" is the set of properties of the designatum of "*x*", the judgment applies the members of "*P*" to the values of "*x*". But, Quine holds, such an application requires a criterion. Let this be "P_1: If this is a *P* then if I do *b*, *a* will happen." However, Quine argues, "*a*" and "*b*" are complex concepts. So their application to the values of "*x*" requires a further criterion, and so we get an infinite regress. Nevertheless, Quine holds that we can abort the regress at some point and then calculate the probability of the singular hypothesis, and he has some fun in showing how to calculate it. This paper also contains Quine first statement of the goal of scientific inquiry, "The ideal conceptual unity at which such a study aims may be described as a system of concepts and interconceptual relations of the maximum simplicity compatible with its accommodating every item of experience falling within the limits of the study."[40]

Lewis wrote a two page comment on this paper. He does not accept Quine's regress argument, pointing out that in his theory, "*a*" and "*b*" are qualia rather than complex concepts, but the commentary is friendly and encouraging.

In May of 1931, Quine wrote a paper for Lewis on "Conceptual Pragmatism." This was the name that Lewis gave to his theory of knowledge in *Mind and the World Order*. "Conceptual pragmatism," Quine wrote, "relies on future verification and is for the purpose of attaining future goals." But, Quine argues, the value of a belief is not identical with its truth. Thus Newton's mechanics may be false, but it is extremely useful. Lewis's theory holds, Quine says, that judgments are two dimensional; they involve the given and the a priori. The given is in no sense pragmatic. It is defined as that in experience that is not altered by our interests or concepts. "Here we have the brute fact element of experience." It is the a priori that is pragmatic – the conceptual system that we seek to apply to experience. "One has a certain latitude as to where he may make his readjustments in the event of an experience recalcitrant to his system; and correspondingly there is some subjective opinion as to whether a chosen concept or a working hypothesis is to be branded as the point of error in the antecedent system." Judgments are pragmatic decrees whose purpose is the partial control of the future given; concepts are pragmatic devices whose office it is to serve as the vehicle for judgments. For an analysis to be correct, the theory must imply the testable consequences, and this is the task of logic. But from whence comes the validity of logic? It comes from definitions, and a "definition is valid in its own right as a convention of usage." But of course everything depends on the correctness of the deduction, and that could go wrong; hence there must be empirical judgments regarding the correctness of deduction. "But it is among futures, essentially, that resides the meaning of truth." It is impossible to know whether Quine accepted Lewis's conceptual pragmatism, but it is clear that he knew Lewis's theory and works thoroughly.[41]

Quine had come to Harvard largely because Whitehead was there, and mathematical philosophy was the field in which he wanted to work.

> Whitehead radiated greatness, and seemed as old as the hills. He was sixty nine. He was short and stooped, with heavy shoulders and a bald round head. Mrs. Whitehead was tall and thin, wore flowing garments, and smoked through a long silver cigarette holder... Mrs. Whitehead spoke disapprovingly of Russell. In the old days, he had more money than was good for him, she said, and he would loll on the Riviera while "Alty" grubbed away at Principia.

Whitehead was more genial. "Bertie thinks I'm muddle-headed" he said, "but I think Bertie is simple-minded." On another occasion, (Quine writes,) "Whitehead told me that he believed Russell to be the greatest analytic thinker the world has ever known, not excluding Aristotle."[42]

Neither Whitehead nor C. I. Lewis, Harvard's most distinguished logicians, were teaching logic; Sheffer was, and so Quine took his course. Sheffer was an able logician; his principal claim to fame was his invention of the stroke function that permitted all of the standard constants of the propositional calculus to be derived from a single primitive, and that made possible Nicod's reduction of the axioms of the propositional calculus to a single one. In the second edition of *Principia*, Russell wrote

> The most definite improvement resulting from work in mathematical logic during the last fourteen years is the substitution ... of the one indefinable "p and q are incompatible" (or, alternatively, "p and q are both false") for the two indefinables "not-p" and "p or q." This is due to Dr. H. M. Sheffer. Consequentially, M. Jean Nicod showed that one primitive proposition could replace the five primitive propositions used in part one of *Principia*.[43]

Sheffer also had a gift for the apt phrase. During the Great Depression, Federal agencies tried to employ unemployed academics. The Department of Agriculture employed a number of philosophers to teach philosophy to farmers. Sheffer remarked that "they had put Descartes before the horse."

Quine fulfilled the requirements for the M.A. degree, not because he had any intention of stopping short of the Ph.D., but because, given the state of the economy at that time, it was a hedge in case he did not complete his doctoral work. Whitehead was his dissertation director, and while Quine found him useful, Whitehead would respond to his questions with comments that drifted into metaphysics and were not much help. Fortunately, Quine needed little direction.[44]

At 9 p.m. on April 1, 1932, Quine handed his completed dissertation to his director, Whitehead – just 3 h before the deadline for the degree.[45] Although Quine subsequently disparaged his dissertation,[46] it was nonetheless a substantial achievement. Despite his admiration for PM, Quine found problems in it. Particularly, he objected to the use of the term "implication" as a name for the material conditional; this was a confusion of use with mention that he said he debated with Lewis and Huntington. He also found the notion of propositional function unclear and ambiguous as between properties and classes. So far as the actual proofs and theorems were concerned, Quine thought all that was needed was classes. He further thought that the limitation of relations in PM to the dyadic was a shortcoming that he could remedy. The system of the dissertation was purely extensional. Instead of focusing on some specific problem in logic, Quine undertook to create a more general logistic system from which that of PM was derivable. That is, he followed the model of PM but sought to improve it. The key idea was the concept of *sequence*. In PM, classes and dyadic relations are treated separately, and n-adic relations, for $n>2$, are not treated at all. Borrowing Sheffer's idea of the "degree" of a relation as the number of arguments that it took, Quine sought a treatment of relations of degree n, for any finite n. Classes then became monadic sequences, and the treatment of classes and relations could be unified. Quine entitled his dissertation "the Logic of Sequences."[47]

Quine borrowed Lewis's terms "quids" and "quods" from *A Survey of Symbolic Logic:* a "quid" is that which is operated upon; a "quod" is the operator.[48] Quine has two quids and four quods. The quids were sequences and functions. By a "function"

Quine meant a propositional function, but he took them extensionally, so that two functions were identical if their extensions were identical. By a "sequence" he meant any n-ad including the monad.

The quods were concatenation, superplexion, predication and assertion. Concatenation meant the combining of elements to form a sequence or of sequences to form a longer sequence. Predication is used in PM but never formally defined. Quine defines it as meaning the endowing of "φ" with an argument "x." Assertion, which is also used in PM, means simply the assertion of an expression, and is symbolized as in PM by "⊢." The most original of Quine's primitives is "superplexion" – a triumph of ingenuity over common sense. The superplex of "α" and "β" symbolized by "α # β," is "the relation of γ to δ where the inclusion of γ in α materially implies the inclusion of δ in β."[49] This definition is heuristic only, since material implication is formally defined in terms of superplexion.[50]

With these primitives, Quine states eight axioms and three rules of inference. He then shows that the system of Part I of Volume I of PM can be derived from his logic of sequences. For almost all of his theorems, he cites the corresponding theorems in PM. But he stresses the points where he considers his system (hereafter LSeq) superior to that of PM. In particular, he stresses the ability of his system to deal with relations of any degree, and the fact that in his system certain operations are capable of iteration although the corresponding operators of PM are not. He adopts Russell's theory of types to deal with the paradoxes and type considerations play an important part in the construction of his system. It is PM that dominates Quine's work, and it is his improvements on PM that he considers to be his contribution.[51]

Quine received his Ph.D. from Harvard in 1932.[52] He also received a Sheldon Traveling Fellowship which allowed him to spend a year in Europe.[53] But Quine also wanted to publish his dissertation. From April to August of 1932 he worked on the revisions, and when he sailed for Europe he left the manuscript with C. I. Lewis with the understanding that Lewis would start it through the press. But Lewis did not send the manuscript to the press, writing Quine in December that there were decisions to be made about the notation that Quine would have to make himself.[54]

The year abroad was a revelation to Quine. He later reflected on his early views.

> When I turned doctor in 1932, I was still wholly under the spell of *Principia Mathematica*. To me this great work, by my teacher Whitehead and his pupil Russell was the [acme] of mathematical logic. Improvement on *Principia Mathematica* was what I thought progress in mathematical logic consisted in. Progress in that sense was indeed no great challenge. The confusion of use and mention cried out for correction, along with the resulting muddles over propositional functions, implication, and ramified types. Uneconomical definitions and notation lay open for pruning. Great pioneer works are particularly easy to improve.[55]

That Quine's views in 1932 were parochial is not a surprise. Before World War II, the intellectual leadership of the world was largely in Europe. The great scientists of the time – Einstein, Bohr, Heisenberg, Schrödinger, Dirac, Fermi, etc. were all European. The same was true of philosophy; Whitehead and Russell were English, even though Whitehead was then at Harvard, and the most important philosophical movements were in Europe. The great painters – Picasso, Braque, etc. were European; so were the great writers – Mann, Proust, Yeats, Joyce, Kafka, etc, were on the other side of the Atlantic, and the leaders in most other fields were in England or on the continent. Today, we are prone to believe that the leadership in the intellectual world is in America. We forget that Hitler and Nazism drove the intellectual leaders of Europe to flee to the refuge of the United States. Many of the world's intellectual leaders were Jewish and would have been liquidated in Hitler's "final solution" had they not escaped. But Nazism did not tolerate freedom of thought by anyone and gentiles and Jews alike were targets of the fascist movements that swept Europe. To a very significant degree, the intellectual rise of the United States in the post-war world was due to the flood of intellectual refugees that Hitler drove to our shores.[56] Nor was Quine oblivious to the growing danger even in 1932. As he later wrote

> Everyone I knew was horrified from the very start, before the Holocaust, before any of the unthinkable things happened. The antisemitism in Vienna was evident before the Nazis took power. There were Nazi demonstrations, graffiti on the walls against Jews, and when the Nazis took over there were scandalous moves, like when they dismissed Einstein from the Prussian academy . . . They were allowed to re-occupy the Rhineland: it was so clear that we should take a firm stand about the gravity of the situation. In 1938 I was in Portugal for half a year, and my friends there talked about their discouragement with the British attitude of appeasement, and with Chamberlain. I didn't know anybody, even back then, who wasn't worried and outraged by the Nazis. When I hear that people didn't realize the extent of deprivation under the Nazis, I don't believe it. Of course we couldn't foresee how bad it was going to be, and we didn't have details about extermination camps. But you could feel an extreme nationalism, and there were many who were very sympathetic to the idea of a reinvigorated Germany. There were a lot of stupid people and there were a lot of rather evil people. I was eager for the United States to get into the war.[57]

As the Nazi menace grew, Quine was to play a significant role in helping European philosophers find sanctuary in the United States.

Quine had the good fortune to have met Herbert Feigl at Harvard. As he later wrote, "Herbert Feigl, a young member of the Vienna Circle, had come to Harvard on a fellowship before Nazi times, and had been instrumental in my choosing Vienna in 1932."[58] Feigl had been a student of Moritz Schlick's and a charter member of the Vienna Circle. When Quine met him, Schlick, with the generosity that characterized him, invited him to attend these meetings. There he met Menger, Godel, Hahn and a number of the others. At a party at Schlick's apartment he met Riechenback, who was a member of the Berlin Society for Scientific Philosophy. Neurath was then in Moscow, and Quine did not meet him until 1939.[59] He also met Philip Frank, who told Quine that he had never before met an American. Schlick invited Quine to lecture to the Circle, which he did, describing his work in his forthcoming book. One cannot but be impressed by the warmth and kindness of the Europeans toward a newly arrived American of whom they had no previous knowledge.

Leaving Vienna, Quine went to Prague where he spent thirty-seven days with Carnap, who had recently moved there from Vienna to take up an Assistant Professorship at the University of Prague. It was Carnap who became Quine's "great teacher." As he subsequently wrote,

> It was my first experience of sustained intellectual engagement with anyone of an older generation, let alone a great man. It was my first really considerable experience of being intellectually fired by a living teacher rather than a dead book.[60]

The timing of the Quine's arrival was perfect. Carnap was writing *Logical Syntax of Language* and Quine read the book "as it issued from Ina Carnap's typewriter."[61] Quine and Carnap became lifelong friends and Quine became for a time a disciple of Carnap's.

When Quine left Prague, he went to Warsaw where he met Lukasiewicz, Lesniewski, Tarski, and others of the Warsaw circle. Lukasiewicz particularly went out of his way to make Quine welcome, but for Quine the central figure was Tarski.[62] Many years later, Quine wrote to Tarski

> When I came to Warsaw you were 32. Logic in America, and in England, had been at a standstill since *Principia Mathematica*. On the continent of Europe it was thriving. Poland was in the forefront, and you were already the leading logician of the Poles.
>
> I came to Warsaw from Prague, a struggling young instructor. It had been through Carnap in Prague that I began to catch up with latter-day continental logic, and it was you who opened the whole bright scene.
>
> I was an unknown neo-doctor of 24, but you asked your seminar students to use German or French for my sake, instead of Polish. It was an impressive seminar, a research center. You were already a great teacher and trainer of research logicians, as you have been now for fifty years.
>
> Also you had already laid up a logical treasure trove on your own account. Even the epoch-making *Wahrheitsbegriffe* already existed in its Polish version.
>
> Six weeks of your seminar and our conversations, and your published papers, and I came away a happier and wiser man.[63]

One should note particularly that Quine knew Tarski's famous paper on the concept of truth when he returned to the United States in 1933. By the time his year was up, Quine had caught up with European logic and philosophy, and had met the two greatest logicians of that era – Godel and Tarski.

Quine's debts to Carnap were very great; many of his doctrines either derived from or were developed in opposition to Carnap's. One of the former sort was Carnap's argument in *The Logical Syntax of Language* that all intensional languages are reducible to extensional ones. Like Carnap, Quine regarded the task of philosophy as the study of science, and subscribed to the positivist idea of the unity of science. A further idea that he clearly got from Carnap was the distinction between the material mode and the formal mode, which in Quine's writing took the form of "semantic assent."

But the influence of Carnap was more profound than this suggests. In those thirty-seven days that Quine spent with Carnap, more than the *Logical Syntax* was discussed.

Quine also read Carnap's earlier work and discussed his ideas on other subjects. Carnap's first major publication was *Der logische Aufbau der Welt*[64] The book was heavily indebted to Russell, not only for the logic it used, which was that of PM, but also for Russell's *Our Knowledge of the External World* where Russell had proposed a logical construction of the world based on sense data.[65] Russell only proposed the project; Carnap tried to carry it out in the *Aufbau*. The book is complex and only certain features of it can be discussed here. Carnap took as primitive only one empirical relation – "remembered as similar." From this base, using the logic of PM, he sought to reduce the world of science to its basis in experience. By "reduce," Carnap meant the following: "If for each propositional function which is explicitly about objects *a*, *b*, ... (where *b*, *c*, ... may be absent) there exists a co-extensive propositional function exclusively about *b*, *c*, ... then *a* is said to be *reducible* to *b*, *c*, ..."[66] By an "object" here, Carnap means "anything about which a statement can be made," including such things as concepts.[67] The domains to be treated are the autopsychological – i.e., the sensory experience of the individual, but without the assumption of any person whose experience this is, the domain of physical objects, the domain of other persons (called the heteropsychological), and the domain of cultural objects, such as values, art, etc. What then is to be reduced to what? Carnap could have started with physical objects, but instead he started with the autopsychological.[68] The reason for this choice, Carnap says, is "epistemic primacy."[69] "An object (or an object type) is called *epistemologically primary* relative to another one, which we call *epistemologically secondary*, if the second one is recognized through the mediation of the first."[70] Carnap begins with the autopsychological, then shows how the physical objects are reducible to the autopsychological, and then having established other persons as physical objects, he shows how psychological states can be imputed to them, using only behavioristic psychology. Finally he shows how cultural objects are reducible to the heteropsychological. For his basic autopsychological units Carnap takes our total conscious sensory experiences at a time.[71] These are unanalyzable, but they can be compared, so that parts of one may be recognized as similar to parts of another, giving rise to the relation of part-similarity.

It is important to understand that Carnap's terms and propositions are taken strictly in extension; intensional statements he says are reducible to extensional ones. It is also fundamental to his enterprise that all objects of the system are defined only by their structural relations.

> Thus, our thesis, namely that scientific statements refer only to structural properties, amounts to the assertion that scientific statements speak only of forms without stating what the elements and the relations of those forms are.[72]

Carnap's example is a railroad map without station names. One can describe the map in terms of the relations among raillines only: i.e., at point 6, eight lines intersect, or between points A and B there are four stops. Thus Carnap attempts to develop his entire system in terms of structural relations only; other properties of objects are used only after being structurally defined.

The result, if it were fully developed, would be a genealogy of concepts where any high level concept, such as values, could be traced back through the steps of its construction to its base in the autopsychological. Carnap gives a careful and brilliant construction of the autopsychological realm, but for the others he gives only an outline of how such a construction might be done. But the ingenuity of his autopsychological constructions seemed to promise that the full system could be constructed.

All this Carnap built on the basis of logic and mathematics, and one primitive empirical relation – remembered as similar. But having gone this far, he seeks to eliminate the basic primitive relation. This can be done by taking some high level theorem about the primitive relation, Rs, and then taking Rs as the relation that satisfies that theorem. Any relation, the relational extension of which is structurally isomorphic to Rs, could be so used.

> It follows that, to each originally constructed object, there corresponds precisely one new one with the same formal properties. Then all statements of the constructional system continue to hold since they concern only formal properties.[73]

But Carnap says that this could destroy the link to sensory experience. He therefore defines a "founded" relation as one which corresponds to extensionally "natural relations." "The various member pairs of the founded relation extension have something in common that can be experienced."[74] "Founded" is a new primitive: it is independent of logic and independent of the constructional system. Accordingly, Carnap proposes to add it to logic as a basic concept.

> That this concept is concerned with the *application* to object domains is not a valid objection to introducing it as a basic concept of logic. The same is true for another basic concept of logic, namely, generality: "(x)Fx" means that the propositional function of Fx has the value true for every argument of an object domain in which it is meaningful.... From this it follows that it [logic] must concern itself precisely with those concepts which are applicable to any domain whatever. And foundedness, after all, belongs to those concepts. In view of these reasons, let us introduce the class of *founded relation extensions* as a basic concept of logic ....[75]

If this is granted, then Carnap has achieved a complete formalization of his system, entirely in structural terms provided by logic in which "founded relation extensions are [taken as a] basic concept of logic."[76]

The *Aufbau* was an exciting work, and it was much discussed within the Vienna Circle. But not everyone agreed with it. Otto Neurath particularly objected to Carnap's choice of the autopsychological domain as epistemologically primary, arguing that the domain of physical objects should have been chosen. There thus developed a running debate between Neurath and Schlick, with Carnap in the middle. As the debate progressed, both Neurath and Carnap gradually changed their positions. Carnap's autopsychological view was phenomenalistic; Neurath's was realistic. Neurath moved toward a physicalist and naturalistic holism. Quine was well aware of Neurath's work as his frequent reference to Neurath's ship simile testifies.

> We are like sailors who must rebuild their ship on the open sea, never able to dismantle it in dry-dock and to reconstruct it there out of the best materials.[77]

Carnap moved slowly, and reluctantly, toward physicalism, but still held in the summer of 1932 that the phenomenological language was epistemologically primary. The problem for Carnap was that the language of science must be intersubjectively understandable, whereas the phenomenalist (autopsychological) language is not sharable. The issue focused on "protocol sentences" – sentences that report the results of observations and experiments. Neurath held that they are physicalist statements and corrigible. Carnap agreed that for the language of science, protocols must be intersubjectively understandable, and therefore must be couched in the physical language vocabulary. But he also held that there was a further level – sentences in the phenomenalist language that could be conclusively verified. It was these, Carnap held, that were basic for epistemology.[78] This was not a position that Carnap held for long, but it was his view when Quine spent thirty-seven days with him in 1932. As we will see, Quine absorbed Carnap's view, although in his hands it developed in a very different direction than that which Carnap followed. Quine's epistemological phenomenalism very likely came from his study of Carnap's *Aufbau*. And his early nominalistic leanings may well have been strengthened by his European sojourn. He would not have gotten that from Carnap, but he may have been persuaded in Warsaw. Lesniewski was a nominalist, though Mancosu doubts his influence on Quine.[79] More interesting is the possibility of Tarski's influence. Little is known about Tarski's early beliefs about nominalism because virtually all of his early papers were lost in the Nazi invasion, but Mancosu has shown that in 1953 he was in full agreement with Quine on nominalism. Both spoke at the Amersfoort conference, and both spoke favorably regarding nominalism, though they did not ignore the problems involved.[80] But whether the Poles influenced Quine on this matter or not, the issue was one that could hardly be avoided. Set theory involved abstract objects; unless one was to adopt Carnap's position that metaphysics was meaningless (which Quine did not), the issue of the status of such abstract entities as classes and sets could not be avoided. As Mancosu has shown, Quine's early writings on nominalism show that while on the one hand he inclined toward it, on the other he fully recognized that adopting nominalism would cost a large part of classical mathematics. But Quine had a further reason for wanting nominalism to be a viable position. One of Quine's foremost traits was his desire for simplicity and elegance in logic and in mathematics. He found the paradoxes of set theory offensive and was skeptical about the theory that led to them. This is an issue that was present throughout his life.

Quine returned to Harvard as one of the charter members of the Harvard Society of Fellows, which guaranteed him three years of freedom to work on logic and philosophy.[81] He was in fact the second person chosen for membership in the society, the first being the behavioral psychologist, B.F. Skinner. These two were to become fast friends – a friendship that endured throughout their lives. Quine wrote to Skinner in 1984 about their friendship that its

> origins go back a full fifty years and more, to when you became for all time the first-born Junior Prize Fellow and I the second... We were kindred spirits, already indoctrinated by Watson in behaviorism, and already imbued with a philosophy of experimental reality and no nonsense. We were both hipped on language, too; I mostly on concrete details of etymology and you on basic linguistic theory. I remember your putting me on to Jesperson; we were sitting on the grassy verge of Belmont Hill, which was then open country. It was probably through you also that I got onto Leonard Bloomfield; and then you were responsible for the pleasure I have had from that fine old edition of John Horne Tooke.[82]

Skinner gave Quine "a first American edition of Tooke's work."[83] Tooke became one of Quine's favorites. John Horne Tooke was an English political figure who had strongly criticized the British actions at Lexington and Concord and supported William Pitt. He wrote a treatise on philology entitled *The Diversions of Purley* in which he argued that instead of talking of "ideas," Locke should have talked about "words."[84] This perfectly matched Quine's view.

By March, Quine had completed the revision of his dissertation, and in1934 the revised manuscript appeared under the title of *A System of Logistic*.[85] This was Quine's first major publication. He revised the list of primitives he had used in LSeq. Superplexion disappeared, predication was kept, assertion was dropped since "as Professor Carnap has pointed out, the assertion sign expresses no primitive idea of logic, but serves rather merely as a device for the presentation of the system."[86] Concatenation is replaced by ordination. Abstraction is introduced as a new primitive, symbolized as in PM, and "congeneration" appears as a new primitive: the congenerate of a class α, symbolized by "$[\alpha]$", is the class of all superclasses of α – i.e., the class of all classes in which α is included. Hence, "$[\alpha], \beta$" is the proposition "$\alpha \subset \beta$", since "β" is a member of "$[\alpha]$" and therefore is a superclass of "α". The system has four rules of inference and six postulates.

Despite these changes, the system is that of the dissertation; like LSeq, it is a rewrite of Part I of Volume I of PM. It is the improvements over PM that occupy Quine; he shows that the logic of PM can be derived in his new system, and it has the same advantages over PM that LSeq had.

The *System of Logistic* shows Quine still under the spell of *Principia*. Looking back later, Quine remarked that he had concentrated on logic and the revision of his dissertation when he returned from his European trip, and only then "got back to Carnap's ideas" two years later.[87]

On the 8th, 15th, and 22nd of November, 1934, Quine gave three lectures on Carnap before the Harvard Society of Fellows. The first lecture, Quine said, was intended to set the stage for the presentation of Carnap's *The Logical Syntax of Language*; the second would describe the book's content, and the third would show the implications of Carnap's work for philosophy generally. In the first lecture, which he entitled "The A Priori," Quine begins with "the analytic character of the a priori," and so with the notion of the analytic. Quine remarks that "The development of foundational studies in mathematics during the last century has made it clear that none of mathematics, not even geometry, need rest on anything but linguistic conventions of a definitional kind."[88] Accordingly, Quine turns to the subject of definition. There are, he says, two sorts: explicit definitions in which the definiens is simply an abbreviation for the definiendum, and implicit definition in which a word occurring undefined in a set of rules or postulates, accepted by convention as true, is given a meaning by showing how it is to be used.[89] Which terms ought we to define first? Following the lead of Whitehead and Russell, Quine gives priority to the terms of logic, and lays down two "rules" or postulates which provide an implicit definition for "neither-nor" – i.e., the Sheffer stroke. It is then possible to give explicit definitions of the standard logical operators for truth functions. "All such sentences become *analytic* – direct consequences of our conventions as to the use of words."[90] By the same technique, Quine introduces universal quantification, and identity. Then, following the logistic thesis, he argues that all of mathematics, including geometry, can be defined in terms of logic and thus shown to be analytic.

But how far should this procedure of creating analytic sentences by definition be carried? Quine notes that it could be extended to empirical terms such as "event" and "time." Clearly, it could be carried through the whole of physics, thus making all the sentences of physics analytic. Why not?

> If *all* empirical generalities are transformed into analytic propositions by redefinition of terms, we shall find ourselves continually redefining and then retrodefining; our definitions will not only be in an unnecessarily extreme state of flux, but there will be no immediate criterion for revising one definition rather than another.[91]

New findings in science continually force such changes, and if every sentence is analytic, the mental gymnastics required to revise our science will prove needlessly exhausting. Hence Quine says "We will do best to render only such sentences analytic as we shall be most reluctant to revise when the demand arises for revision in one quarter or another."[92] However, Quine does not believe that this criterion excludes all propositions of physics. Einstein, Quine says, by his definition of simultaneity made "'the Michelson-Morley law' analytic"[93] – i.e., the invariance of the velocity of light regardless of the motion of the observer.

Quine had titled this lecture "the a priori," but he does not take up that subject until near the end of the lecture. Traditionally, the a priori is that which is true independently of experience. Analytic sentences are a priori since their independence of experience is guaranteed by the syntactic rules. Quine's approach here follows from the grounds he had already developed.

> The more firmly accepted sentences we choose to modify last, if at all, in the course of evolving and revamping our sciences in the face of new discoveries. And among these accepted sentences which we choose to give up last, if at all, there are those which we are not going to give up at all, so basic are they to our whole conceptual scheme. These, if any, are the sentences to which the epithet "a priori" would have to apply. And we have seen ... that it is *convenient* so to frame our definitions as to make all these sentences analytic, along with others, even, which were not quite so firmly accepted before being raised to the analytic status.[94]

There is a subtle but important distinction here. The sentences that we will surrender last, if at all, are those he calls a priori. And since *any* sentence can be made analytic by redefinition, he says that the a priori sentence should be made analytic "along with others." To put it differently, the a priori are a subclass of the analytic distinguished by our determination never to give them up. But "all this," Quine says, "is a question only of how we choose to systematize on language." And so he concludes "Carnap's thesis that philosophy is syntax is thus seen to follow from the principle that everything is analytic except the contingent propositions of empirical science."[95] He is then ready to take up *The Logical Syntax of Language*.

In the second lecture, which Quine entitled "Syntax," he gives a description of the content of Carnap's book. Quine notes, really for the first time in these lectures, that Carnap constructs a specimen language for study (Language I). He does not mention that Carnap also constructs a metalanguage (Language II) in which to discuss Language I. Quine points out the role of formation and transformation rules (rules of inference), and makes Carnap's choice of a formalized language a consequence of the complexity of these rules in natural languages.[96] Carnap's use of numerical operators and descriptive operators is summarized, and his method of identifying physical objects by numerical coordinates – a device Quine would later use himself. The primitive signs of the language are implicitly defined; the formation rules generate the basic sentences, and the transformation rules permit the derivation of their consequences.

> The entire syntactic structure of a language is determined once we do know what passes for a sentence and what passes for an immediate consequence within that language... It is in this sense that *sentence* and *immediate consequence* are concepts of syntax. They are strategic *syntactic functions* of a language.

A further syntactic function is being analytic. Further, "a sentence is *analytic* if it is a consequence of every sentence."[97] Quine emphasizes that the terms "analytic" and "consequence" are relative to a language; what is analytic or a consequence in one language need not be so in another. Similarly, the "contradictory" is defined for a sentence or a set of sentences as that from which every sentence can be derived. A sentence which is neither analytic nor contradictory is synthetic.[98] Similarly, "synonymy" is a syntactic notion. "Two signs are *synonymous* if, when we replace either sign by the other in any given sentence, the resulting sentence is a consequence of the given sentence."[99] Quine apparently found no problems with the notions of analyticity and synonymy at this time. In fact, he salutes Carnap's success in giving precise meanings to these terms. But the concept of truth is not defined in Carnap's model language; truth is not a syntactic concept but a semantic one.

Quine shows how, using the technique devised by Godel, Carnap arithmetizes the entire syntax of his specimen language – "Language I." Carnap further shows that the syntax of Language I can be expressed within Language I itself. Quine is careful to point out that in doing so Carnap does not run afoul of any of the well known paradoxes. Quine urges "that the arithmetization of syntax greatly increases our powers of syntactic investigation."[100] As proof of this, he cites the example of Godel's proof of the incompletability of arithmetic, which Quine calls "the most famous recent discovery in the foundations of mathematics."[101]

Quine entitled his third lecture "Philosophy as Syntax." The purpose is to show the implications of Carnap's work for philosophy generally. Referring to his previous lectures, Quine holds that we can divide all sentences of philosophy into the analytic a priori and the synthetic a posteriori, meaning by "synthetic" the statements of empirical science. But it should be noted that the concept of analyticity that Quine employs in these lectures is a mixture of his own with the classical one.

Quine lays great stress on the distinction between "use" and "mention," although he does not employ the word "use" in doing so. Language, he says, is made up of signs. Syntactic properties are properties of signs, not of the referents of signs. Thus "Boston is populous" is a sentence about the object Boston. "'Boston' is disyllabic" is a sentence about the word "Boston." All syntactic properties are sign-properties, although the converse is not true.[102] He then introduces Carnap's concept of the quasi-syntactic. He illustrates this by the following two sentences:

1. Roosevelt mentioned Boston.
2. Roosevelt uttered a synonym for "Boston."

The two sentences are consequences of each other, but he says that (2) is a purely syntactic sentence whereas (1) is not. Sentence (1) predicates the quasi-syntactic property of having been mentioned by Roosevelt. Any sentence that is equivalent to a syntactic sentence is, if not purely syntactic, then quasi-syntactic. But Quine also stresses the distinction between formal syntax and empirical syntax.[103] The latter sentences of syntax involve the history of the use of words. (1) is therefore a sentence of empirical syntax but nevertheless synthetic. But statements such as "The negation of an analytic sentence is contradictory" belong to formal syntax.

The point of this is to draw the distinction between three kinds of sentences; syntactic, quasi-syntactic, and synthetic.

> Roughly, a syntactic sentence may be characterized as a sentence which treats both ostensibly and actually of a sign, while a quasi-syntactic sentence treats actually of the sign but ostensibly of the object of the sign. Clearly the quasi-syntactic is an indirect idiom, and should be eliminated in favor of the syntactic translation when we are concerned with a logical analysis of what is being said.[104]

Quine then proceeds to give four examples of what such a program of purification involves. His first target is the concept of meaning, which he seems here to equate to denotation. "When the quasi-syntactic idiom is eliminated we find ourselves working within the syntactic level quite independently of the meaning-relation."[105] The second example is modality. Quine argues that the modal term "logically impossible" is translated into syntax as being contradictory. Thus

> The sentence "It is logically possible that —." becomes in the syntactic idiom "The sentence '—' is not contradictory." . . . The sentence "It is logically necessary that —" becomes in the syntactic idiom "The sentence '—' is analytic."[106]

Obviously, all model terms can be translated into syntax in this way. But there is also the problem of empirical possibility. "Empirical impossibility is impossibility in view of accepted empirical laws."[107] This however is a quasi-syntactic sentence which Quine calls "empirically contradictory," since the conjunction of the sentence with other accepted sentences, including the physical laws, will be contradictory. This will be a sentence of empirical syntax but "none the less a syntactic property."[108] Third, Quine holds that "we see that the philosophical difficulties of the *universals* – properties and relations – can be reduced similarly to syntax." And finally, he argues that most of the problems of metaphysics can be reduced to syntax by methods similar to those he has already spelled out.[109] Quine emphasizes that all such reductions are relative to a language, but he thinks that similar results are obtainable in any language sufficiently rich. Here Quine seems to have misunderstood Carnap, since as Creath points out "Carnap rejected both the assertion that there are [ontological] entities and the denial of their existence; both were metaphysical nonsense."[110]

So then what does all this come to? "All this, assuming it to be valid, points to the dissolution of much philosophic controversy."

> [Carnap's] concern is . . . to clear away confusion and lay the foundations of a rigorous and fruitful study of the logic of science: for it is the logic of science, in the broadest sense of the phrase, the analysis, criticism, and refinement of the *methods* and the *concepts* of science,

> that Carnap regards as the defensible province of philosophy. And the medium for all such studies is, according to Carnap, syntax.[111]

This view of the subject of philosophy was one that Quine endorsed and would take to more extreme form in his later writings.

It is important to recognize that Quine is not criticizing Carnap in these lectures; rather, he is expounding and defending what he takes to be Carnap's position. Quine accepts the notions of analyticity and synonymy, but he does interpret these in a non-Carnapian way. According to Creath

> Quine does recount that in these first days [in Prague] he read the pages of *The Logical Syntax of Language* as they poured out of Ina [Carnap's] typewriter. Astonishingly, Quine's very first reaction (preserved in a brief shorthand note by Carnap) contains in embryonic form his whole view of the matter. Might not, he wondered, the difference between the (analytic) axioms of arithmetic and (synthetic) empirical claims about physical bodies be a difference of degree? Might not these degrees reflect our relative willingness to abandon the various beliefs under consideration?[112]

This strongly suggests, not that Quine was communing with his future self, but that he already held this view at the time he met Carnap. And there is every reason to believe that he did. Quine had been a student of C. I. Lewis at Harvard, and he had read Lewis' *Mind and the World Order*, so he was well aware of Lewis's views. In response to PM, Lewis had created an alternative logic based on the modal notion of strict implication. He was then confronted with the question of which system of logic was the correct one. Lewis regarded all statements of logic as analytic and a priori; hence he did not believe that any empirical test could decide the matter. Instead, he took the position that the choice between alternative logics was pragmatic; we choose whichever system best suits our needs. The problem, he thought, was analogous to that of choosing among the metrical geometries. But in rejecting one logic for another, the a priori analytic character of the rejected system was not compromised. Further, Lewis held that it is the principles of logic that we are the least willing to alter. Thus Lewis.

> The whole body of our conceptual interpretations form a sort of hierarchy or pyramid with the most comprehensive, such as those of logic, at the top, and the least general, such as "swans" etc., at the bottom; that with this complex system of interrelated concepts, we approach particular experiences and attempt to fit them, somewhere and somehow, into its preformed patterns. Persistent failure leads to readjustment; the applicability of certain concepts to experience of some particular sort is abandoned, and some other conceptual pattern is brought forward for application. The higher up a concept stands in our pyramid, the more reluctant we are to disturb it, because the more radical and far-reaching the results will be if we abandon the application of it in some particular fashion. The decision that there are no such creatures as have been defined as "swans," would be unimportant. The conclusion that there are no such things as Euclidean triangles, would be immensely disturbing. And if we should be forced to realize that nothing in experience possesses any stability – that our principle, "Nothing can both be and not be," was a mere verbalism, applying to nothing more than momentarily – that denouement would rock our world to its foundations.[113]

Further, in his famous article on "The Pragmatic Conception of the A Priori", Lewis wrote of the laws of arithmetic

> Those laws and those laws only have necessary truth which we are prepared to maintain, no matter what. It is because we shall always separate out that part of the phenomenon not in conformity with arithmetic and designate it by some other category – physical change, chemical reaction, optical illusion – that arithmetic is *a priori*.[114]

Quine knew Lewis' views, and his response in 1932 looks very much like an application of Lewis' doctrine. Quine was more indebted to Lewis than he ever acknowledged.

Quine does emphasize Carnap's doctrine that different languages have different syntaxes, and that the choice among languages is purely pragmatic. And while he makes it clear that Carnap worked with formalized languages, Quine takes these doctrines as applying to natural languages. Certainly this is true in Lecture one and in much of Lecture three. But Quine never mentions Section IV of *The Logical Syntax of Language* in which Carnap attempted to create a general syntax – "that syntax which relates not to any particular individual language but either to all languages in general or to all languages of a certain kind."[115] Why the omission? Carnap's position was not as relativistic as Quine's. It is Quine who is the extreme relativist here, rather than Carnap who believed that syntax for all languages was possible.

Quine claimed in his first lecture that we can make any sentence analytic if we so wish, including the whole of physics. He rejected the idea of doing so because of the problems of redefinition that scientific progress would cause. But this did not deter him from holding that Einstein's principle of the constancy of the velocity of light was analytic. Einstein would not have agreed, and Quine did not repeat this blunder in his later work. But note that Quine's claim that we can make any sentence analytic trivializes the notions of analyticity and definition, whether Quine intended it that way or not.

In 1936, Quine published an article entitled "Truth by Convention."[116] The purpose of the article is not to question the validity of the notion of truth by convention but "its sense."[117] He first examines various proposals for founding logic and mathematics on definitions. He draws the distinction between terms that occur vacuously in sentences and those that occur essentially. The vacuous terms of a sentence can be replaced by any terms of the same linguistic category without changing the truth value of the sentence; the essential terms are those that must occur in all substitution instances of the sentence obtained by replacing the vacuous terms. Quine then considers at length four different ways in which definitions might be said to establish the truth of mathematics, and finds all four to be wanting. He does however accept the logistic thesis that all of mathematics (including geometry, transformed into algebraic form by analytic geometry) is reducible to logic; that the primitive terms of mathematics are definable in logical terms and the axioms of the former are derivable from those of logic. The question then becomes whether the truths of logic are true by convention. In his first lecture on Carnap, Quine had trivialized the notion of analyticity by showing that virtually any statement can be rendered analytic by redefinition. He follows the same strategy here, showing that any statement can be made true by convention. He then shows that upon examination this method of creating truths leads to a regress. "The difficulty is that if logic is to proceed *mediately* from conventions, logic is needed for inferring logic from the conventions."[118] Quine does consider briefly the possibility that these conventions may have arisen in our behavior and only subsequently been formulated into explicit statements, but he rejects that: "It is not clear wherein the adoption of the conventions, antecedently to their formulation, consists; such behavior is difficult to distinguish from that in which conventions are disregarded."[119]

Historically, this claim is surely false. Logic did not spring fully formed from the mind of Aristotle like Athena from the mind of Zeus. Peirce drew a distinction between "logica utens," meaning the "classification of arguments, antecedent to any systematic study of the subject" and "logica docens," which was "the result of scientific study."[120] Human beings have been arguing ever since they came down from the trees, and probably before that, and the distinction between good arguments and bad ones goes back well before Aristotle. The pre-Socratics carried on extensive arguments about metaphysics, science, ethics, etc; Socrates himself was no mean debater, and Plato likewise. From any historical point of view, it must be obvious that Aristotle's work is the culmination rather than the beginning of serious concern with valid arguments. It does not diminish the luster of Aristotle's achievement or his originality to point out that his logic was in part a codification and clarification of the lessons learned from his forebears. Thus, Quine's "knock-down" argument against Carnap ignores the history of how logic actually began.

Why does Quine dismiss this possibility out of hand? He does so because he wants to clear the way for his own theory.

> Viewed behavioristically and without reference to a metaphysical system, this contrast [between the a priori and the a posteriori] retains reality as a contrast between more and less firmly accepted statements; and it obtains antecedently to any *post facto* fashioning of conventions. There are statements which we choose to surrender last, if at all, in the course of revamping our sciences in the face of new discoveries; and among these there are some which we will not surrender at all, so basic are they to our whole conceptual scheme. Among these latter are to be counted the so-called truths of logic and mathematics.[121]

Quine is here paraphrasing what he had said in Lecture I. What is different here is that Quine takes this as a *behavioral* criterion for statements being a priori. Thus he says "We may wonder what one adds to the bare statement that the truths of logic and mathematics are a priori, or to the still barer behavioristic statement that they are firmly accepted, when he characterizes them as true by convention in such a sense."[122] Looking back years later, Quine said that this article "showed already the beginnings of my misgivings over analyticity"[123] What Quine denies in the article is that the truths of logic and mathematics are true by convention, and to do so he uses Lewis's doctrine. But what Quine questions in the article is the standard view of the a priori, and it is for that that he finds in Lewis's doctrine a behavioral criterion. Since he held all analytic statements to be a priori, their definition applies to both, but his focus in the above passage is on the a priori. Lewis's doctrine of the pragmatic a priori has here been converted into a behavioral criterion of a priority.

Presumably Quine is here rejecting Carnap's view of conventionality. But nowhere does Quine mention formalized languages or the fact that Carnap saw conventions as specific to such formalized languages. Quine is talking about natural languages, as is made very clear by his insistence that, as Russell had held,[124] our mathematical terms must accord with everyday usage. But Quine was clearly moving in a direction different from Carnap's.

Quine's job at Harvard was teaching logic, and his interest in Carnap's work did not diminish his productivity in logic. In 1932, he published a brief article correcting two errors in Joergensen explication of Nicod's postulate.[125] The next year, he published a short paper entitled "A Theorem in the Calculus of Classes,"[126] showing that for two classes, the product of the sum of pairs equals the sum of the products. But for n classes combined into k groups, the sum of the products of the groups is n – k+1. This paper draws its inspiration from his undergraduate thesis at Oberlin.

In 1934 he published "Ontological Remarks on the Propositional Calculus."[127] His target is the belief that in the propositional calculus, sentences denote propositions, usually conceived as the meanings of the sentences. Quine's argument is that such "propositions" play no essential role in deduction and should be eliminated.

> Without altering the theory of deduction internally, we can so reconstrue it as to sweep away such fictive considerations; we have merely to interpret the theory as a formal grammar for the manipulation of sentences, and to abandon the view that sentences are names. Words occurring in a sentence may be regarded severally as denoting things, but the sentence as a whole is to be taken as a verbal combination which, though presumably conveying some manner of intelligence (I write with deliberate vagueness at this point), yet does not have that particular kind of meaning which consists in denoting or being the name of something.[128]

The schemata of the calculus now are counted as standing in for sentences, so that for example "$(p .\supset (q \supset p))$" expresses the fact that if declarative sentences are substituted for "p" and "q", the result is an "abbreviation of any sentence of the form" "If so and so, then if such and such then so and so."

There are three important points here. One of the arguments for taking propositions as the meanings of sentences was that they offered a way of explaining how translation from one language to another was possible. To say that if the English sentence "The dog is big" is translated into French as "Le chien est grand," then they must have the same meaning was taken as asserting that they refer to the same proposition. In throwing out propositions, Quine left it unclear just what it was that translation between languages leaves invariant. Second, Quine's "deliberate vagueness" masked the fact that at that point he had no alternative theory of meaning to offer, and third, one can see here his dislike of metaphysical entities. One should bear in mind that this paper was written in the same year that he gave the three lectures on Carnap. It is not surprising that he had omitted any mention of Carnap's attempts at a general syntax.

But there is another point that needs noting. Quine here takes sentences as having no denotation; their terms denote but not the sentences themselves. This is a change from the position he had held earlier when he had taken sentences as names of ordered pairs. Thus he had taken "Socrates is a man" as naming the pair "Socrates" and "mankind." He dropped this interpretation for Wiener's definition of the ordered pair.[129] Why did Quine make this change? Why rule sentences devoid of denotation?

There is, I think, a reason beyond the attractiveness of Wiener's redefinition. Quine already knew Tarski's theory of truth. If one says with Tarski

"Snow is white" is true ≡ snow is white.

it appears that it is truth that ties sentences to the world, and Quine does say this repeatedly. What then is there for "Snow is white" to denote? For Quine, to say that it denotes a fact or a state of affairs would seem to be the introduction of a redundant entity. This argument is found very explicitly in Davidson, and I think it is what led Quine to say that sentences have no denotation. But this simple and seemingly plausible decision was to have enormous consequences as Quine developed his system.

When Quine's Junior Fellowship ended, he received a three year appointment as an instructor in the Philosophy Department at Harvard. Mathematical logic was still regarded as an esoteric field within philosophy, even though it was well established at Harvard, but the times were changing. In 1935, the Association for Symbolic Logic was founded, and in 1936 the *Journal of Symbolic Logic* was established with Alonzo Church of Princeton as its editor. Church, whose famous proof that there can be no mechanical decision procedure for quantification theory appeared that year, was the foremost American logician. He had studied under Hilbert, and was a major power in the new movement. He was also a very successful teacher. Among his more distinguished students were Barclay Rosser, Stephen Kleene, and Allen Turing, the inventor of the Turing machine that played an important part in the development of machine computing. Quine was of course an enthusiastic supporter of this movement, and did yeoman service as a reviewer for the *Journal of Symbolic Logic*.[130]

Quine also had to adjust to two new roles: those of a father and of a faculty member. His daughter Elizabeth was born in the early fall of 1935. The delivery was difficult, and took place in the midst of their move from one apartment to another – not ideal planning, but they all survived it.[131] As a new faculty member, Quine was fortunate in being allowed to teach courses in his specialty instead of being required to teach the survey course on the history of philosophy and similar fare that are the lot of most newly minted instructors. In 1937 he was even allowed to teach a course on Logical Positivism.[132] Lewis was chairman of the Philosophy Department from 1937 through the first half of 1939, and was probably instrumental in allowing Quine to specialize. Also, Lewis got him to join the American Philosophical Association. He also joined the Harvard Teachers Union, but he soon dropped out of it, "unsympathetic to the intense and strident politics" he found there.[133] This result was not surprising. Quine was politically conservative, and the Teacher's Union was the scene of some very bitter wrangling between the left wing radicals and the conservatives. This was a time when war was in the air; Fascism was on the march in Europe, England and France seemed paralyzed, and many pined their hopes on the Soviet Union, only to be bitterly disappointed.

In 1936 Quine published seven articles in addition to "Truth by Convention". One of these was an article dealing with Schoenfinkel's method of developing logic without variables.[134] Quine shows that Schoenfinkel's "unitary operators" can be reconstrued in terms of ordinary logic. While nothing Quine says about Schoenfinkel's system is either negative or positive, it is clear that he was intrigued by the idea of a system that eliminated variables.[135] Quine was later to revisit Schoenfinkel's work in a somewhat more positive vein.

Quine's work did not go unnoticed. During his years as a Junior Fellow, he received offers from Oberlin, Michigan, and Princeton. He turned them down, wanting to complete his fellowship and to enjoy the freedom that it gave him, and also because he was promised that on the completion of his fellowship, he could count on "something" at Harvard, which turned out to be a three year appointment as an Instructor.[136] To have passed up such opportunities, even in 1936 when the Great Depression seemed to be waning, showed an extraordinary self-confidence on his part.

What shows particularly in Quine's articles is his dislike of Russell's theory of types. In PM, Russell proposed the theory of types as a way of eliminating not only the paradox he had discovered but a raft of paradoxes that he enumerated. But the type hierarchy was encumbered with a hierarchy of orders that differentiated propositional functions within types.[137] In 1926, Frank Ramsey separated the strictly logical paradoxes, such as Russell's, the Burali-Forti paradox, Cantor's paradox, etc., from the semantic paradoxes such as the lair paradox, and urged that the latter category be dropped from the theory of types, thus eliminating the hierarchy of orders and greatly simplifying type theory.[138] In 1936, Quine published an article entitled the "Axiom of Reducibility."[139] This axiom had been introduced by Russell to offset the complexity of his theory of types. Quine argues that the axiom of reducibility, together with what he called "the partial extensionality principle for propositional functions" amounted to adopting Ramsey's proposed simplification.

In a more extended article on "Set-Theoretical Foundations for Logic,"[140] Quine suggested an amplification of Zermelo's system. One of the great attractions of Zermelo's theory was that it offered an alternative to the theory of types yet avoided the paradoxes. The axiom that led to Russell's paradox was the axiom of abstraction

$$(\exists y)(w)(w \in y \equiv Fw)$$

from which we have

$$(w)(w \in x \equiv Fw)$$

$$x \in x \equiv Fx$$

$$x \in x \equiv -(x \in x)$$

$$[(x \in x) \supset -(x \in x)] \cdot [-(x \in x) \supset (x \in x)]$$

$$[-(x \in x) \vee -(x \in x)] \cdot [--(x \in x) \vee (x \in x)]$$

$$-(x \in x) \cdot (x \in x)$$

$$(\exists z)[(z \in z) \cdot -(z \in z)]$$

Zermelo blocked this derivation by substituting for the axiom of abstraction the axiom of separation. In its modern form due to Skolem, the axiom is

$$(\exists y)(x)(x \in y \equiv. \ x \in z \cdot \ Fx).$$[141]

In developing his system, Zermelo had presupposed standard logic. Quine wanted "to frame the Zermelo system as a deductive system in its own right"[142] by including standard logic within it. This led him to create a system of his own, which he called "Γ" and which he said is similar to Zermelo's but "more economical."[143] He took for primitives truth functions, quantification, and membership – what he called the "neo-classical primitive notation that Tarski and Godel had settled on in 1931."[144] He adds a revised form of the axiom of separation "$(\exists y)(x)(x \in y) \equiv ((y \subset z) \cdot P)$" where "P" replaces "Fx" and "$(y \subset z)$" replaces "$(y \in z)$." He also adds the axiom of extensionality and truth functions and quantification, but he does not add the axiom of infinity or the multiplicative axiom, preferring to treat them as hypotheses to be invoked when necessary. The system "Γ" has the peculiarity that it contains no null class and no types. But Quine then shows that from " Γ" he can derive Tarski's delta system which he identifies with "standard logic," and which brings both types and the null class back again. It also brings the rule that, if the variables carry indices indicating type, then "the elementary Δ -formulae are found by replacing □x□ and □y□ in □(x∈y)□ by Δ-variables whose indices are consecutive and ascending."[145] This device was to become Quine's method of stratification. And it also permits the identification of individuals with their unit classes, which was also to become standard for Quine.[146] What Quine is doing here is exploring alternatives to type theory that will still avoid the paradoxes.

In 1935, Quine was writing a paper entitled "Toward a Calculus of Concepts" that was published the next year.[147] In a notebook, he reflected on the question of what the nouns and sentences of the calculus denoted. Even if sentences are taken as denoting truth values (in the manner of Frege) and nouns to denote classes and relations, they should not be regarded as denoting anything real.[148] Singular terms may denote concrete entities, and general terms like "cat" can be taken distributively as referring to each cat. "The ontology on which the conceptual calculus may be regarded as ultimately based comprises concrete individuals, better, simply concrete objects, which is all I envision for an individual." This, Quine remarks, is "nothing more nor less than a logical validation of nominalism."[149] Does this mean that Quine was a nominalist? Not exactly. Quine would have liked to be a nominalist, but he was well aware that doing so would cost much of mathematics, since the classes of mathematics, in terms of which numbers are defined, are abstract objects. He thought that all the known ways of dealing with the paradoxes were ad hoc and unintuitive, and he would have been happy to abandon set theory, had he been able to do so without giving up so much of mathematics. But why did Quine want to be a nominalist? The reason, I believe, is that he thought the paradoxes of set theory called the entire theory into question. Quine had a strong desire for simplicity and elegance. He found the paradoxes offensive, and all of the known ways of avoiding the paradoxes he considered ad hoc and unseemly. He would, I think, have been happy if he could do without set theory entirely, but he saw mathematics as too vital for science to be abandoned. This was a problem that would occupy him for a decade at least.

In a 1937 lecture to the Harvard Philosophy Club, Quine made it clear that the price to be paid for nominalism was very dear. "Nominalism, then, in any sense such as has here been considered, is incompatible with ordinary logic and mathematics; possible only if we are prepared for the intuitionist sacrifices."[150]

In 1936, Carnap was invited by Harvard University to the celebration of the university's 350th anniversary. Quine of course played a significant role in arranging this, and given the situation in Europe Carnap was glad to accept. That year the American Philosophical Association met in Baltimore. Nelson Goodman, then a student of Quine's as well as a Boston art dealer, borrowed his parent's car, and he, Carnap, Quine, and David Prall drove to the meeting. It was on the way back from this meeting that Carnap became aware that Goodman was working on a project that paralleled the *Aufbau* in many respects, and which would later issue in *The Structure of Appearance*. The Carnaps did not return to Europe; he was offered a position at the University of Chicago and later became a U.S. citizen.[151]

In May of 1937, the Quines' second child was born. She was named Norma, and apparently the delivery was uneventful, since a few weeks later Elizabeth and Norma were left with Naomi's aunt while their parents took off with Naomi's sister and her boyfriend on a motor trip through the southwest.[152] Quine's passion for travel was a lifelong addiction that never abated.

In 1937, Quine published an article entitled "Logic Based on Inclusion and Abstraction."[153] The point was to show that, taking inclusion and abstraction as primitive, he could develop the whole of logic. He constructs the system, defining the usual logical constants including membership, so class theory is also a part of the system. One interesting detail is Quine's introduction of the notational device of corners. Thus $\ulcorner(\zeta \subset \eta)\urcorner$ is the "expression which is formed by combining the terms 'ζ' and 'η', whatever they may be, within corners." "In general, an expression beginning and ending in corners is to denote the expression which we obtain, from the expression between the corners, by replacing all Greek letters by the expressions which those Greek letters are intended to denote."[154] This is I believe the first appearance of this notation in Quine's writing.

Also in 1937, Quine published an article entitled "New Foundations for Mathematical Logic."[155] This was another attempt to formulate a logical system of his own that would be superior to its competitors and establish his position in the field. The system was based on three primitives: the Sheffer stroke, universal quantification, and membership. All of the standard truth-functional constants are defined in terms of the Sheffer stroke. The existential quantifier is defined in terms of the universal quantifier. He also introduces identity by "$(\alpha = \beta)$" for "$(\gamma)((\alpha \epsilon \gamma) \supset (\beta \epsilon \gamma)))$" so "x = y" means "x" and "y" must belong to exactly the same classes, including unit classes. He introduces the iota operator; his theory of descriptions is a somewhat modified version of Russell's. Relations are defined as classes of ordered pairs, following the Wiener-Kuratowski definition, and classes and relational abstracts are defined. Quine further introduces the requirement of stratification: it must be possible to put numerals for the variables in such a way that "ϵ" comes to occur only in contexts of the form "n ϵ n + 1". The sole postulate is

$$((x \subset y) \supset ((y \subset x) \supset (x = y)))^{156}$$

He also states three rules and two inference rules. If "/" is the Sheffer stroke, and "φ," "ψ", "χ" and "ώ" stand for arbitrary formulae, then the first rule as Quine gives it is

R.1 $((\varphi/(\psi/\chi))/((\acute{\omega} \supset \acute{\omega})/((\acute{\omega}/\psi) \supset (\varphi/\acute{\omega}))))$ is a theorem

which, as Quine says, answers to Lukaseiwicz's reduced form of Nicod's postulate, i.e.,

$$((\varphi/(\psi/\chi))/((\acute{\omega}/(\acute{\omega}/\acute{\omega}))/((\acute{\omega}/\psi)/((\varphi/\acute{\omega})/(\varphi/\acute{\omega})))))$$

since '$(\acute{\omega}/(\acute{\omega}/\acute{\omega})) = \acute{\omega} \supset \acute{\omega}$" and "$((\varphi/\acute{\omega})/(\varphi/\acute{\omega}))$" reduces to " $(\varphi/\acute{\omega})$."

R.2 If ψ is like φ except that β occurs in ψ as a free variable whenever α occurs in φ as a free variable, then $((\alpha)\ \varphi \supset \psi)$ is a theorem.

R.3 If "x" does not occur in φ, $(\exists x)(y)(y \in x \equiv \varphi)$ is a theorem.

R.4 If φ and $(\varphi/(\varphi/\chi))$ are theorems, χ is a theorem.

R.5 If $(\varphi \supset \psi)$ is a theorem, and α is not a free variable in φ, then $(\varphi \supset (\alpha)\ \varphi)$ is a theorem.

Rule 3 of course would lead straight to Russell's paradox if it were not for the requirement of stratification. With stratification R.3 becomes

R.3′ If φ is stratified and does not contain "x", $(\exists x)(y)((y \in x) \equiv \varphi)$ is a theorem.

The effect of stratification is to render the whole theory of types superfluous. Variables can then have unrestricted ranges that are not limited to specific types. There is only one null class instead of a different one for each type, and similarly there is only one universal class, and although the existence of some classes corresponding to unstratified formulae can be proven, the existence of classes which will produces the paradoxes cannot be proven.

"New Foundations" was to have a long and controversial history. One shortcoming of the theory was pointed out by Quine the year it appeared. In "New Foundations," Cantor's Theorem cannot be derived. It is possible to show that if "x" is a class, then the class of subsets of "x" is larger than "x", but the theorem that for any class of "n" members, the cardinality of its power set is "2^n" and that "$2^n > n$," for all values of "n," cannot be derived. On the other hand, in a 1939 article, Rosser found no evidence that the system led to contradiction. He did not prove "New Foundation" consistent, but he did say "All efforts to produce relations, which are unstratified in such a way as to enable one to prove either the Russell paradox, the Cantor paradox, or the Burali-Forti paradox, have so far failed."[158] Moreover, in his 1953 work *Logic for Mathematicians*, Rosser adopted Quine's method of stratification from "New Foundations."[159] Specker proved in 1953 that the axiom of choice is not consistent with NF, but that the axiom of infinity can be derived in the system.[160] Subsequently, Jensen proved that by a slight change in the Extensionality principle for NF, yielding a modified system NFU, he could prove that NFU was consistent; further, it remained consistent when the axiom of choice was added, even though that axiom is not consistent with NF.[161] Further, the axiom of infinity, which Specker had shown followed in NF, is independent of NFU.[162] Thus research on NF has continued over the years, but not, it is important to note, by Quine himself.

But in 1938, after NF was published, and he had shown that Cantor's theorem could not be derived in NF,[163] Quine was led to reconsider the possibility of nominalism.[164] Cantor's theorem is a prolific generator of transfinite sets, though it is not the only way that Cantor had for generating such sets. In the lecture he gave for the Congress for the Unity of Science at Harvard that year, Quine took a stronger stand. The issue of nominalism is an issue of ontology. Its resolution requires a clear criterion of what constitutes an ontological commitment. It is here that Quine first gave his famous dictum "To be is to be the value of a variable." (The dictum appears in the portion of the lecture that was published as a "Logistical Approach to the Ontological Problem.") Quine now frames the problem as follows.

> If, as is likely, it turns out that fragments of classical mathematics must be sacrificed under all such [nominalistic] constructions, still one resort remains to the nominalist: he may undertake to show that these recalcitrant fragments are inessential to science.[165]

One can see here how Quine's emphasis on the importance of ontology grew out of his desire to embrace nominalism, and what he believed would be required if he was to do so.

Meanwhile, the situation in Europe grew ever more grim. The Munich agreement in September of 1938 left many wondering if there was any hope of stopping Hitler; whether all of Europe would simply capitulate to the Nazis. In November of 1938, Carnap wrote to Quine

> The mess in Europe is incredible. In the last days of September it looked as though Hitler was due for a showdown and the forces of democracy still had a good chance; but now the hope of frustrating Hitler's world domination begins to look like wishful thinking.[166]

All those who, like Quine, had friends in Czechoslovakia and Poland were desperate to get them out of Hitler's reach before they were killed. The mood in Europe and among those in the United States who had a realistic view of the Nazi menace, was one of deep apprehension and dismay. (The mood of the times is well reflected in Vincent Sheehan's *Not Peace but a Sword.*)

The next year (1938), Quine was given a full term off with pay. Naomi had been ill with bronchitis during the summer, and they decided to spend the leave in a warmer climate. They chose Ponta Delgada in the Azores. Quine was working on his book on mathematical logic; the warm climate suited his wife and the two little girls, and Quine also began learning Portuguese. That linguistic acquisition was to have consequences.[167]

In 1938, Quine published a paper entitled "On The Theory of Types"[168] which made explicit his reasons for wanting to eliminate the theory. As Quine viewed it, the theory of types contained two aspects: "we must distinguish between the *metaphysical* or *ontological* aspect of this theory and the *metalogical* or *formal* aspect." Viewed from the metaphysical perspective, type theory imposed a hierarchical ordering upon all classes that involves an infinite reduplication of all logically definable classes, while requiring all members of a class to be alike with respect to type. From the metalogical point of view, the theory denies meaningfulness to relations of membership that are not between objects of "consecutively ascending types," and to all relations of identity and inclusion where the subjects involved are not of the same type. The metaphysical aspect of type theory Quine rejected outright as irrelevant to logic. But the metalogical aspect requires retention in some form if the paradoxes are to be avoided. Quine takes as primitives abstraction and inclusion, which he had shown in his article "Logic based on Inclusion and Abstraction"[169] the year before provided adequate primitives for the construction of logic. But even the metalogical aspect involved problems. Quine points out theorems of arithmetic that are only partially provable in type theory. Similarly, if a term is held to be meaningless, then formulae in which it occurs are meaningless even though they concern inclusion among genuine classes.[170] Further, formulae may be rejected as meaningless even though they have substitution instances that are meaningful. And there are further problems as well. Quine's remedy is to assign numbers to all terms, to limit inclusion to cases where the classes involved have like numerical assignments, and to require that whatever the number assigned to an abstract, the next lower number is assigned to each free occurrence of the variable in each occurrence of the abstract. There cannot then be a meaningless truth function of meaningful terms,[171] and the anomalies of substitution also disappear. What Quine's procedure amounts to is making *stratification* a requirement. But even that he regards as too restrictive. He is willing to admit as meaningful some unstratified terms, e.g., " $V \epsilon V$," but believes that his rule of inference will not lead to trouble.

Quine's use of stratification seemed to have solved the problem of avoiding the paradoxes – at least for the moment. But Quine disliked it, as he did type theory; he considered such devices as ad hoc, unintuitive, and inelegant. His passion for simplicity and elegance was offended by such devices, yet there seemed to be no other way to avoid the paradoxes. Russell, Quine said, followed common sense in taking every term to determine a class, and then had to admit ad hoc procedures to eliminate the ones that led to paradoxes; Zermelo followed the opposite course by admitting only the classes that caused no trouble. Neither solution suited Quine.

There are several further points to be made here. Throughout this and his earlier papers, Quine has taken "ϵ" as a symbol of logic. He had not yet reached the view that the term "logic" should be restricted to a first order calculus. Also evident here is Quine's passion for simplification; he had always considered Russell's type theory overly complex and cumbersome. His antimetaphysical and nominalistic bias accounted for some of his hostility toward it but even in the metalogical aspect he wanted a simplified version, which he hoped he had provided in "New Foundations."

In 1939, Quine's three year appointment as Instructor was up; given his production of publications, he hoped for promotion. But Harvard decided at that point to extend the Instructorship term to five years, which left him still an instructor. He was not happy about it.[172] Even in 1937, he had written to Skinner

> Day after day I yearn for *offers* – I haven't had one since my second year in the Society of Fellows. I don't want an offer merely as a means to jack them up here – the way I feel now a good job outside would be extremely welcome. Things contribute to this feeling: heavy teaching load, poverty, lack of interest in my field on the part of the rest of the department, and lack of good friends and intellectual conversation outside the department. I think our department chairman, Lewis, is something of an obstacle ... It is significant that of those institutions that approached me in 1934-35, *none* filled their openings until last spring; and one of the places at least, Michigan, was known to have written Lewis last spring unofficially reexploring their prospects of getting me. But none approached me again; two already of those jobs, Michigan included, have now been filled, and the report has seeped back to me through two channels that "Michigan wanted me but I wouldn't leave Harvard."[173]

Quine clearly felt ill-used by Harvard and by Lewis and was asking friends for help in finding another job. He understood how the academic game is played; nobody loves you until somebody else loves you. Outside offers are one of the main routes to promotion.

In August of 1939, Harvard hosted the fifth International Congress for the Unity of Science, which was basically the Vienna Circle, plus some kindred spirits such as J. H. Woodger, Earnest Nagel, and Charles Morris.[174] This was the device used to get Tarski out of Poland, and just in time. Harvard gave him "a meager makeshift appointment" until something better could be found. In 1940–1941, Russell came as William James Lecturer, and Carnap was a visiting professor for the year. The collection of star power was impressive.[175] But this was also the occasion of a major confrontation over analyticity. Carnap delivered a lecture that brought the issue sharply into focus.

> My main thesis was that mathematics has no factual content, and, therefore, is not in need of empirical confirmation, but that it nevertheless has a very important function in empirical science as an instrument of deduction. I thought that this was an old story and at any rate a purely academic question. But to my great surprise, the audience responded with vehement emotions. Even before I had finished my lecture, excited objections were raised. Afterwards we had a long and heated discussion in which several people often talked at the same time. Richard von Mesis stated bluntly that the sentence "2 + 2 = 4" (if taken not as a theorem in an uninterpreted axiom system, but in its customary interpretation) was just as much of an empirical nature as the sentence "Solid bodies expand when heated." I thought: are we now back with John Stuart Mill? The attacks by Tarski and Quine were even more spirited, but also more disconcerting.[176]

Tarski's view was not Quine's, but the fact that Tarski rejected the analytic/synthetic distinction is important. I have already suggested that Quine's views on this matter owed much to Lewis, but he may well have been influenced also by Tarski. Quine regarded Tarski as a friend,[177] and he and Carnap made strenuous efforts to find an appropriate position for him here in the United States. In February of 1951, Quine wrote to the publisher Malone, "Tarski ranks with Godel as one of the two greatest living logicians, and he is a very distinguished mathematician on other counts as well."[178] Tarski's views on the analytic/synthetic distinction are probably best laid out in his letter to Morton White on Sept. 23, 1944,[179] but, as John Corcoran has pointed out to me, they also appear in his 1935 paper in Vienna that was published the next year under the title of "On The Concept of Logical Consequence"[180] where he wrote

> Underlying our whole construction is the division of all terms of the language discussed into logical and extra-logical. This dividion is certainly not quite arbitrary. If, for example, we were to include among the extra-logical signs the implication sign, or the universal quantifier, then our definition of the concept of consequence would lead to results which obviously contradict ordinary usage. On the other hand, no objective grounds are known to me which permit us to draw a sharp boundary between the two groups of terms. It seems to be possible to include among logical terms some which are usually regarded by logicians as extra-logical without running into consequences which stand in sharp contrast to ordinary usage. In the extreme case we could regard all terms of the language as logical.[181]

One should also note Carnap's record of his discussions with Tarski in the early 1930s.[182]

There is no suggestion here that Quine copied his ideas from others. In the case of Lewis, Quine absorbed Lewis's theory of the pragmatic a priori from his courses with Lewis, and developed it in his own way into a behavioral criterion. No doubt he and Tarski discussed the issue of analyticity, and Quine certainly found corroboration of his views in their similarity to Tarski's. On the other hand, Tarski described his own position as similar to that of John Stuart Mill, which is not how Quine saw his position. Nevertheless, having the support of one of the two greatest living logicians must have bolstered his conviction that the analytic/synthetic division was untenable.

But why did Quine reject analyticity? The standard notion of analyticity was that an analytic statement was one in which the meaning of the subject term includes the meaning of the predicate; this is essentially Kant's notion, and it was also the view that Carnap held. On this view, there was a sharp line between analytic and synthetic statements. But on the view that Quine had derived from Lewis – that those statements are analytic that we are least willing to surrender – no such sharp line exists. And we have already seen in his first lecture on Carnap that Quine trivialized the notion of analyticity by holding that any statement can be made analytic by redefinition. Here of course Carnap and Quine talked past each other since they were using different definitions of analyticity.

In 1940–1941, Quine published two books: *Mathematical Logic* and *Elementary Logic*[183] They require separate discussion. Quine had often complained about the lack of suitable textbooks for a beginning logic course; *Elementary Logic* was intended to fill that need. It deals with the statement (propositional) calculus and quantification, with only some brief remarks about membership and classes at the very end. Quine's principal concern in this book appears to be teaching students how to translate ordinary English into a logical language. He takes as primitives for the statement calculus conjunction and negation; no other statement connectives are used in the logical language, and where he needs them in his metalogical discussions, they are given in terms of conjunction and negation only. The standard symbols for disjunction, if…then, equivalence, and identity nowhere occur. In dealing with quantification, he introduces only the existential quantifier "∃" and the universal quantifier appears only as " –∃– ". Granted that this keeps his primitives down to three, it still makes for some very complicated expressions. He introduces as usual some new terminology. The term "frame" is introduced as follows: "The letters 'p', 'q', 'r', and 's', also with subscripts in the fashion 'p_1', 'p_2', 'q_1' 'q_n', will be called *statement variables*, and these, together with all expressions thence constructible by conjunction and denial, will be called *compositional frames*."[184] The term "matrix" is introduced as "an expression which is not a statement but can be turned into a statement by applying one or more quantifiers."[185] "Frames" are extended to quantification by including "∃" among the operators used in construction, though certain restrictions are placed on the permissible constructions.[186]

Quine does not introduce truth tables. The validity or invalidity of formulae are demonstrated by reduction of the formulae to the frames " –(p · –p)" or "p · –p " respectively. Nor does he use modus ponens. Instead he gives rules for substitution. For example, he gives

I "p · q · –q" is equivalent to "p · –p"[187]

and

II Given that F implies G, we can infer that F is equivalent to the conjunction of F and G[188]

One should note that Quine has no hesitation in speaking of "predicate variables" – e.g., "We call 'f', 'g', etc., themselves predicate variables."[189] Schematic letters are not mentioned in the volume. It may not be amiss to give an example of a proof to indicate how the system works. Quine proposes as a theorem the following.

Given that F implies G and G implies H, we can always conclude – again only by transformations - that F implies H.[190]

The proof is as follows. The hypothesis to be proven is a conditional; it will therefore be true if the consequent is true. Quine's strategy is to assume the consequent as an hypothesis and to show that its denial is contradictory. That is, using Quine's notation, to show that the denial of " –(F · – H)" is contradictory, or equivalently to show that

(F · −H) is equivalent to p · −p.

1.	F · −H	Hyp.
2.	F ⊃ G	Given
3.	F ≡ F · G	2, II
4.	F · G · −H	1, 3, II
5.	G ⊃ H	Given
6.	G = G · H	5, II
7.	F · G · H · −H	4, II
8.	p · −p	7, I

Quine does not symbolize equivalence, as I have done here, but states it in English. All the proofs follow this general design, using substitution of equivalents.

At the end of the book, Quine briefly mentions membership and classes. He remarks that "ϵ" can be thought of as a symbol of logic or as an extralogical symbol of mathematics. If the one course is followed, then logic includes mathematics and "this book is limited to what we may call *elementary logic*. According to the other version logic stops short of mathematics and this is a general work on logic. The question is wholly one of terminology, and we may leave it undecided."[191] But in fact Quine chose to entitle the book *Elementary Logic*; that would seem to indicate that in 1940, he still held to the logistic thesis.

Why did Quine choose this peculiar presentation of logic? What he tells us in his autobiography is this.

> We saw reasons a few pages back for minimizing logical notation, as was my way in *Elementary Logic* .... But there are reasons also to the contrary, if certain algorithms are to be used. If we are to exploit duality, we need alternation [disjunction] as the dual of conjunction and we need both kinds of quantification. If we are to use alternational or conjunctional normal forms, in which the negation signs are driven inward so as to govern only atoms, again we need both alternation and conjunction. If we are to prove implications by proving chains of conditionals, then a conditional sign is perspicuous. In my mimeographs of 1946 and 1948, for such reasons, I reverted to the redundant notation of *Mathematical Logic* and other literature.[192]

The reasons given led Quine to alter his approach in his revised versions a few years later, yet one wonders why he did not follow them in 1941. So far as the logical constants are concerned, PM and his own *Mathematical Logic* were models that he chose not to follow. Tarski's *Introduction to Logic*, published in 1941 defined the constants straight away[193] and used truth tables. Grentzen's work appeared in 1934, and Quine surely knew it but chose not to use it in 1941. *Elementary Logic* remains an odd book. One doubts if Quine's insistence on the economy of primitives proved very effective in instruction. Furthermore, the book provoked some adverse comment from others. Carl Hempel wrote Quine "I do not think this is really an approach suitable for beginners.... It is somehow like introducing high school students who never before heard about number variables, negative numbers, etc., to the study of algebra by means of an extremely elegant condensed presentation of abstract algebra."[194] Hempel thought it far too difficult for an elementary text; Hempel was right. Quine's teaching load usually involved an undergraduate course on logic and a graduate seminar. He was not a particularly good undergraduate teacher; he disliked expository teaching[195] and was not an exciting lecturer. Moreover, he liked teaching technical material more than non-technical.[196] He was, however, much better in teaching a subject such as set theory. But he was a brilliant teacher of graduate students. The list of graduate students that studied with him is a distinguished collection of the next generation of logicians and empirical philosophers, and includes George Berry, William Craig, Donald Davidson, Burton Dreben, Dagfin Follesdal, Nelson Goodman, Jakko Hintikka, Henry Hiz, Saul Kripke, Hughes Leblanc, John Myhill, Charles Parsons, Barry Stroud, Hao Wang, and Morton White, among others. As of 1940, he had not adjusted his sights to an undergraduate audience.

In 1940, Quine published *Mathematical Logic*[197] (hereafter ML). This was to be his major work in logic, though he changed and refined it in subsequent editions. In order to define what logic is, Quine specifies its vocabulary – the standard logical constants. Thus for example any statement of the form

If every -------is--------and---------is a --------then--------- is ----------

is true no matter what (categorematic) terms are substituted for the blanks, so long as the first and fourth are filled alike, the second and last are filled alike, and the third and fifth similarly. The logical terms here occur essentially, in that any variation of them is likely to lead to falsehood, whereas the ones substituted for the blanks occur vacuously – the truth of the statement does not depend on them. All statements true by virtue of their skeleton of logical particles are true statements of logic. But Quine is quick to point out that if the vacuous terms were replaced by terms of geology, or any other science, logic would thereby be incorporated into those sciences. Logic is, he says, "the common denominator of the special sciences."

This however is not a sufficient description of logic. Quine includes also metalogic, the study of logic. And he further holds that "mathematics reduces to logic," so that mathematics itself and metamathematics fall within the domain of logic. The future, Quine holds, will provide many applications of logic in the sciences. This has been true of mathematics, particularly where measurement is possible, but "to the scientist longing for non-quantitative techniques, then, mathematical logic brings hope." Quine stays short of saying that logic is a science itself, but he comes close; later he would go all the way.

In this work, Quine has no hesitation about introducing the full panoply of logical constants. Conjunction, if–then, negation, alternation (which is Quine's name for disjunction) and the biconditional are all defined and the truth table method of testing logical formulae is extensively used. Having defined truth-functions, Quine remarks that all statement composition is truth-functional. In discussing subjunctive conditionals, Quine uses the term "disposition," in the sense that applies to terms such as "soluble," "malleable," "hard," "sensitive," "intelligent." etc. Of course he places great emphasis on the distinction between use and mention, and the problems that the confusion of the two has caused. In an extended footnote on pages 31–33, he criticizes Whitehead and Russell for their confusion of the material conditional with implication, and takes Lewis' strict implication as the product of the same confusion – a change Lewis denied.[198] It should be noted that here and elsewhere, Quine has no hesitation is talking about the meaning of an expression.[199]

Having introduced the standard logical constants, Quine introduces the Sheffer stroke function and shows how negation, conjunction, alternation, the conditional, and the biconditional can be defined in terms of it. He then adds one of his famous dicta: "to define a sign is to show how to avoid it." But unlike his practice in EL, he continues to use the standard constants, taking them to be shorthand symbols for the elaborate formulae in the Shefffer notation. He also introduces Wittgenstein's term "tautology." A statement is tautological if it proves true by the truth-table method for every assignment of truth values to its ultimate truth functional components.[200]

Quine then shows that conjunction, alternation, and the biconditional are commutative, associative, and transitive. Conjunction and alternation are idempotent. Conjunction is distributive into itself and alternation; alternation and the conditional are distributive into conjunction, alternation, the conditional, and the biconditional. He then gives DeMorgan's laws – that

$$-(\varphi \cdot \psi) = -\varphi \vee -\psi$$
$$-(\varphi \vee \psi) \equiv -\varphi \cdot -\psi$$

Similarly, he gives the law of transposition

$$\varphi \supset \psi \ .\!\equiv\!. \ -\psi \supset -\varphi$$

and of double denial. $- - \varphi \equiv \varphi$

Quine then turns to quantification. With quantification comes the serious business of logic. Quine introduces the notions of variables, freedom, bondage (complete with his famous diagrams), matrices, formulae, and closure – all clearly explained. Then come the axioms of quantification theory

100. If φ is tautologous, $\vdash \varphi$.
101. $(\alpha)(\beta)\,\varphi \supset (\beta)(\alpha)\,\varphi$
102. $(\alpha)(\varphi \supset \psi) \supset (\alpha)\,\varphi \supset (\alpha)\,\psi$
103. If α is not free in φ, $\vdash \varphi \supset (\alpha)\,\varphi$
104. If φ' is like φ except for containing free occurrences of α' wherever φ contains free occurrences of α, then $\vdash (\alpha)\,\varphi \supset \varphi'$
105. If $\varphi \supset \psi$ and φ are theorems, so is ψ

The concepts of theorem and of metatheorem are introduced. So are those of "ponential," meaning the theorem derived from "$(p \supset q)$" and "p", and of "stacked conditionals," meaning a chain of conditionals in which each antecedent is the consequent of the preceding conditional and is inferred by modus ponens. Next come universal and existential quantification.[201]

Chapter 3 introduces membership and classes. Quine discusses briefly the difference between properties and classes, and opts for classes because the criterion of identity for classes is clearer than that for properties. This of course means the introduction of abstract entities. He comments

> One may prefer to regard abstractions as fictions or manners of speaking; one may hope to find a method whereby all ostensible reference to abstract entities can be explained as mere shorthand for a more basic idiom involving reference only to concrete objects (in some sense or other). Such a nominalistic program presents extreme difficulty, if much of standard mathematics and natural science is to be really analyzed and reduced rather than merely repudiated; however, it is not known to be impossible. If a nominalist theory of this sort should be achieved, we may gladly accept it as the theoretical underpinnings of our present ostensible reference to so-called abstract entities.

But lacking such a nominalistic theory, Quine says that we must admit classes as abstract entitles. That Quine's preference would be for the nominalist theory is obvious; so are his doubts about the possibility of achieving it. Classes therefore remain, and Quine defines "atomic formulae" as having the form " $x \in y$" where "y" is a class. What if "y" is not a class but a concrete individual? Then Quine construes " $x \in y$" as "x = y." The definition of atomic formulae allows the characterization of logical formulae as formulae whose atomic parts are atomic formulae. With the introduction of membership Quine notes that the three primitive notions of joint denial, universal quantification, with its variables, and membership are sufficient for the totality of logic and mathematic Quine then discusses Russell's paradox and points out that its avoidance requires restrictions on what can be a member of what. The course Quine takes is that of von Neumann; he restricts the realm of membership eligible entities, or "elements," by declaring certain classes ineligible for membership in any class. Hence " $(\exists x)(x \in y)$" can mean only that "x" is an element of "y." Using the expression "$\hat{x}$ (---)" for the class of statements meeting the condition "---," Quine gives the definition

$$\text{“}(\beta \in \hat{\alpha}\ \varphi)\text{” for “}(\exists \gamma)(\beta \in \gamma \cdot (\alpha)(\alpha \in \gamma \supset \varphi))\text{”}$$

Identity is then introduced as

$$\text{“}(\zeta = \eta)\text{” for “}(\alpha)(\alpha \in \zeta .\equiv. \alpha \in \eta)\text{”}$$

Quine deals with the problem of identical individuals by redefining "individual" to mean unit class. Thus the concrete entity "x" vanishes into its unit class, allowing all relations to be stated in terms of classes without having to make exceptions for concrete non-classes. The usual theorems of identity are then proven. The occurrence of the variable "x" in the context "$\hat{x}$(---)" binds all occurrence of "x" in "---" unless they are already bound to an occurrence of a quantifier within "---." Thus Quine says, "An abstraction prefix '$\hat{\alpha}$' operates in the same way as a quantifier '(α)', so far as bondage and freedom are concerned."

Quine introduces the universal and null classes with the usual symbols, "V" and "Λ" respectively.[202] He then introduces descriptions, using the PM notation of "(ιx)φx" for the one and only thing that is "φ." Should it be the case that there is nothing that is "φ," then "(ιx)φx = Λ." And if there are two or more things that are "φ," Quine also equates "(ιx)φx" to the null class, rather than considering it meaningless, as is the case in PM.[203] This machinery enables Quine to deal not only with descriptive phrases such as Russell's "author of Waverley" but with names as well. The underlying issue here is what to do with a name that does not denote, such as "Pegasus." Quine's solution is to convert them into descriptive phrases or terms, such as "Pegasizes," and then to nullify them as

$$(\exists x)(x \text{ Pegasizes}) = \Lambda.$$

We thus escape the multitude of primitive names and all the complexities that they would involve.[204]

The addition of membership as a primitive notion requires further axioms, and also a method of testing whether or not one can avoid Russell's paradox, and the related paradoxes of Cantor and Burali-Forte. As the test, Quine introduces stratification, as he had in NF, namely

> Let us speak of a formula as *stratified* if it is possible to put numerals for its variables (the same numeral for all occurrences of the same variable)in such a way that "ϵ" comes to be flanked always by consecutive ascending numerals ("n ϵ n + 1").

By admitting only stratified formulae, Quine believed he had ruled out the possibility of the vicious circle paradoxes.

Quine is then prepared to state his axioms of membership.

200. If φ is stratified and has no free variables beyond $\alpha, \beta_1 \ldots \beta_n$ then $\vdash (\beta_1 \ldots \beta_n \in V .\supset \hat{\alpha}\varphi \in V)$ $|\vdash \hat{\alpha}\varphi \in V$ when $n = 0|$

201. If φ is atomic, and φ' is formed from φ by putting α' for an occurrence of α, then $\vdash(\alpha - \alpha' .\supset. \varphi \supset \varphi')$

202. If β is not α nor free in φ

$\vdash (\exists\beta)(\alpha)(\alpha \in \beta .=. \alpha \in V \cdot \varphi)$

In an extended footnote on pages 163–166, Quine compares ML's method of avoiding the paradoxes to those of Russell, Zermelo, and von Neumann, and to his own NF. The objective of these safeguards is not only to avoidance of contradiction but also simplicity and as near an approximation to common sense as possible.

Quine then defines logical product, logical sum and complement, and proves a number of theorems concerning them, including double negation and DeMorgan's laws. He defines inclusion as "(ζ⊂η)" for "(α)(α∈ζ .⊃. α∈η)" where α is "foreign" to "ζ" and "η". Quine carefully distinguishes inclusion from membership. Given his extensional views, he gives

$$(x)(y)(x = y . \equiv . x \subset y \cdot y \subset x)$$

Mutual inclusion of classes therefore yields identity. Quine also defines unit classes, using iota, as "ι ζ" for "$\hat{\alpha}(\alpha = \zeta)$." If "y" is a member of "ι x" and an element, then "x = y."[205]

In the next chapter Quine takes up relations. Here too Quine imposes the restriction that relations are to hold only among elements, since otherwise contradictions can be derived. Quine uses the Wiener-Kuratowski theory of relations as classes of ordered pairs. He symbolizes such pairs as "ζ; η", understood to mean that the sole members are "ι x" and "ι x ∪ ι y." The two members must be distinct since the relation is a class of ordered pairs. Two pairs, "x;y" and "z;w", are identical only if for all "w", "x", "y", and "z",

$$(w)(x)(y)(z)\ [x, y, w \in V . \supset :\ x; y = z; w\ . \equiv .\ x = z \cdot y = w]$$

Ordered pairs are of course dyadic. A triadic relation among "x", "y", and "z" is defined as a dyadic relation of "ι ιy" to "z;w."

Relations, being classes of pairs, admit of inclusion, and similarly sum and product apply to relations as does abstraction, and all relations have a converse, so converse "x" is symbolized "$\breve{x}$". Quine coins the term "projection" of a class "y" by a relation "x", symbolized by "x " y," as the class of all the elements that bear "x" to one or more members of "y." The relative product of the relations "ζ" and "η" is

$$\text{“}(\zeta \mid \eta)\text{” for “}\hat{\alpha}\ \hat{\gamma}\ (\exists\beta)(\zeta(\alpha, \beta) \cdot \eta(\beta, \gamma))\text{”}$$

The relative product of "brother of" and "father of" is "uncle of."

Quine then takes up the ancestral. He first defines "closed" for a class with respect to a relation as "x " y ⊂ y" – i.e., whatever bears "x" to a member of "y" belongs to "y". The class of even numbers is closed with respect to the relation "square of" since all squares of even numbers are even numbers. The name "ancestral" comes from the fact that, if we regard a person as one of his own ancestors, then that person and all his ancestors form a class closed with respect to the relation "ancestor of." Quine symbolizes the ancestral of a relation "x" as "*x." The great importance of the ancestral lies in its role in defining successor.

Functions then occupy Quine. Functional application is defined as "(ζ ' η)" for "(ι α) ζ (α, η)" – i.e., a function is a relation such that a unique element α bears the relation "ζ" to "η". The "x of y" is given by

$$(\exists w)(z)(z = w. \equiv x(z, y))$$

The class of elements "y" with respect to which "x" is a function is termed the range of functionality of "x" and is symbolized by "rx".

$$\text{"r }\zeta\text{" for "}\hat{\beta}\ \ (\exists\gamma)(\alpha)(\alpha = \gamma. \equiv \zeta(\alpha, \beta))\text{"}$$

Thus Quine says "the range of functionality of the relation *auther of e.g.,* is the class of his uncollaborated writings." Quine also points out that there are functions that apply to more than one argument at a time; his example is the power function which, for arguments "x" and "y", gives "x^y" Like other relations, functions admit of abstraction, symbolized by "λ." The "double of" is "$\lambda_x(2 \times x)$." The formal definition is

$$\text{"}\lambda_\alpha\zeta\text{" for "}\hat{\beta}\ \hat{\alpha}\ \ (\beta = \zeta)\text{"}$$

With this machinery, Quine defines the identity function as

$$\text{"I" for "}\lambda_x x\text{"}$$

This is the function whose value for any element as an argument is that element itself. "I" is included in every ancestral. Membership is also a relation, which is defined as

$$\text{"€" for "}\hat{x}\,\hat{y}\,(x \in y)\text{"}$$

"€" holds only among elements.[206]

Quine now turns to number. He defines "0" as "ι Λ" He can then define "1" as "$\hat{x}\,(\exists y)(y \subset x \cdot x \cap \overline{\iota y} \subset 0)$". That is, "x" has a member "y", and the intersection of "x" with the complement of the unit class of "y" (i.e., "x" without "y") is "0." Then "2" can be similarly defined as a class x which, if one member is dropped, is 1, and so forth. Thus "1" is the class of all one member classes of elements. "2" is the class of all two member classes, and so forth. These are basically Russell-Whitehead numbers. Generating further numbers requires the relation of successor, "S"

$$\text{"S" for "}\lambda_z\ \hat{x}\,(\exists y)(y \in x \cdot x \cap \overline{\iota y} \in z)\text{"}$$

The class of natural numbers, Nn, can then be defined as 0, S'0, S'(S'0), etc. This can be done by using the ancestral. The class of natural numbers is

$$\text{"Nn" for "(*S“}\iota 0)\text{"}$$

All successors of natural numbers are then natural numbers. This allows Quine to prove the principle of mathematical induction.

If ψ, φ' and φ_o are like φ except for containing free occurrences respectively of ζ, S□α, and "0," wherever φ contains free occurrence of α, then $\vdash (\alpha)(\varphi \supset \varphi') \cdot \varphi_o \cdot \zeta \in Nn . \supset \psi$

Quine then defines "counter sets." If "x" is a class then " $x \cup \iota x$" has one more member than x, namely, its unit class. Clearly, any class can thus be increased by 1 by the addition of its unit class. Quine calls such classes "counter sets". This is a function whose value for any element "x" as argument is " $x \cup \iota x$", and is called Sa

Thus we have

"Sa" for " $\lambda_x (x \cup \iota x)$"

The class of counter sets, "C", comprises "Λ", "Sa□Λ", "Sa□(Sa□Λ)", etc. Hence "C_s" is "(*Sa" ι Λ)" where "*Sa" is the ancestral of "Sa". "C_s" contains one member from each natural number, and thus guarantees that for every natural number, there is a class having that number of elements.

A class is finite if it is a member of the natural numbers. Hence the class of all finite classes, which Quine calls Fin, can be defined as

"Fin" for "(Є"Nn)"

A class is infinite only if it belongs to no natural number. "V" is such a class. Hence where as usual the vinculum signifies the negation of any expression above which it is placed,

$$V \in \overline{Fin}$$

Hence "V" is an infinite class.

The ancestral *x, being a set of pairs, contains the relation "z;w" such that "z" is identical with "w". This relation is "I". Then there are the pairs "z;w" such that "z" bears "x" to "w". Then there are the pairs "z;w" such that "z" bears "x" to something "y" that bears "x" to "w". Clearly, these relative products can be extended to any number of pairs i.e., "(x|x)", "((x|x)|x)," etc. These relations Quine terms the relative powers of "x", and are symbolized by "x^0", "x^1", "x^2" ....So if "x" is the relation "parent of," then "$x^0 = I$" "x^1" is parent of, "x^2" is grandparent of, "x^3" is greatgrandparent of, etc. Clearly, the powers of a relation yield a series of relative powers that can be extended indefinitely.

Counting off is done by pairing the elements "$z_0, z_1, \ldots, z_y$" to the natural numbers from "0" to "y", as for example "$z_0; 0, z_1; 1, \ldots z_y; y$." The relation which any such pair "z_i; i" bears to its successor "$z_{i+1}; i + 1$" is called "δx". Then the power "x^y" is the relation of any element "z" to any element "w" such that "z;0" bears the ancestral "* δ x" to "w;y".

Quine then defines arithmetical addition among the natural numbers by the following theorem.

$$\text{“}(\zeta + \eta)\text{” for “}(S^{\eta} \square \zeta)\text{”}$$

Similarly multiplication,

$$\text{“}(\zeta \times \eta)\text{” for “}(\lambda_{\alpha}(\zeta + \alpha)^{\eta} \square 0)\text{”}$$

He uses the inverted radical as the symbol for arithmetic powers to avoid confusion with the relative powers. I will use the sign " $\lrcorner$ " for this purpose.

$$(x)x^{\lrcorner}0 = 1$$

$$(y)(x)(x \in Nn) \,.\supset.\, x^{\lrcorner}(S \square y) = x \times (x^{\lrcorner} y)$$

When applied to natural numbers, all these operations yield natural numbers. Quine then derives a series of theorems concerning arithmetical identities, such as the commutative, associative, and distributive laws for addition and multiplication. He then introduces rational numbers and real numbers, and proves a series of theorems for each sort.[207]

Quine has at this point accomplished the objective of his book – he has given an exposition of logic and shown that numbers and arithmetic are derivable within that system. But there is another question – one which in his lectures on Carnap he had called "the most famous recent discovery in the foundations of mathematics" – Godel's proof of the incompletability of arithmetic. Quine therefore devotes the final chapter of ML to the problem of incompletability, but he does so in an original form. He begins with syntax. He has shown that logic requires only a small set of primitives – the Sheffer stroke, universal quantification with its variables, and membership. All of these have been formally defined. Quine now introduces names for the signs used in their definition, as follows,

S_1 = "w"
S_2 = "x"
S_3 = "y"
S_4 = "z"
S_5 "–"
S_6 = "("
S_7 = ")"
S_8 = " ↓ "
S_9 = "ϵ"

To these he adds a new symbol for concatenation, for which I will use the symbol "†", so that "(x ∊ y)" can be written as "$S_6 \dagger S_2 \dagger S_9 \dagger S_3 \dagger S_7$." Logical notation, supplemented by these nine signs, provides the means of translating all the preceding theorems and metatheorems into syntax.

But Quine then defines what he calls "protosyntax" which is syntax without S_9 – i.e., without membership. And within protosyntax, he is able to define logical formula, the axioms of quantification theory, matrix, and tautology. He then applies protosyntax to itself – i.e., to protosyntax. The attempt to define "theorem" in protosyntax then leads to the conclusion that there is a protosyntactic statement that "is thus true if and only if it is not a theorem." In other words, protosyntax is protosyntactically incompletable. That leads to the further demonstration that logic is protosyntactically incompletable. Quine remarks

> Apprised only of the protosyntactical incompletability of logic, we should probably have blamed the difficulty on the connective "∊" of membership and hence questioned the admissibility of that connective. The notion of membership is a natural object of suspicion; for it is this notion that imports the whole realm of classes of higher and higher orders of abstractness, and even calls for *ad hoc* measures such as the distinction between element and non-element for the avoidance of contradiction. Protosyntax itself, on the other hand, is wholly independent of the notion of membership and the theory of classes; it calls for no non-elements, indeed no entities whatever beyond an infinite domain of finite expressions each of which is nameable within the notation of protosyntax in systematic fashion ... When protosyntactical incompletability reasserts itself in as simple a field as this, we cease to regard such incompletability as a ground of suspicion and come rather to expect it in every fairly untrivial field.[208]

This paragraph is interesting not only for the incompletability result, but also for what it tells us about Quine's motivation. Quine had a strong nominalist bent; if it had been possible to construct mathematics without classes, he would have been delighted. He accepted the existence of classes because he could see no way to do mathematics without them. But he never showed any great interest in problems such as those of transfinite arithmetic; unlike Godel, Hilbert, Sierpinski, and others, he seems to have found issues such as the continuum hypotheses of little significance. I suspect that he had thought the problem of incompletability might be due to the membership relation and that that is why his investigation took the form it did. But having found that even protosyntax, which does not contain "∊," was protosyntactically incompletable, he dismissed the possibility that incompletability was due to classes and membership and recognized that it was an inevitable feature of what he called any "untrivial field."

Looking back on 1941, Quine wrote, "the free world was collapsing before the onslaught of the Nazis." Then, on December 7, the Japanese attacked Pearl Harbor. Quine wrote "For me, the shock was combined with relief; at last we would get into the war."[209] Quine had been involved for some years in efforts to bring refugees to the United States; Carnap had made it, Tarski had too but only just, arriving here only days before the German invasion of Poland, and there were many others. But of course not all of them did make it. Grelling was shot by French Nazis while trying to escape over the Pyrenes, Wajsberg died in a concentration camp, and a number of others also died at Nazi hands. What is remarkable is how many did escape and reach the Unites States. Some like Reichenbach had to follow devious courses to get here; he taught for a while at the University of Istanbul before managing to reach America.

In the summer of 1941, Quine was promoted to Associate Professor with tenure.[210] He now knew that he could stay at Harvard as long as he wanted; his third book was published and his twenty-third article. All of this was of course very good news. But in September he received a letter from Barkley Rosser informing him that Rosser had been able to derive the Burali-Forti paradox (the paradox of the greatest ordinal) in the system of his ML.[211] Quine remarked that he was surprised, "having taken precautions against the analogous paradox of cardinals. It was less analogous than I thought."[212]

Cesare Burali-Forti discovered the paradox that bears his name in 1895, and it appeared in print 2 years later. As Whitehead and Russell describe it

> The ordinal number of the series of numbers from 0 (including 1) to any ordinal α is $\alpha+1$; hence $\alpha+1$ exists, and is therefore $>\alpha$. But the ordinal α is similar to the segment of the series of ordinals consisting of the predecessors of α, and is therefore less than the ordinal number of all ordinals. Hence the ordinal number of all ordinals is greater than every ordinal and therefore than itself, which is absurd; moreover, though the greatest of all ordinals, it can be increased by the addition of 1, which is again absurd.[213]

The paradox is really a double paradox. First, the ordinals are serial numbers. For any series, if one counts the series from 1 to n, then the number of numbers in the series is n. But if one begins the count with 0 instead of one, then the last number used in the count is n-1 although the series still has n members. Suppose that there is a series that contains *all* ordinals. Let the series of ordinals in the series, counting from 0, be α, so α must be the greatest ordinal. But the number of the series itself is $\alpha+1$, and $\alpha+1 > \alpha$. But since the series from 0 to α contains all ordinals, $\alpha+1$ must be a member of that series. But then it would have to be equal to or less than α, which is a contradiction. Second, any ordinal can be increased by adding 1. Hence there cannot be a greatest ordinal.

The difficulty, as Rosser showed, lay in Quine's axiom *200. The damage was not irreparable, but it did require some important revisions. Quine's immediate response to Rosser's proof was to try to save a much of ML as possible. He undertook to do this in "Element and Number" that appeared in the *Journal of Symbolic Logic* in 1941. The problem in ML, as Rosser had shown, lay in the elementhood axiom *200. Quine commented

> The argument that leads to Cantor's paradox *is* obstructed in [ML]; for the argument depends on showing that $\hat{y}\,(y \in x)$ always has more members than x, and it is obstructed by the fact that $\hat{y}\,(y \in x)$ comprises in general only certain of the subclasses of x, viz. those that are elements. My own inattention to Burali-Forti's paradox came, indeed, of confidence in the parallelism between that paradox and Cantor's.[214]

There were in ML nineteen theorems that depended on *200 for their proofs. Quine's strategy was to delete *200, and then to try to deal with the nineteen by other means. He was able to do this for the first five chapters of the book by introducing the following.

400. $(x)(\iota x \subset V)$
401. $(x)(y)(x, y \subset V .\supset. \bar{x} \cap \bar{y} \subset V)$

By deleting ten of the nineteen consequences of *200 and adding 400 and 401, he could prove the remaining nine, and so keep the first five chapters of the book substantially intact. For the chapter on number, Quine also found it necessary to alter

610. $(z)(z \in V .\supset \hat{x}\, (\exists y)(y \in x \cdot x \cap \overline{\iota y} \in z) \in V)$
614. $(z)(z \in V .\supset. S \,\bar{\underline{\;}}\, z = \hat{x}(\exists y)(y \in x \cdot x \cap \overline{\iota y} \in z))$

to the weaker forms

!610 $(x)(x \in V .\supset. x \cup \iota x \in V)$
!614 $(x)(x \in V .\supset. S'x = x \cup \iota x)$

The weaker axioms !610 and !614 are adequate for the book.[215] These changes were the substance of the Errata slip that Quine put in the later editions of ML during the 1940s.[216] But he went on in the article, "Element and Number," to propose a radical revision of his theory of number. The theory of number in ML is basically the Russell-Whitehead theory of number: the number "n" is the class of all n-membered classes. But Quine had also used what he had called "counter sets." Quine now proposed to delete his previous definition of number and to take the basic idea of the counter sets in its place. When one counts the members of a class, one matches the natural numbers to the class members one to one and the number of numbers used is given by the last number in the count; thus the number of numbers used in counting is "n". But if we take "0" as the first number used in the count of a class of "n" members, the number of numbers *preceding* "n" will be "n", even though the last number used in the count is "n-1". So one may construe the number "n" as the class of all numbers preceding itself. This method was developed by von Neumann, and Quine proposes to adopt it. He uses the successor relation as in !614, "$0 = \Lambda$, $1 = S\Box 0, 2 = S\Box S\Box 0$", and so on, for all finite "n". This method has the advantage of assuring the existence of a class corresponding to any finite "n" – namely, the class of the predecessors of "n", and it identifies the natural numbers with the ordinals.

In ML, the application of natural numbers was by membership; to say that "x" has "n" members is to say that "$x \in n$". In the proposed new theory, application is by correlation; one needs to establish a one to one correspondence between the members of the class being counted and the sequence of natural numbers starting with "0." The cardinal number is then simply the number of ordinals less than "n". A cardinal number, Quine says, "is any ordinal that has more predecessors than has any of its predecessors."[217]

In the infinite, cardinals and ordinals diverge. Although "$\aleph = \omega$" where "$\aleph$" is the cardinal of the natural numbers and "ω" is their ordinal, the successor of "ω" is "$\omega + 1$," whereas "$\aleph + 1 = \aleph$". The series of ordinals does not form a class greater than "$\aleph$" until "ω_1," so "$\aleph_1$" is taken as the cardinal corresponding to "ω_1." The series of transfinite ordinals and the series of transfinite cardinals are both infinite, but they do not correspond one to one. Quine is not here addressing problems in ML as it stood, but outlining what would subsequently become his theory. Meanwhile, he was content to let ML be republished with the Errata slip.

Even so, ML was by no means an ideal system. Rosser had proved that the class of natural numbers cannot be proven in that system to be a set, unless the system is inconsistent. "One must assume that it is a set, in order to found the theory of real numbers but the addition of such an ad hoc postulate is unwelcome technically" Quine remarked.[218] But the great virtue of ML is its clarity and simplicity of exposition. Quine was a gifted expositor and stylist, not only in logic but in other matters as well. Here it is worth noting Hao Wang's comment

> What has struck me for many years and seems to have drawn little attention is the formal perfection of ML. It has often been remarked that in a strictly formal system, a machine should be able to check the proofs. I have seen this goal stated and aimed at. But so far as I know, *ML* is the only extended development which satisfies this stringent requirement.[219]

But, as Quine himself noted, ML is still modeled on PM. Further, Quine did not pursue the correction of ML beyond "Element and Number"; it was not Quine who found and corrected the basic error, but Wang.[220]

Chapter 2
Semantics and Ontology

In 1941, Quine was invited by the Office of the Coordinator for Inter-American Affairs to go to Brazil as a visiting lecturer. Quine accepted, and in May of 1942 he flew south – his first experience of air travel. With his extraordinary gift for languages, and his knowledge of Portuguese, Quine was able to lecture in the local language, and the lectures became a book, *O Sentido da Nova Logica*, that was published in 1944.[1]

Before going to Brazil, Quine contributed an extended essay entitled "Whitehead and the Rise of Modern Logic" to the Library of Living Philosophers volume on Whitehead.[2] Quine began with a very brief survey of developments since the time of Boole, and then moved at once to Whitehead's *Treatise on Universal Algebra* of which he gives a brief overview. He then describes Whitehead's subsequent paper, "Memoir on the Algebra of Symbolic Logic." It was at this time that Whitehead discovered Peano's work, and began working with Russell. In his "The Logic of Relations, Logical Substitution Groups, and Cardinal Numbers," he used Peano's formal notation to formulate Boolean algebra and further explore its possibilities. Whitehead had projected a second volume of the *Universal Algebra*, and Russell had projected a second volume of his *Principles of Mathematics*; neither ever appeared. The collaboration of the two men soon convinced them both that a new work was required. Whitehead remarked, "I believe that the invention of the Peano and Russell symbolism ... forms an epoch in mathematical reasoning." But neither of them recognized how great and how difficult the task before them would be; for eleven years they labored almost continuously. Russell has somewhere remarked that so great was the effort involved that after the completion of PM, he turned away from logic with a feeling of nausea. But the result, Quine says, was that "there appeared, in 1910–1913, one of the great intellectual monuments of all time: the three volumes of *Principia Mathematica*."[3]

Quine treats the three volumes in some detail, and in doing so lays out his criticisms of them. It is these criticisms that are relevant here. Their treatment of the propositional calculus, which Quine prefers to call the "calculus of statements," draws his criticism particularly on the confusion of use and mention. Whitehead and Russell took the material conditional as the *implication* of one statement by another,

M.G. Murphey, *The Development of Quine's Philosophy*, Boston Studies in the Philosophy of Science 291, DOI 10.1007/978-94-007-2424-2_2,

and the biconditional as asserting the *equivalence* of two statements. But Quine argues that "implication" is a relation between the names of statements, whereas the material conditional is a device for statement composition. The same obviously holds for "equivalence" in relation to the biconditional.[4] Quine holds that the truth functional character of the statement calculus is essential to the simplicity and clarity of logic, and that it should not be encumbered by imputing meanings to its contents that belong properly to relations between their names. Further, Quine argues, the schematic letters in logical formulae are not names; they do not refer. To take them as referring is to adopt a theory of attributes.

With respect to quantification theory, Whitehead and Russell generally follow Frege. But the use of "propositional functions" draws Quine's criticism.[5] First, the expressions such as "φx" are taken as predicating "φ" of "x", but the relation of predication itself is never defined in PM. Second, Whitehead and Russell speak of both attributes and classes, with the result that "φx" is taken both as predicating an attribute "φ" of "x" and as marking the membership of "x" in the class "φ". In fact, Quine holds, it is classes that they actually employ in deductions; attributes are a gratuitous distraction that ought not to have been introduced. Third, Whitehead and Russell take predicates such as "φ" as variables for quantification. By doing so, they commit themselves to the existence of attributes, and so to Platonism.[6] Quine also criticized Whitehead and Russell's treatment of relations. They should, he claims, have taken them as classes of ordered pairs, following the example of Peano.

Quine describes the theory of types in its simple and also in its ramified form. Ramsey's work got rid of the ramification, but Quine is no happier with the remaining theory of types. There are other devices that could have been used to block the paradoxes. Quine agrees that some such device is necessary for this purpose, but he remarks "that none of these proposals, type theory included, has any intuitive foundation." So far as the paradoxes are concerned, he says "Common sense is bankrupt, for it wound up in contradiction."[7] But no matter how unintuitive the methods employed, one of them has to be used to avoid falling into contradiction.

In PM, Whitehead and Russell adopt Frege's notion of cardinal number. They also adopt the axiom of infinity.[8] Quine describes briefly the cardinal and ordinal numbers, finite and infinite, and the concept of relation number. Finally, Quine describes the last section on measurement.

Quine's critique of Whitehead's work contains few surprises. He does introduce the argument that predicate letters do not denote but are dummy letters that are parts of schemata. His criticisms of the confusion of use and mention, and the practice of quantifying over predicates as if they were variables are what one would expect from him, and his claim that statements do not refer had been made before. But in an article on Whitehead, it is surprising that he has so little to say about Whitehead's books on natural science that followed soon after the *Principia*. Perhaps Quine thought them irrelevant to logic.

Quine had applied for a commission in the Navy before going to Brazil. In October of 1942, he began his service as a lieutenant; he was assigned to Naval Intelligence in Washington, where he was part of the anti-submarine operation in the Atlantic. His job was translating and analyzing deciphered German U-boat radio messages which were then passed on to the high command and back to cryptanalysists. It was congenial work in which his fluency in German served him well, and his unit included a number of bright and congenial people.[9]

But also during this period his marital difficulties became serious. Quine had married Naomi Clayton when he entered Harvard; they had become engaged at Oberlin. But Naomi turned out to involve problems. One was her college debts, which Quine found himself obliged to pay, at a time when they were just scrapping by financially. The Quines were constantly short of money, and he considered his income "meager." A second was her health. Naomi had serious physical problems which resulted in a number of operations. Just what the problem was Quine does not say, but after one operation he remarked "tumors again." Quine believed she was manic-depressive, and in her manic phases she would buy recklessly – something they could ill afford. Finally, Naomi took the children to Boston, leaving him in Washington. Quine filed for divorce on grounds of desertion. The divorce was bitter; Naomi called his parents, the Philosophy Department, etc., complaining about her treatment. They tried a reconciliation which did not last and finally they were divorced. Meanwhile, Quine was courting Marjorie Boynton, a young woman who served in his Navy department as a Wave. She took a job in Boston teaching nursery school when he went back to civilian life, and in 1948 they were married. (It should be born in mind that this account is based on Quine's account of his divorce and remarriage; what Naomi's account would have been we do not know.)[10]

Prior to the war, Quine's work had been primarily in logic, and had been, as he himself said, under the spell of PM. The systems he had constructed – LSeq, SL, NF, and ML – had all been efforts to rewrite PM in simpler and clearer terms. But his interests were changing. He would continue to work on and teach logic for years to come, but it became less a field of research for him and more a tool to be employed in dealing with other subjects, particularly semantics and ontology. Even before the war, the discussions between Quine, Tarski, Carnap, and Goodman in 1940–1941 focused not only on the analytic/synthetic issue but also on the possibility of a nominalistic version of science. Quine, Tarski, and Goodman were all strongly inclined toward nominalism, Carnap less so since he considered metaphysics meaningless, but all of them understood the problems of constructing a nominalistic mathematics and were uncertain whether or not enough mathematics could be so constructed to meet the demands of science.[11]

In 1943, while still on active duty, Quine published "Notes on Existence and Necessity" in the *Journal of Philosophy*.[12] The paper, he tells us, was taken from his Brazilian book, *O Sentido da Nova Logica*.[13] Accordingly, the article should be dated 1942–1943. This is a major paper that explored new ground. Rather than being a paper on logic, it is mainly concerned with semantics. He first offers the following definition: "The relation of a name to the object whose name it is, is called *designation*; the name 'Cicero' designates the man Cicero."[14] An occurrence of a name in a context in which the name refers simply to the object designated, he calls "purely designative." One would think that identity is a relation such that if two things are identical, then either can be substituted for the other in any context without change of truth value. But Quine then provides examples that show this is not true. In the statement "Cicero denounced Catiline," we cannot substitute "Cicero" for Cicero – that is, we cannot substitute the name of the word <<Cicero>> for the purely designative occurrence of the word "Cicero". To do so would be equivalent to saying that Cicero's name denounced Catiline, which is nonsense. Similarly, "Giorgione" and "Barbarelli" designate the same man, but "Giorgione was so-called because of his size" is true, whereas "Barbarelli was so-called because of his size" is false.[15]

Contexts of this sort Quine says are not purely designative. They are what he would later term "opaque" contexts, in contrast to the purely designative, or what he later called "transparent" ones. He then goes on to argue that "the ontology to which one's use of language commits him comprises simply the objects that he treats as falling within the subject-matter of his quantifiers – within the range of values of his variables."[16] Whatever is true of a substantive is true of something. Hence, if the occurrence of "Cicero" in the sentence

Cicero denounced Catiline

is purely designative, it follows that

(∃x)(x denounced Catiline).

One can quantify into purely designative contexts, but not into those which are not purely designative. It is therefore important to determine what contexts are not purely designative. One set of such contexts are those of propositional attitude. Thus it may be truly said of Jones that

Jones believes that Cicero denounced Catiline.

but it does not follow that

Jones believed that Tully denounced Catiline

for Jones may not know that "Tully" and "Cicero" name the same person.[17] Similarly, Quine holds that modal contexts are not purely designative. For his example, Quine picks the sense of "modality" by which to say "p is necessary" is to say "p is analytic." Then if one holds that "$9 > 7$" is analytic, and therefore necessary, and if "the number of the planets $= 9$," which it was before Pluto was demoted from planetary status, then one can derive the consequence that "Necessarily the number of the planets is > 7", which is clearly not true. Modal contexts are thus not purely designative. Quantification into such contexts can produce unacceptable results, e.g.,

(∃x)(necessarily the number of planets is greater than x)[18]

which is clearly false. Similarly, Quine holds that quantification over attributes is mistaken because there is no criterion for the identity of attributes.

The failure of the substitution of identicals in not purely designative contexts is not the only such failure; synonyms have the same fate. This leads Quine to a discussion of synonymy and meaning. He writes

> Just what the *meaning* of an expression is – what kind of object – is not yet clear; but it is clear that, given a notion of meaning, we can explain the notion of *synonymity* easily as the relation between expressions that have the same meaning. Conversely also, given the relation of synonymy, it would be easy to derive the notion of meaning in the following way: the meaning of an expression is the class of all the expressions synonymous with it. No doubt the second direction of construction is the more promising one. The relation of *synonymity,* in turn, calls for a definition or a criterion in psychological and linguistic terms. Such a definition, which up to the present has perhaps never even been sketched, would be a fundamental contribution at once to philology and philosophy.[19]

It is quite clear here that as of 1943 Quine had by no means abandoned hope of finding satisfactory definitions of "meaning" and "synonymy" and that he was looking to psychology and linguistics for the answers. Thus although Quine had found problems with analyticity three years before, he had not as of 1943 yet rejected the notions of meaning, synonymy and analyticity. When he republished parts of this paper in "Reference and Modality" in *From a Logical Point of View* in 1953, these comments about meaning and synonymy were not included; by then Quine had made up his mind. But it is clear that even before the war, his interests were turning to semantics and ontology.

In August of 1943, while he was still in the Navy, Quine wrote a manuscript that he entitled "Foundations of a Linguistic Theory of Meaning."[20] He makes it clear at the start that he wants to avoid mentalism in semantics, and therefore requires a behavioristic definition of synonymy. The definition of synonymy in terms of substitutions that leave the truth of a sentence unchanged, is rejected since "truth" involves the world. Synonymy requires a linguistic definition for an individual at a time that is implicit in the individual's or the community's behavior. But Quine notes that we cannot have a single definition for the community because individual differences are too great. To take the common denominator for the community would leave us with too meager a result; to accept all the definitions of the community members would lead us into contradictions.

Quine then proposes to reduce the problem to one of belief; beliefs he thinks can be tested by direct questions that require yes or no answers. How would we establish what constitutes a yes or a no response? Quine says that that can be done, but he will not attempt it here. But he thinks reducing the problem to one of beliefs narrows the problem. Hence, "what we are concerned with in the present study, then, is the reduction of synonymy to field data on belief." The belief sentences to be tested must be univocal so there is no chance of confusion of different meanings of a word. How is that to be established? "The univocally believed sentences are the largest class of sentences such that the individual in question believes them not only singly but also believes all constructions (joint assertions) of them, in all combinations." This requires a criterion for determining when a word is used with one or a different meaning. After several attempts at finding a criterion, Quine narrows the problem to an individual and immediate synonymy – one statement directly transformable into another with no intermediate transformations. He then defines syntactical uniformity: "S is syntactically uniform with respect to f if there is no form whose substitution for one occurrence of f turns S into a statement, while its substitution for another occurrence of f turns S into a non-statement." He then defines a "semantic split" as follows: "There is a statement D such that BD is believed (univocally, as usual) and CD not, and every such statement contains f." By "a proper 'belief context of a form f' will be meant any *belief which is syntactically uniform with respect to f and contains as parts no two beliefs which constitute a semantic split of f.*" (emphasis in original). Then the definition of "immediate synonymy" for an individual is "f is an immediate synonym of a form g under a context C if, for every belief [B] such that CB is a proper belief context for f, the conjunction CB (B: g/f) is believed" – i.e., in the context C, the substitution of g for f results in a sentence believed by the individual.

> What has been defined is immediate synonymy *under* (or in) *a context* C. But we can derive the absolute notion of immediate synonymy: *Two expressions are immediate synonyms (simpliciter) if one is an immediate synonym of the other under a proper belief context of the one.* A *synonymy sequence* (for an individual) is any sequence f_1, f_2, f_3, etc. of forms such that each is an immediate synonym with the next (for the individual) under some one proper belief context of f. Forms are *synonymous* (for an individual) if they belong to one synonymy sequence (for him). A meaning (for an individual is *any class whose members comprise the terms of some one synonymy sequence form (for him) and do not all belong to any longer synonymy sequence (for him)* (emphasis in original).

A "synonymy sequence for a community is any set of forms f_1, f_2, f_3 such that each is immediately synonymous with the next, for some individual of the community, under some one proper belief context C of f." What will break the chain is failure to believe, not disbelief. But this procedure, Quine says, will produce a "gerrymandered selection from the whole community." Quine adds a note saying that this procedure will not do.[21]

This paper is interesting in several respects. It shows that in 1943, Quine was trying to find a suitable definition of synonymy and of meaning. The method adopted here is a development of that suggested in his earlier paper "Notes on Existence and Necessity." What is particularly interesting is that Quine thought he could develop a behavioristic criterion of belief by asking respondents to give yes or no answers even though he remarked in the paper that belief is a psychological state and is not identical with affirmation. Nevertheless, he proceeds to employ affirmation and denial as tests of belief. In view of his subsequent use of affirmation and denial in *Word and Object* one may well wonder just what affirmation and denial do signify for Quine. But since he declines to discuss just what responses constitute affirmation and denial in this paper, the problem remains unanswered. Nevertheless, there are suggestions of WO here, particularly in the proposal to reduce the problem to field data.

Quine's published output was understandably reduced during the war since his Naval duties took up most of his time. The "Notes on Existence and Necessity" was a translation of something he had written in 1942, and he published nothing in 1944. Late in that year, Quine wrote to Goodman "In the matter of logic and philosophy,

I am more at a standstill than I have been for half a generation." But, Quine added, "I am still dickering with the Introduction of a book on ontology."[22] Quine continued to write papers on logic, but from here on he was chiefly concerned with clarifying and deepening his understanding of the logical notions he had already developed,[23] while at the same time there is a discernable shift of emphasis in his work away from pure logic and toward other philosophical matters – especially ontology, semantics, and epistemology. In August of 1945, he wrote to Goodman that he was jotting down ideas for a new logic book (probably the revision of *Elementary Logic*) but added "I'm more looking forward to the long postponed ontological and semantic book."[24]

Why this change of direction? The 1930s had been a period of explosive development in logic. Quine's major contributions to this era had been NF and ML, and the use of "stratification" to block the paradoxes. He had been productive – astonishingly so, turning out a flood of articles and reviews for the *Journal of Symbolic Logic*. For a man of thirty-six, he had achieved much. But although he was a very good logician, he was not a great one. His work, significant though it was, did not rival that of Russell or Godel or Tarski, or even Church. There was no "Quine theorem" or "Quine proof," no result comparable to those of the pace setters of the field. Quine was not a humble man; he aspired to eminence, and it was, I think, clear to him that it would not come in logic. It was time to look to other fields. In 1945 semantics and ontology looked like greener pastures. Quine did not abandon logic; he continued to work on it, but it became for him increasingly a tool to be used in the investigation of other subjects.

But what was this book on semantics and ontology that Quine projected? What was the "Introduction" with which he was dickering? Since this book was never written, we have only fragments from which to determine its nature, but there are fragments enough to give us some idea. In December of 1943, Quine wrote

a) Ontology as metaphysics: bootstrap difficulty. Realistic. is trivially true for the one who adopts it. But false for other languages, yet not linguistic in content.
b) Epistemology: bootstrap difficulty. Resolution. Definition of epistemological priority in child psychology.
c) Theory of value. Validity of values argument from oneself. Bootstrap difficulty. Yet values *not identifiable* with individual taste.

On March 3, 1944, Quine projected a book as follows.

> One is epistemologically an idealist and ontologically a realist. There is no contradiction: on the contrary it is as it should and the traditional issue between idealism and realism thus turns merely on a confusion between ontology and epistemology.
>
> Epistemology is the philosoph[ical theor]y of knowledge.
> Ontology is the philosoph[ical theor]y of being.
>
> Write a chapter on epistemology, carrying it far enough to show how it related to psychology. In the same chapter bring out the relation with ontology; how both are in a sense all-inclusive.
> Next chapter, on ontology, show *ultimacy*: how ontology is trivially true for him who adopts a science of the kind that presupposes it. But false for other languages, yet not linguistic in

content [footnote: draw the analogy from the theory of value: ultimate values aren't provable, yet "good" [here part of the paper is missing] ... meaning "I like" (characteristic of that which people call metaphysical). This is to be general and elementary, still (criterion based on whole or identity is still to come.) How even the gross opposites as high and low are two rather than one for a (phylogenetically) new-fangled point of view, according to Freud; but yet aren't subjective generalities. Also primary qualities as those of touch, learned in the womb (Henderson).[25]

A few days later (March 27, 1944), Quine wrote

Ontology and epistemology: how are they distinct and how are they mutually inclusive. Ontology is realistic, epistemology idealistic: but no contradiction.

On October 4, 1944, in a fragment entitled "Introduction," – perhaps the document he mentioned to Goodman – Quine wrote

Sign and Object: or, the Semantics of Being.

The Reality of the External World.

Things are *physically* composed of atoms, but are epistemologically (or: ideas of things are epistemologically) composed of qualia etc. (Mill. Two senses of the real. Reason there seems to be a rival claim for reality in an absolute sense is that an argument can be made for reducing one to the other, and a counter-argument can be made for the opposite reduction.) But both reductions are right. Depends on purpose but each purpose is good: realistic basis for clarifying the basis of epistemology (e.g., that only light rays etc. are given; nature of epistemological proceeding etc. and vice versa) for as Carnap says, physicist retreats in epistemological direction when statements are challenged). Needless, though, to use the word "reality" here. Epistemological primacy for the one; and what there is, simply – "reality", if you like for the other.

And in another fragment, Quine writes "So the epistemologically real is the phenomenal. But this I think is a tautology."[26]

What is one to make of these fragments? Several points stand out. First, the relation of epistemology and ontology is just that with which Carnap was struggling when Quine met him in 1932. Carnap abandoned the phenomenal (autopsychological) for the purely physicalistic. But evidently Quine did not and apparently for the same reasons Carnap found the choice difficult. Second, when Quine speaks of idealism, it seems clear that he means subjective idealism of the Berkeleyan sort or phenomenalism – probably the autopsychological of Carnap's *Aufbau.* Third, Quine tends to equate "language" with "theory." It seems clear that Quine had already accepted the notion that science defines ontology; what there is is what science says there is. Further, he sees science as realistic. But Quine also feels, as Carnap did, that some immediate contact with experience is necessary to preserve the empirical character of knowledge. But why are Idealism (phenomenalism) and Realism both right? Science determines what is – hence it determines ontology. But we project sense qualia onto the world. We see "green"; we know this is due to light rays that are not themselves green, but that make things look green. The notion that ontology and epistemology are both all-inclusive is there but not explained. But the relation of epistemology to psychology, particularly child psychology, is already seen.

Quine's concept of epistemology here is phenomenalistic, and apparently involved perceived qualia as a sensory base. External objects would then be some form of logical construct from immediately given qualia. This will involve him in all the problems Carnap saw – particularly the issue of subjectivity. If the world is constructed from sensory qualia, how is intersubjective knowledge to be possible? Furthermore, Quine was a committed behaviorist. It is not easy to see how he could have reconciled such sensory qualia with his behaviorism. Unfortunately, these papers are too fragmentary to provide answers to these questions, if he had answers.

It is also interesting that Quine sees a role for a theory of value here. Value theory is not an area in which Quine had ever published, and we know little about his theory, if he had one. But his point that ontology is trivially true for one whose science assumes it, but not for one who adopts a different science [Aristotle's perhaps?], appears to fit his view of values. One automatically adopts the values of one's community as one does its science. But note that Quine has the problem of different notions of reality – one ontological and one epistemological, and needs a way to resolve this division. There are here a number of hints of what was to come, but they are only that – hints. Quine still has a long way to go. Meanwhile, he had other problems to deal with.

Near the top of the list was the need to revise EL into an adequate text for his undergraduate class. In 1945, Quine published "On the Logic of Quantification" in the *Journal of Symbolic Logic*.[27] In this article he presented a decision procedure that he had first published in *O Sentido da Nova Logica* in Portuguese, and which is a simplification of earlier results by Lowenheim and Hilbert and Ackermann. This procedure was to figure prominently in his revisions of EL. The article consists of several sections. First, he defines his terminology. By "predicates" he means schematic letters such as "p", "q", "f", "g", etc. The important thing about these schematic predicate letters is that they do not refer; rather than standing for something, they stand in for genuine predicate expressions such as "is red" or "walks." The letters "x", "y", etc. are used to represent variables, which of course do refer. By a "matrix," Quine means an expression with some free variables that would become a statement if the variables were bound. The "closure" of a schema consists in the attachment of quantifiers that bind all the free variables in the expression, and the result of closure is a "closed" expression. A "medadic" expression is one built up from the letters "p", "q", etc. by truth functions with no free variables. A "monadic" schema is one in which the predicate letters occur with only a single variable, as e.g., "fx", "gy", etc. A polyadic schema is one in which one or more predicate letters are followed by an n-ad of variables ($n > 1$).

Quine states a generalization of modus ponens as follows

> If a conditional is valid, and its antecedent consists of zero or more quantifiers followed by a valid schema, then its consequence is valid.[28]

Quine then defines a "basic quantification" as consisting of a universal quantifier followed by a truth function built up of components each of which "consists of a predicate letter and a single occurrence of the variable of quantification." He then presents a method whereby any closed monadic schema can be converted into an equivalent closed schema that is basic. The method consists of working outward from the innermost quantification of the schema, reducing that to a closed form and exporting it beyond the scope of any other quantifier of the schema, and continuing this process until the schema has for components only closed schemata; in other words, the schema then consists of a sequence of closed schemata such that the scope of no quantifier extends beyond its own component. In this process, all existential quantifiers must be converted into universal ones by the identity "(∃x) = −(x)−." Then all the variables of the schema must be relettered as a single variable. Since no quantifier's scope extends beyond its own component, there is no danger that the relettering will lead to an illegitimate capture of a variable by a quantifier. Quine then describes the method of testing a basic schema, labeled ψ, as follows.

> First, set up a truth table under ψ, assigning T's and F's in all combinations to the statement letters and quantifications of ψ and calculating, for each row of the table, the value of ψ. Keep only the rows that yield "F" for ψ (If there are none, so that ψ is medadically valid, the test is already at an end.) Then determine whether each of these rows meets at least one of the following conditions:
>
> (a) It assigns "F" to a quantification whose scope is medadically valid.
> (b) It assigns "T" to one or more quantifications whose scope, or the conjunction of whose scopes, is medadically contravalid.
> (c) The scope, or the conjunction of scopes, of one or more quantifications assigned "T" medadically implies the scope of a quantification assigned "F".
>
> As soon as a row is found to meet one of the conditions (a)–(c), we can drop it and go on to the next. As soon as a row is found to meet none of (a)–(c), we can stop altogether, having found ψ non-valid. If every row meets one of the of the conditions (a)–(c), ψ is valid.[29]

In the next section of the paper he proves that this decision procedure is itself a valid one, and gives illustrative examples.

Quine then turns to the polyadic case. Given a valid monadic schema, one can substitute polyadic schemata for the mondadic schemata, the same polyadic schema obviously being put for each occurrence of the monadic components for which it is substituted. Then, by the generalized rule of modus ponens, we can infer the consequent. This method gives a decision procedure for some polyadic schemata, but not for all, since by Church's theorem it is known that there can be no general decision procedure for quantification theory. Nevertheless, it significantly expands the domain of schemata for which there is a decision procedure.

Quine had long had an interest in concatenation theory, not only for its linguistic implications, but also for its use in logic. He had employed it in the last chapter of ML to prove the incompletability of protosyntax, and in 1946 he used it again to establish the following:

> It will be shown not only that the elementary arithmetic of natural numbers can be embedded in the elementary theory of concatenation, but that it can be so embedded as to exhaust the

> latter, rendering the elementary theory of concatenation and the elementary arithmetic of natural numbers identical.[30]

The idea, Quine says, was suggested by Tarski. Elementary concatenation theory, Quine describes as dealing with finite sequences of objects that are themselves sequences, and are termed "atoms." The concatenation of these sequences consists of joining them end to end to form new sequences. Elementary concatenation theory includes identity, truth functions, and quantification, the values of whose variables are sequences. One of the attractions of concatenation theory for Quine was that its ontological commitments included neither classes nor attributes; the objects of which the sequences are formed can be of any sort, including marks or inscriptions. Since numbers are taken to be classes in PM and ML, the reduction of the arithmetic of natural numbers to elementary concatenation theory is more than just a formal feat; it has significant ontological implications. As Quine remarked, "elementary number theory *is* protosyntax."[31]

By the late forties, Quine was again looking for offers. As he wrote to Goodman, then well established at Penn,

> I am very grateful for your good offices in arranging for the talk at U.P. and the fifty bucks. By putting in appearances publicly thus in southeastern Pa and at Princeton and at the ASL meeting I might conceivably encourage a good offer, which, as you know, is an enduring interest of mine.[32]

In February of 1947, Quine wrote Goodman in greater detail.

> I'm increasingly anxious to get a job elsewhere. There are only two important things in favor of Harvard: occasional good students, and a good library ... Against Harvard there is the philosophical sterility of the place (apart from good students) and the slowness of advancement. This latter point has come to loom large again in view of the recent promotion of [John] Wild and [Donald] Williams; this makes five full professors, so there won't be room for another for a very long time. Scholarly productivity counts for nothing here, as you know; nor, considering the make-up of the department, could it be expected to count for much. Seniority of tenure is the single guiding principle. So I should take a lot of pleasure in giving up the department as a bad job; Lewis would be the only man left in it to all practical purposes, and he will soon retire. Nothing would please me more than a professorship in a lively place with a serious thinker or two among my colleagues and a moderate teaching load. But it is hard to know how to go about it. There is a silly idea abroad that people don't want to leave Harvard. And I must proceed, of course, with much discretion; frank talk like the above is for few ears.[33]

Goodman wrote back

> Your desire to move is quite understandable in view of those two slaps in the face of you and Harvard both; those were wild and wily promotions.[34]

Quine's timing was good. West Churchman was chair of the Penn philosophy department and could be counted on to support acquiring Quine; Glen Marrow, a distinguished philosopher and Plato scholar, was dean. Goodman asked Quine for a complete list of his publications. He promised to push hard for the appointment but warned that money would be a problem; he thought the highest salary possible would be $7000. Penn had attractions for Quine; not only was Goodman there, but Morton White was there as well, and Goodman mentioned [Francis] Clark and

[John] Adams as good colleagues. Quine responded warmly: "Pennsylvania would please me on many counts; proximity of congenial minds, proximity of New York, and mild climate."[35]

Goodman had his work cut out for him. Morrow was very skeptical that Quine would leave Harvard. Goodman wrote to Quine that "I have practically had to swear that you would [leave Harvard]. So if you do get the offer, you had damn well better accept it or you will be leaving me away out on a long limb."[36] But Quine was also working on the Harvard administration. Harvard provost Paul Buck promised him that in fifteen months he would be a full professor with a $900 dollar raise. Quine told Goodman that if Penn topped the Harvard offer, he was still interested, but it was clear that he was not, and he apologized to Goodman for leaving him in the wind. Goodman wrote back

> OK you bloated plutocrat. As they say, money yells. Now I know what those Harvard shirts are stuffed with – raw and shameless dough. We are depressed by our failure, but had our fingers crossed anyway. Have a good time rolling in it!! You can dig in for life now, although someday maybe you will be able to climb the final step to Yale.[37]

But Morrow did not give up. He went to see Quine in Cambridge. Goodman wrote Quine, "Morrow has in mind that when you are gray and hoary you will follow Whitehead's course and depart more and more from technical matters."[38] Morrow did come up with the necessary money, but Quine used that to jack up his Harvard salary. Actually, the salary that Penn offered was higher than Quine's Harvard salary. But Quine wrote that for the year he would have to wait for his Harvard promotion, "my pay would be higher by a fifth at Penn than at Harvard, but the distant salary prospects at Harvard were better than at Penn. Also, Harvard had other attractions, and there was the trouble and expense of moving. So I stayed."[39]

Would Quine have left Harvard? There is no way to be sure, but probably not. Quine knew very well how the game is played. Your home institution rarely recognizes your value unless some other institution makes you an offer. But this game requires a delicate touch and nice judgment. If you go to your dean with an offer from elsewhere, he may simply wish you well in your new job and tell you to close the door on your way out. If you are at a school like Harvard, your offer has to be credible: it must be from an institution of approximately equal status and one your administration will believe that you will move to if they don't respond. Penn met the requirements, and Paul Buck clearly did take the offer seriously. Quine was not shy; he had a high opinion of his own worth and was willing to threaten his administration with leaving. This is a tactic Quine had used before and would use again. No doubt he would have moved if the Harvard administration had refused to respond to his threat, but they did respond, and one suspects Quine was fairly sure they would. Whatever his real intentions, the gambit worked and Quine stayed at Harvard.

In 1947 Quine published a paper entitled "The Problem of Interpreting Modal Logic."[40] The difficulties Quine finds with modal logic are several. First, leaving aside quantification, one might assume that a necessary statement is the same thing as an analytic statement. But on this tack one has the problem of defining analyticity. That, Quine notes, could be done if we had a definition of synonymy. Such a definition would have to be, like other linguistic concepts, of the form "the expression x is synonymous with the expression y for a person z at a time t," and it would have to be behavioral, like those of "general linguistics." "I should like," Quine writes, "as a service both to empirical semantics and to philosophy, to offer a satisfactory definition, but I have none."[41] Nevertheless, he considered analyticity clearer than the modal notions, and he therefore tries to use analyticity to explain modality.

The second problem is that "is analytic" attaches to the name of a statement to form a statement about the named statement, whereas "necessarily" attaches to a statement to form a statement containing the prior statement. Hence, "necessarily" (symbolized by "□") can be iterated whereas "is analytic" cannot. But Quine holds that the result of prefixing "□" to any statement is true if and only if the statement is analytic.[42] To avoid that problem Quine proposes a hierarchy of "intentions": a statement devoid of modals is of first intention; a statement is of (n + 1)st intention if "□" occurs in it in application to statements of the nth intention and no higher. The phrase "is analytic" can then be applied by a statement of intention n only to statements of intention (n – 1). So construed, Quine considered modality relatively harmless.

The third problem arises with quantified modal logic. A statement beginning with "(∃x)◇" can be transformed into one initiated by "◇(∃x)", thus keeping the modal operator prefixed to the whole statement. But where one has a statement such as

$$(\exists x)(x \text{ is red} \cdot \Diamond(x \text{ is round}))$$

the interpretation is unclear. Quine proposes that it be interpreted as follows:

> A. An existential quantification holds if there is a constant whose substitution for the variable of quantification would render the matrix true.[43]

But the result is to rid the universe of concrete objects. How so? Assume three entities Venus, the Evening Star, and the Morning Star that bear to each other, and to themselves, the relation "C". "C" is a stand-in for identity, since all three refer to the same object, which, as Tycho Brah discovered, is the planet Venus. Thus we should have

Morning Star C Evening Star · □(Morning Star C Morning Star)

Therefore, by A

1. (∃x)(x C Evening Star · □(x C Morning Star))
   But we also have
   Evening Star C Evening Star · – □(Evening Star C Morning Star)
   Hence 2. (∃x)(x C Evening Star · – □(x C Morning Star))

Quine says: "Since the matrix quantified in (1) and the matrix quantified in (2) are mutual contraries, the 'x' whose existence is affirmed in (1) and the 'x' whose existence is affirmed in (2) are two objects." This shows, Quine says, that the version of modal logic so described cannot have physical objects as its ontology, but only attributes or concepts. Obviously no such result is acceptable to Quine.[44] This paper is part of his campaign to banish modality, attributes, and mental entities from science

Quine had a strong nominalist bent. We have seen that after Cantor's theorem turned out to be unprovable in NF, Quine's interest in nominalism revived. Goodman had an even stronger bent toward nominalism. As early as 1940, the two of them were discussing the possibility of a joint paper on nominalism.[45] Goodman was working on his calculus of individuals that he hoped would take the place of classes.[46] The war interrupted whatever plans they had made, but by late 1946 the idea had revived. A letter of December 30, 1946 shows Goodman working on nominalism; a letter of January 6, 1947 shows Goodman worrying about how to define the ancestral without Platonism. By that time they were both engaged in the project, with apparently an agreement that each was to write certain sections. Goodman was worrying about how to say there are more cats than dogs. This exchange brought forth a bit of poetry from Quine.

> The unrefined
> Inchoate mind
> Of *Homo javanensis*
> Can only treat
> Of things concrete
> And present to the senses.[47]

Goodman responded in kind. "To refer to expressions that ain't there is just as reprehensible as to refer to anything that ain't, and you have shown that it is not only reprehensible in itself but you have shown that its wages are those of the blackest sin."

> Nothing is wrong with being
> Quiet or very small;
> The only sin is being
> Not anything at all.[48]

Goodman remarked "I tend to compare our advances with nothingness; you compare then with everythingness."

By March, Quine was working on a nominalist theory of syntax, and a system based on the Sheffer stroke and quantification, nominalistically defined. The problem, Quine says, is substitution. He says that he has worked out how to translate the axioms of membership into nominalist language, and he can get a nominalist theory of proof, but not of "theorem" since there can be theorems for which we have no proof. Quine adds that the nominalist claim that any two sequences have a concatenation is false since the result could be too long to fit into the spatio-temporal universe. By April, Goodman is pushing to get the paper done; he has a Guggenheim Fellowship for the next year and needs to prepare for that. On April 26, 1947, Quine writes that he likes Goodman's ideas on substitution and thinks they solve the problem. On April 29, Goodman writes that he wants his name to appear first, which is what happened. Quine sent the article to Church for publication on July 15, and it was published in the December issue of the *Journal of Symbolic Logic* with the title "Steps Toward a Constructive Nominalism."[49]

But during the process of research that led to "Steps," Quine gave a talk and published a paper drawing on the work that would become "Steps." The first, in 1946, was a talk that he gave before the Harvard Philosophy Colloquium that he entitled "Nominalism."[50] He makes it clear from the start that he is drawing on work done in collaboration with Goodman, and he presents, not a defense of nominalism but a discussion of the problems involved. "I should like to be able to accept nominalism," Quine says, thus making his view clear.[51] What then is nominalism? Quine defines it as follows: "Discourse adequate to the whole of science can be so framed that nothing but particulars need be admitted as value of the vbls [variables]."[52] Quine considers two versions of nominalism – the mental version according to which particulars are "concrete, specific mental events," and the physical version according to which particulars are "physical events." He chooses to talk here only about the physical version. So his doctrine is that "the only things there are are spatio-temporally extended physical objects." He cites Eddington as his authority for the claim that there are only a finite number of physical objects in the universe. The nominalist rejects attributes and classes, which Quine says are universals. Such an ontological claim, he notes, would be dismissed by Carnap as meaningless, but Quine holds that it is meaningful and important. The problem then is to show that science can get along on a nominalistic basis. He first urges that not all nouns denote; some can be taken as syncategorematic. And using his criterion of ontological commitment – to be is to be the value of a variable, he points out that it is the variables, not the nouns, that carry such commitment. But mathematics requires quantifying over classes, and the problem is how to deal with that. Some expressions that seem to require such commitment to classes can be avoided by quantification theory; this is his doctrine of "virtual classes." But most cannot, and so the question is, granted that that a lot of mathematics has to be rejected by the nominalist, is there enough mathematics that the nominalist can accept to meet the needs of science? Why insist on trying to do so? Quine points out that the paradoxes of set theory indicate that something is fundamentally wrong with set theory. How then to proceed? Quine cites Goodman's theory of fusions: the fusion of a class of particulars is the smallest particular which has all the original particulars as parts.[53] This method avoids many classes, but not enough. He also cites the traditional nominalist tactic of reducing universals to mere names – the *flatus vocis*. Since words are themselves classes of tokens, this seems at first futile, but Quine proposes to dodge the problem by talking only of inscriptions geometrically similar to one another. But there is a problem here: since there are only a finite number of objects in the universe, the nominalist could find himself with inscriptions so long that all the matter in the universe would not suffice to inscribe them. Contextual definitions offer yet another way to avoid classes such as numbers. But Quine cannot claim that these devices are enough: "the issue of nominalism is still an open problem," and there he leaves it.[54]

There are several issues here that require comment. In this talk, Quine put the nominalist issue as particulars versus universals; on other occasions, he puts it as concrete versus abstract. These are not the same, so why does he conflate them? I think the answer is physicalism; universals and abstracts are not physical entities; particular physical objects are also concrete entities. Physicalism was for Quine a fundamental doctrine, like behaviorism, that he would never abandon. Second, why does Quine want the number of objects in the universe to be finite? In "Steps," he claims that if there are only a finite number of entities in the universe, then it is possible to develop quantification without any classes at all. Further, where the number of entities is finite, a universal quantification is equivalent to a conjunction and an existential quantification to an alternation. Third, in this talk, Quine has no hesitation in using terms such as "a priori" and "analytic": thus he writes "There is no *a priori* ground for throwing out the statement ..." and "since arithmetic is *a priori* ..."; further "... arithmetic of constants becomes *analytic* in a very extreme sense."[55] As these usages indicate, as of 1946, Quine had not yet rejected the use of these terms.

The paper that Quine wrote and published was entitled "On Universals"; he says it was stimulated by his discussions with Goodman. This paper appeared in the *Journal of Symbolic Logic* over Quine's name only.[56] The first part of the article Quine says is drawn from a paper that he read before the Association for Symbolic Logic in February; the second part draws upon the forthcoming collaborative paper, "Steps." We will look first at "On Universals" and then at "Steps." Quine begins "On Universals" by asserting his doctrine that the only things to the existence of which a theory (language) is committed are those that are the values of its bound variables. The doctrine is combined with his view that predicate letters are not variables but mere schemata used to depict the structure of statements. He takes the same line with respect to the "p", "q", ... of what he calls truth function theory – i.e., the propositional calculus. He describes various ways in which signs that are usually taken to refer to universals can be reconstrued so as to avoid actual reference to abstractions and finds most of them inadequate.

Quine then turns to the formulation of quantification theory. He adopts three axioms and six rules of inference, allowing predicate letters to serve as bindable variables. But not only does this formulation bring in universals – that is, classes, by Cantor's theorem it brings in an infinite number of classes, and it leads us to Russell's paradox. To escape the paradox, Quine invokes Russell's theory of types. The result is a system essentially like one of Tarski's, and can serve as a foundation of mathematics. But it is Platonistic, and its defense against the paradox is strictly ad hoc. Such a theory is, from the point of view of a nominalist, unsatisfactory.

> But let us now consider quantifiers whose range is restricted to classes of not more than k members; and for these let us use the notation "$(x^n)_k$" meaning "for all classes x^n of nth type having not more than k members." For any fixed k, and with n fixed at 1, it turns out that we can define this kind of quantification without initially assuming class variables at all. The only foundation needed is ordinary quantification theory ... plus the theory of *identity*.[57]

Quine then proceeds to develop such a system of limited quantification. "What is to all intents and purposes a logic of limited quantification over classes of concrete individuals is thus erected, by conventions purely of notational abbreviation, on a basis involving no classes or other universals." Thus "$(x^1)_k$ Fx^1" would range over all of the k individuals in the universe. But note that Quine requires k to be finite. This seems indeed to show the dispensability of Platonism, but it leads to problems.

> The fact must be recognized, however, that a logic which can be reconciled with nominalism only upon a highly speculative physical hypothesis ... is little better than a logic which cannot be reconciled with nominalism at all. Also there is another respect in which the scrupulous nominalist might find the foregoing reduction of platonistic logic unacceptable; the actual expansion of "$(x^1)Fx^1$" ... [for known k] would be *too long to exist*..... Even if we think of an inscription merely as an appropriately shaped distribution of particles, without requiring that it be delineated visibly by the hand of man against a contrasting background, still there would not be matter enough in the whole of space-time.[58]

The paper, entitled "Steps toward a Constructive Nominalism,"[59] opens with the declaration "We do not believe in abstract entities."

> Why do we refuse to admit the abstract objects that mathematics needs? Fundamentally this refusal is based on a philosophical intuition that cannot be justified by appeal to anything more ultimate. It is fortified, however, by certain *a posteriori* considerations. What seems to be the most natural principle for abstracting classes or properties leads to paradoxes. Escape from these paradoxes can apparently be effected only by recourse to alternative rules whose artificiality and arbitrariness arouse suspicion that we are lost in a world of make-believe.[60]

Goodman and Quine thereupon renounce also the claim that there are infinitely many objects. Since mathematics does assume infinite sets – e.g., the natural numbers, this finitism poses a threat to mathematics. Classical syntax also assumes an infinite realm of objects. But here again the limits of the finite world intervene. "We cannot say that in general, given any two inscriptions, there is an inscription long enough to be the concatenation of the two." "Our problem is solely to provide, where definitions are called for, definitions that are free of any terms or devices that are tainted by a belief in the abstract."[61]

It is instructive to see how Quine and Goodman perform this feat. One concept that had worried them in writing the paper was how to define the ancestral. Frege's definition of "*b* is an ancestor of c" is rendered as

> b is distinct from c; and, for every class x, if c is a member of x and all parents of members of x are members of x, then b is a member of x.[62]

The authors introduce the notion "Part st," meaning that the individual *s* is part (or all) of the individual *t*. We need only replace "class" by "individual" and "member" by "part," provided we also stipulate that *b* is a parent and *c* has a parent. Then we have

$$\mathbf{b \neq c \cdot (\exists u)(Parent\ bu \cdot (\exists w)(Parent\ wc \cdot (x)(Part\ cx \cdot (y)(z)(Part\ zx \cdot Parent\ yz\ .\supset Part\ yx\ .\supset Part\ bx))))}$$ [63]

Goodman's calculus of individuals counted as individuals not only single objects, such as a particular dog, but all dogs, however separated in time and space they may be. With this understanding of "individual" the authors attack the problem of how to say that there are more cats than dogs. They define the predicate "is bigger than"; then they define "is a bit" as applying "to every object that is just as big as the smallest animal among all cats and dogs." "Now if and only if there are more cats than dogs will it be the case that every individual that contains at least one bit of each cat is bigger than some individual that contains at least one bit of each dog."[64]

These are examples of how statements that seemingly involve abstract entities can be converted into a nominalist syntax. But the authors remark "certainly we have not as yet reached our goal of knowing how to deal with every statement we are not ready to dispense with altogether. But there is as yet no convincing reason for supposing the goal unattainable." The most obvious problem is how to deal with mathematics. They propose to deal with mathematical statements formally, as strings of meaningless concrete marks, that is, inscriptions. "Such intelligibility as mathematics possesses derives from the syntactical or metamathematical rules governing those marks."[65] They specify a vocabulary of six marks, together with variables, and take concatenation as the method for combining the marks. Then having defined some auxiliary predicates, they define "formulas" and "proof." This requires them to specify their axioms. They adopt Lukasiewicz's simplification of Nicod's axiom schema as the axiom governing the Sheffer stroke. For quantification they use three standard axioms for quantification theory that are translatable into their nominalistic syntax. To these four they add an axiom of membership. This requires solving the problem of substitution for their system.

> We have to find a way within nominalist syntax of defining "Subst wxyz," meaning that the *formula w is like the formula z except for having free variables like x wherever z contains free variables like y*. Our method of definition depends on the fact that the condition in the foregoing italics is equivalent to the following one: *what remains when all free variables like y are omitted from the formula z is like what remains when some free variables like x are omitted from the formula w.*[66]

From this notion they then derive a formal definition for substitution.

> They give in addition two rules of inference.
>
> (1) From any formula, together with the result of putting a formula like it for "P" and any formulas for "Q" and "R" in "(P/(Q/R))", infer any formula like the one which was put for "Q"
> (2) For any formula infer any quantification thereof.[67]

(1) is Nicod's generalization of modus ponens; that it serves as such is seen by the translation "P/(Q/R)" as "p ⊃. q · r" from which, given "p", "q" follows. (2) amounts to "φx ⊃ (x)(φx)," where "φ x" is a formula.

With this machinery, the authors can give formal definitions of "axiom," "proof," and "theorem." They conclude

> The gains which seem to have accrued to natural science from the use of mathematical formulas do not imply that those formulas are true statements. No one, not even the hardiest pragmatist, is likely to regard the beads of an abacus as true; and our position is that the formulas of platonistic mathematics are, like the beads of an abacus, convenient computational aids which need involve no questions of truth. What is meaningful and true in the case of

> platonistic mathematics as in the case of the abacus is not the apparatus itself, but only the description of it: the rules by which it is constructed and run. These rules we do understand, in the strict sense that we can express them in a purely nominalistic language.[68]

But "Steps" was a turning point for Quine; it showed him that no nominalistic theory was possible that would provide the foundation for science that he had hoped for. By 1948, Quine abandoned hope that a nominalism could be developed that would do for science. In a letter to Woodger on March 22, 1948, Quine remarked "It seems, more than ever, that the assumption of abstract entities and the assumptions of the external world are assumptions of the same sort."[69] From this point on, Quine seems to have given up on nominalism.

When Quine returned to his teaching duties at Harvard after the war, one of his first priorities was the revision of EL into a satisfactory elementary logic text. It should be pointed out that in fact this was no longer necessary. At the time EL was published, there was a genuine dearth of suitable texts for an introductory logic course. But in 1941, Tarski's *Introduction to Logic*[70] was published in English. Quine had urged Oxford to publish the book and had promised that he would use it himself if they did.[71] But Quine wanted his own textbook, and having flubbed his first attempt, was determined to correct the situation. This turned out to be an extended process. Early revisions were mimeographed for use in classes, but not until 1952 was he sufficiently satisfied with his revisions to publish his text, nor was this the end of the revisions – subsequent editions were modified to reflect developments in his thought.

Since Rosser had discovered that the Burali-Forti paradox could be derived in ML, Quine had been content to republish the book with a correction slip that deleted his membership axiom *200 while letting some of its consequences serve as axioms that could be invoked when needed. But in 1951, he published a revised edition of ML[72] that incorporated changes that would solve the problem. In the preface to the revised edition he credits two major changes to George Berry and Hao Wang, both of whom had been his students. Berry's correction made it possible to reduce his axioms for ML from six to five by deleting the second axiom of the first edition – namely,

$$(\alpha)(\beta)\ \varphi \supset (\beta)(\alpha)\ \varphi$$

This correction gives a redefinition of "closure" so that only a statement in which the quantifiers are applied in alphabetic order is to count as the closure of the matrix. Thus

$$\text{(i)}\quad (x)(y)(x < y\ .\!\equiv\!.\ y > x)$$

is not the closure of "$x < y\ .\!\equiv\!.\ y > x$" because the quantifier "(y)" was applied to the matrix before "(x)," while

$$\text{(ii)}\quad (y)(x)(x < y\ .\!\equiv\!.\ y > x)$$

is the closure since "x" is alphabetically earlier than "y"[73] whereas in the first edition the opposite ordering had been used. Slight as this change may seem, it eliminates the need for axiom 2, which is then derived as theorem.

But the most important change in the new edition of ML was that which restored consistency by blocking the derivation of the Burali-Forti paradox. In the first edition, Quine had briefly reviewed the measures that had been taken to free logic from Russell's paradox. Russell's answer had been type theory, the problems of which have already been noted. He had followed common sense in admitting as classes all classes that satisfy the criterion "the class of all entities such that....", and then imposed a stratification of all entities into types such that "each entity is conceived as belonging to one and only one of a hierarchy of so-called types; and any formula which represents membership as holding between entities of other than consecutive ascending types is rejected as meaningless, along with its contexts." Zermelo's scheme, which like Russell's was presented in 1908, began with a limited set of classes, and then admitted as new classes only those which satisfied the criterion "the class of all entities such that **x**ϵ**y** and ... x ..." Zermelo's theory avoided the paradoxes, but was difficult to work with; it did not admit complements of classes, class generation was "laborious and uncertain, and it had no exhaustive class."[74] In NF, Quine says, he tried to take a middle course between Russell and Zermelo. He proposed stratification without types, but unstratified formulae remained meaningful; he had an exhaustive class "V", and included the complements of every class, thus making the universe symmetrical, and class generation was as easy as in Russell's theory. (But NF had its own problems; as we noted above) The von Neumann-Bernays system provided yet another approach by dividing classes into elements and non-elements, the latter being barred from membership in any class. In ML Quine followed the von Neumann-Bernays approach, but with certain differences.

> Whereas the elements of the von Neumann-Bernays system comprise approximately the classes of Zermelo, the elements of the present system comprise all the classes of "New Foundations." For von Neumann and Bernays, no element embraces as members more than an infinitesimal proportion of the totality of elements; consequently the class V of all elements is not an element, nor is the complement $\bar{x}$ of any element x. For the present system on the other hand V is an element (!210) and so is $\bar{x}$ for each element x (!274); the totality of elements is symmetrical as between small and large.[75]

What Wang pointed out was that Quine was wrong in thinking that the classes of ML corresponded to those of NF. Quine's axiom *200

If φ is stratified and has no free variables beyond $\alpha, \beta_1, \ldots \beta_n$ then

$$\vdash \beta_1 \ldots \beta_n \in V \,.\!\supset\!.\; \hat{\alpha}\varphi \in V$$

allows for more classes than NF "because in the defining formulas of the elements of [ML] there may occur bound unrestricted class variables which under the normal interpretation have no counterparts in [NF].[76]" It was the failure to restrict the bound variables of *200 to elements that opened the way for Rosser's derivation of the Burali-Forti paradox. Instead of *200, Wang proposed **200 – "If φ is normal and stratified and" (etc. as in *200), where by a formula being "normal," Wang meant "all the bound variables in it are element variables." Hence in the revised edition of ML Quine changed *200 to read

> If φ has no free variables beyond $\alpha, \beta_1, \ldots \beta_n$ and is formed from a stratified formula by restricting all bound variables to elements,
>
> $$\vdash \beta_1 \ldots \beta_n \in V \;.\supset.\; \hat{\alpha}\varphi \in V$$ [77]

As Wang added concerning his own system P which included the restriction of the bound variables of ML to elements, "In P, the whole of Quine's book [ML] can be developed without any changes except the replacement of *200 by **200 in a few proofs." "Indeed," Wang added, "we may even guess that P is the system Quine originally intended to present but that he made a mistake in his presentation."[78] But it is significant that it was not Quine who found the error. As with NF, Quine's interest in the subject declined as the 1940s wore on.

The issue of analyticity reached the stage of public argument in 1940, as we have seen. It is not easy to say just when the issue was first raised. Quine has suggested that his "Truth by Convention" in 1936 marks at least a very early stage of the argument, but I think his problems with the notion probably go back to his conversion of Lewis's doctrine that our a priori beliefs are those we are least willing to abandon into a behavioral criterion for a priority and analyticity. That was of course not what Lewis meant; his claim was that there are different analytic theories (i.e., logics) and that the choice among them is pragmatic. Willingness to abandon a logic did not alter its a priori analytic character as Lewis saw it; it simply meant that it was not as pragmatically useful as an alternative theory. Quine took it differently; for him, "reluctance to abandon" became a measure of analyticity, or more exactly, we call those statements analytic that we are least willing to abandon. But there were other factors involved here. Quine's nominalistic bent was evident early on; he was not just deeply suspicious about abstract entities, as were the Logical Empiricists; he was suspicious of notions such as "intension," "meaning," and "synonymy," which he regarded as tainted with mentalism. Quine's behaviorism, imbibed in a fairly raw form from Watson and Stetson, was perhaps his most deeply held doctrine. There is a very interesting exchange of letters between Carnap and Quine in early 1938. Quine wrote Carnap about the errors in the English translation of *The Logical Syntax of Language,* but went on to attack Carnap's intensionalism.

> I proceed to inveigh against your recent intensional propensities, as reported by Hempel. First I schematize your motivation, as I understand it:
>
> (1) Intensional languages are legitimate, by the principle of tolerance.
> (2) Therefore our syntax language must be adequate also to treating intensional languages.
> (3) But now we find that the syntactical treatment of languages must in general be supplemented by semantic treatment.
> (4) To treat a language semantically we must be able to translate the language into ours.
> (5) Hence, to treat intensional languages adequately we must be able to translate them into ours.
> (6) It results that our language also must be intensional.

> The course I should prefer is to repudiate (2). As I told Hempel, I fear your principle of tolerance may finally lead you even to tolerate Hitler.[79]

This was not only a rather smart aleck letter, but in the context of 1938, the final comment can only be regarded as rude.

Carnap replied

> Your sermon against my sin of intensionality has made a great impression upon me. But I may say as an apology, I do not indulge this vice generally and thoroughly. I used an intensional meta-language only for certain special purposes and I found it useful and even necessary for these purposes, namely, for the investigation of the relation of translation between an extensional and an intensional language. It seems to me that certain interesting results are found in this way. Although we usually do not like to apply intensional languages, nevertheless I think we cannot help analyzing them. What would you think of an entomologist who refuses to investigate fleas and lice because he dislikes them?[80]

Carnap nailed the point. Quine did want to eliminate intensions, and he considered intensional languages ill advised. Carnap's reply put Quine on the spot. His reply was

> I am glad we agree on wanting to be extensional where extensionality suffices. Your analogy of fleas and lice is forceful; we must study intensional languages as a human phenomenon. But will a syntactical treatment not satisfy us here? – as in the case of *metaphysical* expressions, which are devoid of denotation, truth, falsehood?[81]

Quine did not deny intensions outright, but it was obvious what he thought of them.

Nevertheless, Quine was not yet ready to throw out notions such as intension, meaning, and synonymy, etc. During the 1940s, while still in uniform, he had written a paper on meaning and synonymy that he did not publish, as noted above. Then in 1947 he received a letter from Morton White about the "paradox of analysis."[82] The paradox, as White describe it, is as follows.

> The paradox, propounded by C. H. Langford, is that if true, a statement such as (1) "The attribute of being a brother is identical with the attribute of being male sibling," says or expresses the same thing as the truism (2) "The attribute of being a brother is identical with the attribute of being a brother."

The "paradox" here is that while (1) conveys information, (2) is a tautology and therefore vacuous. Yet if the "attribute of being a brother" is synonymous with "the attribute of being a male sibling," it seems paradoxical that the substitution of one for the other can change an informative statement into a vacuous truism. White remarks on Church's solution which involved postulating a number of additional attributes.

Quine took this occasion to "express my general attitude on this problem."[83] He rephrased the problem without the ontology of attributes as follows:

> An "analysis" has the form $\zeta = \eta$, where ζ and η are synonymous, therefore the whole analysis is synonymous with, or translatable into, the triviality $\zeta = \zeta$.

Quine continues

> Resolution of the paradox. Distinguish between *intensional* and *structural* synonymy. The distinction is made, I believe, by Carnap in *Meaning and Necessity;* also in effect by Lewis, *Analysis of Knowledge and Valuation* p. 199.

Lewis's theory of what Quine calls "structural synonymy" arises in his theory of meaning where he drew a distinction between "holophrastic" meaning and "analytic" meaning. The problem for Lewis is that the holophrastic linguistic meanings of analytic statements, such as those of mathematics, are all alike since on his theory all analytic statements have universal comprehension and zero intension. This would appear to mean that all analytic statements have the same meaning, which obviously they do not. For statements that are neither analytic nor contradictory, identity of intension suffices for synonymy, but for analytic and contradictory statements something more is required – namely, analytic meaning: there must be a one to one correspondence between the components of the statements, such that each component has the same intension as its correspondent, and the components of the two statements should have the same syntactic order, or can be made the same without altering the intension of either whole statement.[84] Thus "2 + 3 = 5" and "all sisters are female" may be analytic but they are not synonymous because they differ in analytic meaning. Carnap's doctrine of "intensional structure" is very similar to Lewis's analytic meaning.

> The two sentences ... must not only be L-equivalent in the whole, but consist of L-equivalent parts, and both must be built up out of these parts in the same way. If this is the case, we shall say that the two sentences have the same intensional structure.[85]

Where the two doctrines differ is that Lewis applies analytic meaning only to statements that are analytic or contradictory and bases the synonymy of synthetic sentences on sameness of holophrastic intension, whereas Carnap applies his concept of intensional structure to all types of sentences.

Quine remarks

> *intensional synonymy* is the basic kind of synonymy; Carnap has called it "L-equivalence," and Lewis "sameness of intension." It is the kind of synonymy that underlies analyticity; also it is the kind that would determine identity of attributes, if there were attributes. It is the kind of synonymy that I have lamented the lack of a behavioral criterion of ... In effect, the definitions of structural synonymy on the basis of intensional synonymy in Lewis and Carnap are substantially alike. No special ontology is really needed, and the definition is straightforward (once intensional synonymy is given), apart perhaps from some minor patching.

Quine goes on to say that the distinction between holophrastic and structural synonymy is needed in indirect discourse and in translation. But he says

> It's bad that we have no criterion of intensional synonymy; still, this frankly and visibly defective basis of discussion offers far more hope of clarity and progress, far less danger of medieval futility, than does the appeal to attributes, propositions, and meanings.[86]

White was intrigued by Quine's suggestion that the paradox of analysis can be generated without invoking attributes, since, "we may solve it along the lines you point out in your reference to Carnap and Lewis."[87] But White is unhappy with the final appeal to "intensional synonymy." He therefore sent the correspondence to Goodman and asked him for his opinion.

Goodman's response was what one might have expected in view of his nominalism.

> My skepticism about Van's hope for a behavioristic criterion of synonymy stems from the fact that I feel that the lack of any behavioristic criterion (or even the dimmest suggestion as to how one might be set up) is a sign that we are not at all clear as to what it is that we have to define.

Even if we had terms that we hypothesized might be synonymous,

> I would not know how to test whether the terms are synonymous because I don't know what the question means – I wouldn't know what the hell I'd be testing for.[88]

Goodman suggests that we might hold two sentences as synonymous if the substitution of one for the other in the context "x believes that ..." leaves the truth value unchanged, but he raises this possibility only to dismiss it. He does however suggest that the controversy is similar to a religious one.

> Church believes devoutly and unquestioningly in the gods; Van finds the conception of God unclear but can't cast him off and hopes to find a meaning for God in human life; I think we'd better recognize that we are going to have to get along without.[89]

Thus from Goodman's perspective, Quine has not renounced synonymy or analyticity and is still hoping for an adequate definition.

Quine replied that he was trying to separate the problem of intensional synonymy from Platonism and intensional logic and get it into a form where it can be solved.

> The problems take the form essentially of a single problem; behavioristic definition of intentional synonymy. I don't say this problem can be solved; what I say is that this is the way that the problems under consideration should be conceived, if at all. And I can't understand such things as the philosopher's (e.g. Carnap's) concept of "analytic truth" unless the problem can be solved.[90]

Quine rejected the idea of defining synonymy as substitution in belief contexts without change of truth value; he had already considered that possibility in the paper he wrote while in the Navy and rejected it. He then adds a note to Goodman concerning their joint paper on nominalism: "A fair typist is already tintinnabulating on nominalism."[91]

This letter provoked a response from Goodman who felt that Quine and White had misunderstood his last.

> When I say I don't understand the meaning of "analytic" I mean that very literally. I mean that I don't even know how to apply the term... Give me a sentence and I can't tell you whether it is analytic because I haven't even implicit criteria.

Quine's remark about the "fair typist" provoked a typically Goodmanesque bit of word play.

> I wish you analyticists would explain how it happens that while all fair typists are fair, many fair typists are not fair. Fiction apparently supposes that all typists are fair and that the large

> majority are fair, even though some are very unfair. I hope that our typist is fair and fair but since you will have all the dealing with her it is not important to me whether she is fair and fair.[92]

The correspondence continued for several more letters without appreciable progress. White attempted to summarize where things stood in of letter of July third. He noted the difference between Quine and Goodman and put himself on Quine's side. This led Goodman to propose that White do a survey article covering the whole problem. Quine seconded this proposal in his July seventh letter, and so White was commissioned.

White's paper, "The Analytic and the Synthetic: An Untenable Dualism" was read before the Fullerton Club at Bryn Mawr College on May 14, 1949, and was published the next year.[93] White dealt only with predicative analyticity, although noting that such statements are often derivable from logical statements by substitution of synonyms, but he focused on analyticity rather than synonymy. He considered two statements that have often been taken as supporting analyticity. (1) "Analytic statements are those whose denials are self-contradictory." But, White points out, the denial of the statement "All men are rational animals" is not self-contradictory, although the statement is usually thought to be analytic. "Some men are not rational animals" does not involve a contradiction unless one substitutes "men" for "rational animals," which takes synonymy for granted and so begs the question.

The second statement usually seen as supporting analyticity is (2) "If we were presented with something that was not a rational animal, we would not call it a man." But who are "we" here? This is an appeal to general usage and its validation would require an empirical study of English speakers, which could only lead to a statistical frequency, and that frequency would in all probability be less than one. Neither statement therefore yields the sort of sharp division between analytic and synthetic that has traditionally been claimed.[94] Here for the moment the matter stood, but not for long.

Meanwhile, in 1948 Quine published "On What There Is" in the *The Review of Metaphysics*.[95] He had previously given it as a paper at Yale and at Princeton. He opens the piece with the problem of non-being. If one says "There is no such thing as Pegasus" one seems to be in the awkward position of talking about something that is not there; that is, in order to deny existence of Pegasus, one seems to be required to assume an object of which one is predicating non-being. Quine deals with this problem by using Russell's theory of descriptions. The sentence "The author of Waverly was a poet" can be analyzed as "Something wrote Waverly and was a poet and nothing else wrote Waverly." Here, Quine points out, "the burden of objective reference which had been put upon the descriptive phrase is now taken over by words of the kind that logicians call bound variables." The theory of descriptions can be applied to the problem of Pegasus by converting "Pegasus" into a descriptive phrase such as "Pegasizes," and then applying Russell's theory.

$$-(\exists x)(x \text{ Pegasizes} \cdot (y)(y \text{ Pegasizes} \supset x = y))$$

Here the existence of Pegasus is denied but without presupposing its existence; in fact it is eliminated from the sentence.

Quine emphasizes the difference between meaning and naming. "Pegasizes" is meaningful but it does not name. This however brings him to the question of universals. Even if "Pegasizes" is not a name, even the name of an attribute, still it has a meaning, and meanings are universals. Quine's way around this is to deny the existence of meanings, at least as any sort of entities. He argues that "I remain free to maintain that the fact that a given linguistic utterance is meaningful (or *significant* as I prefer to say . . .) is an ultimate and irreducible matter of fact."[96] Giving a meaning usually means giving a synonym. Of course the terms "significant" and "synonym" require explanation, but that Quine says is not provided by introducing universals.

The means of reference, Quine holds, is the bound variable, not the name, which, as the example of "Pegasis" shows, can be eliminated in favor of the variable. And here Quine again states his famous formula: "To be is to be the value of a [bound] variable."[97]

In debating ontology, Quine argues for the advantages of operating on a semantic plane. Disagreements on ontology involve basic disagreements in conceptual schemes.

> In so far as our basic controversy over ontology can be translated upwards into a semantical controversy about words and what to do with them, the collapse of controversy into question-begging may be delayed.[98]

This is of course Carnap's strategy of moving from the material mode to the formal mode. Quine had learned a great deal from Carnap.

Quine then turns to the medieval ontological arguments among Realism, Phenomenalism, and Nominalism. The issues that exercised the schoolmen are still with us. Realism, Quine states, "is the Platonic doctrine that universals or abstract entities have being independently of mind." Quine takes Logicism, as represented by Frege, Russell, Whitehead, Church, and Carnap as condoning the use of bound variables to refer to abstract entities. *Conceptualism* is the "doctrine that there are universals but they are man-made;" he sees it emerging in Intuitionism, as represented by Poincare, Brouwer, and Weyl. Formalism is apparently linked to Nominalism, though Quine does not say this explicitly; this was the path he and Goodman had taken in "Steps."[99] "The formalist keeps classical mathematics as a play of insignificant notations" that do not commit him to any ontology. But the argument among systems is better carried out at the semantic level if we are to find enough common ground to argue, which will be found in the rules that govern the manipulation of notations.[100] These alternatives are basic to our conceptual scheme, which is itself chosen to make our world as simple as possible.

Quine then compares phenomenalism with physicalism. "Each," Quine says, "deserves to be developed." "Each may indeed be said to be the more fundamental, though in different senses: the one is epistemologically, the other physically, fundamental."[101] This echoes Quine's statements in 1944 and 1945 that phenomenalism has epistemological primacy. But Quine regards the question of which ontology to choose as open. All these options he says should be pursued with tolerance to see where they lead. But

> From among the various conceptual schemes best suited to these pursuits, one – the phenomenalistic – claims epistemological priority. Viewed from within the phenomenalist conceptual scheme, the ontologies of physical objects and mathematical objects are myths. The quality of myth, however, is relative; relative, in this case, to the epistemological point of view. This point of view is one among various, corresponding to one among our various interests and purposes.[102]

Here one sees the same doctrine that Quine had set forth in his fragments of 1943–1944 and which he had learned from Carnap in 1932.

"Identity, Ostension, and Hypostasis"[103] was published in 1950, and drew on some of Quine's earlier papers. The article is largely a conjectural account of how certain features of our conceptual scheme and language may have come to be. Quine starts with Hericleitus's claim that one cannot bathe in the same river twice. Quine distinguishes between the water at a given ostensively indicated place and time and the river as a process. To reach the latter requires the identification of the river stages as parts of one spatio-temporally extended process – the river. This is an induction, and one that simplifies our world. Quine gives a similar analysis of "red." We can by ostension note many discrete reds, but by a similar induction these can be brought together as parts of a large discontinuous red individual thing. These reflections lead Quine to formulate what he calls the "maxim of the *identification of indiscernibles*" best rendered as "the references to the original objects should be reconstrued for purposes of the discourse as referring to other and fewer objects, in such a way that indistinguishable originals give way each to the same new object."[104]

Spatio-temporal objects and processes serve to unify some of our experiences but not all. Shapes, such as triangular and square, do not fit the previous pattern. If one thinks of a square with one diagonal drawn, then adding the resulting triangles gives not a triangle but the square. (This depends on how you add them) "Triangle" and "square" are not concrete entities that are additive but abstract entities.

> We come to recognize two different types of association: that of concrete parts in a concrete whole, and that of concrete instances in an abstract universal. We come to recognize a divergence between two senses of "is": This is the Cayster [river] versus "This is square."[105]

The ostension that led to the general terms "river" and "red" does involve identity; the ones leading to the singular term "square" do not. Quine sees the move from the general term "square" to the abstract singular term "squareness" as a major evolutionary step because it commits us to having attributes or classes in our ontology. So as Quine sees it, our conceptual scheme has developed bit by bit by steps such as these.

We can change our conceptual scheme, Quine holds, but only a bit at a time; it is like rebuilding Neurath's ship while at sea. But we cannot inquire into the absolute correctness of a conceptual scheme as a mirror of reality, since any judgment must be from within some conceptual scheme. "Our standard for approving basic changes of conceptual scheme must be, not a realistic standard of correspondence to reality, but a pragmatic standard."

One has here another avowal of pragmatism on Quine's part. But one also has the example of a conjectural history of how our conceptual scheme came to be. The genetic account is purely conjectural, and not very convincing; one rather imagines that rivers were recognized before anybody thought about river stages. But Quine was prone to this form of argument, as would become very evident in his later work.

As these papers suggest, Quine was in the process of rebuilding his own philosophical ship. By 1950, he had become convinced that nominalism was not a tenable position. As he wrote to Woodger in March of 1948, "My ontological attitude seems to be evolving rather rapidly at the moment."[106] The failure of "Steps" to open the way to a sustainable nominalism meant that he would have to accept the reality of abstract sets, with all its attendant dangers of Platonism. He was also convinced that analyticity, at least as traditionally defined, was not a tenable notion because he could see no way of defining meaning that was consistent with his behaviorism and physicalism. But abandoning analyticity had consequences; if logic and mathematics were not a priori analytic, what then was the foundation on which they rested? Through most of the 1940s, Quine had subscribed to the logistic view that mathematics was reducible to logic; this was his position in ML, as it had been throughout his earlier work. But by the late forties, logicism seemed questionable to him. By 1947, he was pondering the difference between logic and mathematics. In a talk that year, he said that mathematical logic could be divided into four parts: truth functions, monadic quantification theory, general quantification theory, and set theory.[107] Mechanical proof procedures exist, Quine says, for the first two divisions of logic; for the third division, no such mechanical decision procedure is possible, as Church proved. But in the third division, if proofs are discovered, they can be checked. So *if* a proof of a theorem in general quantification theory can be found, it can be checked, and there is no reason to doubt that all statements of general quantification theory are in this sense provable. But this is not true for the fourth division, set theory. Indeed, Godel has shown that "no matter how carefully we devise a general consistent system of logical proof, there will be logically valid statements of set theory that not merely will not, but *can*not, be proved in it." There is thus, Quine says, a distinction between logic of the first three divisions, and set theory, and if we draw this distinction, "we can no longer say that mathematics is part of logic, or derivable from logic."[108] "My present interests," Quine says, "... cluster around the *boundaries* I have described, and various problems relating to them." But once Quine has come this far, he was faced with the problem of what guaranteed mathematical truth.

In 1949, Quine gave a talk to the Harvard Philosophy Club that he entitled "The Entangled Philosophies of Mathematics."[109] He divides the philosophy of mathematics into logicism, intuitionism, and formalism – the standard divisions at that time. He barely touches here on the grounds of logic; his topic is the foundation of mathematics. He reviews each of these three positions, and concludes that in each case they reduce to pragmatism. "Logicism," he says, "has become irrelevant; justification whether by intuitive considerations or pragmatic ones is no better nor worse for 'logic' than for mathematics outright." The pragmatic justification of mathematics then becomes the same as the justification of science.

> The function of science as a whole may be taken to be prediction of experience. A hypothesis or law in physics or biology is good only insofar as the total system of which it forms a part enables us to anticipate experience; and the same may be said of pure mathematics itself, thought of as a functioning part of the total scientific conceptual scheme ... Despairing of intuitive grounds for general set theory and classical mathematics, we settle for pragmatic grounds.[110]

What of intuitionism? Quine says, "any intuitionism is bound to sacrifice much of classical mathematics – portions surely of the theory of real numbers and theory of functions." What then of formalism? Mathematics would become a system of uninterpreted symbols, the justification of which would be a conformity to rules of proof. But, Quine says, "Godel's proof is the fairest flower of proof theory" and Godel's proof shows that formalism cannot succeed. But

> As pragmatists we can go on with classical mathematics, obeying formal rules of proof as usual; but we must recognize that the previously suggested boundary between mathematics and the rest of science has been rubbed out by Godel's discovery.

So Quine concludes

> I seem to have ended up with an overall pragmatism, embodying much of the formalist point of view; but still with attachment of a special value, a central importance, to that core of mathematics which is acceptable to the intuitionist.[111]

The classical claim was that mathematics was true because all statements of mathematics are provable from axioms, the truth of which can be determined by inspection. Godel's proof of the incompletability of mathematics destroyed this claim. But it left intact the argument that all statements of mathematics were analytic – "tautologies" as Wittgenstein called them – and therefore certain. But if analyticity was now to be rejected, what was left to support mathematics? Quine could at that point think of no better basis than pragmatism – the position Lewis held. But Quine saw something more: he took Godel's proof as eliminating the distinction between mathematics and science generally. But if so, was mathematics a form of empirical knowledge after all?

Quine was also trying to fit some of his other beliefs into his rebuilding system. If what he had said of mathematics holds, what of logic? "I see no non-standard sense in which logic is not as factual as theoretical physics."[112] But he went on: "so the *epistemologically* real is the phenomenal. But this I think is a tautology."[113] Hence

> I can imagine motivations for getting rid of universals, as least the high orders. We are no doubt stuck with the leap to physical objects but would like to leave it at that, plus just extrapolation to the big and small. Hence *materialism* as a goal for a conceptual scheme, post-epistemological, then the *epistemologically* real remains phenomenal. Show how classical mathematics is realistic; hence so is *science* since it includes mathematics.[114]

Here again the problem of a phenomenalistic epistemology and a realistic ontology shows through. So Quine is considering two orders of reality here – the epistemological and the scientific (ontological). But just how they fit together is as yet unclear. Note however that Quine is invoking pragmatism as the basis of mathematical truth, and since the boundary between mathematics and science is called into question by Godel's proof, his justification for science is also pragmatism.

In 1950, Quine read "Two Dogmas of Empiricism" before the Eastern Division of the American Philosophical Association, and in January of 1951 it was published in the *Philosophical Review.* Two years later it appeared in *From a Logical Point of View*[115] but by that time it was already a center of controversy. "Two Dogmas of Empiricism" is beyond question the most famous article Quine ever published, and it is generally regarded as one of the most important philosophical papers published in the twentieth century. It therefore requires some special attention, particularly because it does not quite say what it has been interpreted as saying.

Quine begins by announcing what the two dogmas are:

> One is the belief in some fundamental cleavage between truths which are *analytic*, or grounded in meanings independently of matters of fact, and truths which are *synthetic*, or grounded in fact. The other dogma is *reductionism*: the belief that each meaningful statement is equivalent to some logical construct upon terms which refer to immediate experience.[116]

Quine restates Kant's definition of "analyticity" as "a statement is analytic when it is true by virtue of meanings and independently of fact." He distinguishes sharply between the theory of reference and the theory of meaning, and then remarks

> Once the theory of meaning is sharply separated from the theory of reference, it is a short step to recognizing as the primary business of the theory of meaning simply the synonymy of linguistic forms and the analyticity of statements; meanings themselves, as obscure intermediary entities, may well be abandoned.[117]

This slight of hand by which Quine banishes meaning from the discussion is one of the crucial steps in the argument. So is the implied assertion that meanings play no explanatory role with respect to behavior. Quine had no idea how to define "meaning," at least in the behavioral sense on which he insisted, and so it was important to change the subject to synonymy and analyticity – as if these concepts could be defined independently of meaning. One may recall here his earlier attempts to define meanings as classes or sequences of synonymous terms.

Quine then distinguishes between two types of analytic statements – logical and predicative. He points out that statements that are predicatively analytic are derivable from logically analytic statements by substitution of synonyms, but his focus here is on the predicative type. He describes Carnap's notion of state descriptions but the requirement that all atomic sentences be independent makes this way of proceeding intolerably restrictive.

Definition then draws Quine's fire. Lexographic definitions (dictionary definitions) are, he points out, based on pre-existing synonymy relations in usage and cannot therefore serve as a basis for defining synonymy. Similarly, Quine shows that "explications" assume synonymy as a basis. The only definitional type that does not assume pre-existing synonymies is stipulative definition, but these are far too few to be of much help.

This brings Quine to the subject of substitution.

> A natural suggestion, deserving close examination, is that the synonymy of two linguistic forms consists simply in their interchangeability in all contexts without change of truth value – interchangeability, in Leibniz's phrase, *salva veritate*.[118]

Quine holds "interchangeability *salva veritate* is meaningless until relativized to a language whose extent is specified in relevant respects." But Quine holds that such a language must be extensional, and it is easy to show that interchangeability *salva veritate* in an extensional language does not yield synonymy – e.g., "creature with a heart" and "creature with a kidney" have the same extension but are clearly not synonymous. But why restrict our language to an extensional one? Quine points out that if we admit adverbs such as "necessarily," so construed as to hold only for analytic sentences, then interchangeability *salva veritate* will give us synonymy, but at the cost of assuming that we already understand analyticity, so the argument becomes circular. This assumes that "necessarily φx" is equivalent to "'φx' is analytic," a point that Quine had made in a previous paper. Quine takes this as sufficient to show that no nonextensional language can be employed in our investigation of synonymy and analyticity. But not all intensional languages need include modals, and one may wonder why Quine thought this argument conclusive.

Having as he believed shown that attempts at defining "synonymy" did not solve the problem of analyticity, Quine turns directly to the analysis of "analyticity." Echoing Goodman's statement from their triangular correspondence, Quine says: "I do not know whether the statement 'Everything green is extended' is analytic." The problem Quine says is not "green" or "extended" but "analytic."[119]

> The notion of analyticity about which we are worrying is a purported relation between statements and languages; a statement S is said to be *analytic* for a language L, and the problem is to make sense of this relation generally, that is, for variable "S" and "L." The gravity of this problem is not perceptibly less for artificial languages than for natural ones.[120]

This leads him to Carnap's work. Let L_0 be an artificial language, and let this language contain a semantic rule stipulating that all statements of a certain class are analytic. Obviously, this assumes a prior understanding of analyticity. But suppose a semantic rule specifying all the true statements of L_0. Then it can be held that "a statement is analytic if it is (not merely true but) true according to the semantical rule." But this Quine says avails us nothing, for it assumes the notion of "semantic rule" which is no clearer than "analytic."[121]

Quine then turns to the verification theory of meaning and reductionism. The theory is summarized as follows: "the meaning of a statement is the method of empirically confirming or infirming it. An analytic statement is that limiting case which is confirmed no matter what." What the verification theory says is that statements are synonymous if and only if they are alike in point of method of empirical confirmation or infirmation. But what constitutes the confirming relation between statements and the experiences that confirm them? Quine reviews the various forms that this doctrine has taken from Locke to Carnap. He then states what he takes to be the basic assumption underlying verificationism.

> The dogma of reductionism survives in the supposition that each statement, taken in isolation from its fellows, can admit of confirmation or information at all. My countersuggestion, issuing essentially from Carnap's doctrine of the physical world in the *Aufbau*, is that our statements about the external world face the tribunal of sense experience not individually but only as a corporate body.[122]

One should note here that Quine attributes his holism to Carnap's *Aufbau*; this is one more indication of Carnap's influence on Quine. At the time he wrote "Two Dogmas," Quine did not know of Duhem's work.

Quine draws the relation of reductionism to analyticity as follows: analyticity makes sense "as long as it is taken to be significant in general . . . to speak also of a limiting kind of statement which is vacuously confirmed, *ipso facto* come what may; and such a statement is analytic." All statements have both a linguistic component and a factual component. The idea that these can be separated so that the truth of some statements can be determined by their linguistic component alone is what underlies analyticity. Quine rejects the possibility of such a separation, as he does the possibility that statements are confirmed singly. "The unit of empirical significance is the whole of science."[123] One should note here that the empirical testing of statements concerns their relation to "experience." What "experience" means is not explained. But it cannot mean physical objects, for Quine says that physical objects are part of the theory. "Experience" then must consist of some sort of sense data, though of what sort Quine does not say. Note that the rejection of the second dogma is introduced only as a support for the rejection of the first.

Quine then compares our knowledge to "a man-made fabric which impinges on experience only along the edges." Or, varying the figure, he compares it to "a field of force whose boundary conditions are experience." At the center of this fabric or web lie the statements of mathematics and logic, which Quine contrasts to statements about brick houses on Elm Street. "These [latter] statements are felt, therefore, to have a sharper empirical reference than highly theoretical statements of physics or logic or ontology. The latter statements may be thought of as relatively centrally located within the total network." The whole body of our knowledge (meaning our scientific knowledge) is underdetermined by experience. Any statement in the theory can be revised if the result is a better fit to experience. Quine says

> Revision even of the logical law of the excluded middle has been proposed as a means of simplifying quantum mechanics; and what difference is there in principle between such a shift and the shift whereby Kepler superseded Ptolemy, or Einstein Newton, or Darwin Aristotle?[124]

In the total system that is our science, some statements are more centrally located in the system – in the sense that they are the ones we are most reluctant to change in modifying the theory. Others are particularly germain to particular experiences, "but in this relation of 'germainness' I envision nothing more than a loose association reflecting the relative likelihood, in practice, of our choosing one statement rather than another for revision in the event of recalcitrant experience."[125] Statements of mathematics and logic lie near the center; the statement "There are brick houses on Elm Street" near the edge.

Then, in a now famous paragraph, Quine extended his doctrine further.

> As an empiricist I continue to think of the conceptual scheme of science as a tool, ultimately, for predicting future experience in the light of past experience. Physical objects are conceptually imported into the situation as convenient intermediaries – not by definition in terms of experience, but simply as irreducible posits comparable, epistemologically, to the gods of Homer. For my part I do, qua lay physicist, believe in physical objects and not

> in Homer's gods; and I consider it a scientific error to believer otherwise. But in point of epistemological footing the physical objects and the gods differ only in degree and not in kind. Both sorts of entities enter our conception only as cultural posits. The myth of physical objects is epistemologically superior to most in that it has proved more efficacious than other myths as a device for working a manageable structure into the flux of experience.[126]

The same standard, Quine holds, applies to classes as entities, to the real and rational numbers, and to all those things we take to be existent. Quine wraps it up by declaring for pragmatism.

> Carnap, Lewis and others take a pragmatic stand on the question of choosing between language forms, scientific frameworks; but their pragmatism leaves off at the imagined boundary between the analytic and the synthetic. In repudiating such a boundary I espouse a more thorough pragmatism. Each man is given a scientific heritage plus a continuing barrage of sensory stimulation; and the considerations which guide him in warping his scientific heritage to fit his continuing sensory promptings are, where rational, pragmatic.[127]

The reception of "Two Dogmas of Empiricism" was all that Quine could have hoped and far more than he had expected. Hempel wrote him

> I find your arguments re analyticity fully convincing, and am, of course, all with you on the issue of reductionism. My expectation, if I may venture a prediction, is that this article of yours will prove to be very influential, and that its first publication will surely not remain its last.[128]

Lewis's response was more complex, and perhaps not what Quine had expected. "I have just been reading your Two Dogmas of Empiricism; it gives me hope." Lewis ignores the first dogma and goes at once to the second. Quine has, he says, three choices: (1) is to repudiate the offending datum. "But you still have to deal with it even if you call it an illusion." (2) There is the conceptual scheme that can be altered. (3) There is the relation of datum to scheme. If you repudiate the datum, you lose contact with the world. This leads to idealism. (3) regards how the datum is interpreted. "So I am thinking that positivism could turn out to be such left wing Hegelianism, with a coherence theory of truth as its epistemological implicate – when it finally faces such issues as you seem to me to raise. I think pragmatism cum radical empiricism represents the only plausible alternative to such coherence theory idealism – or –positivism. And I don't see that you have chosen yet."[129] Lewis clearly saw Quine as a positivist and "Two Dogmas" as bringing Quine closer to the pragmatic camp, and he sought to encourage Quine to become a full fledged member.

Equally interesting is Carnap's response to "Two Dogmas." He is not at all phased by Quine's attack. Quine had said that he did not know whether the sentence "Everything green is extended" is analytic, and that the problem was not "green" but "analytic." Carnap's reply is "It seems completely clear to me, however, that the difficulty here lies in the unclarity of the word 'green.'" In its everyday use, the word "green" is too vague to settle the question of its application to a point. Carnap holds that precision regarding meaning is not to be found in natural languages but only in formal languages whose terms are precisely defined. The point of formal languages, Carnap says, is to provide clear explications of concepts that are only vaguely defined in the natural language. “Our rules are meant ... [as] the explication of an inexact concept already in current use.” More precisely, we advance the claim that the defined concept embraces what philosophers have meant, intuitively but not exactly, when they speak of “analytic sentences” or, more specifically, of “sentences whose truth depends on their meanings alone and is thus independent of the contingency of facts.” So, Carnap concludes, “it follows from this clarification that the analytic-synthetic distinction can be drawn always and only with respect to a language system, i.e., a language organized according to explicitly formulated rules, not with respect to the historically given natural language.”[130] It is I think clear that here Carnap and Quine talked past each other. Carnap assumes a concept of meaning that Quine denies, so the two really do not confront the same issue.[131]

These are only three of the many replies, denunciations, and celebrations that greeted the paper. Quine himself was surprised by the reaction to the paper. But it is important to realize that Quine was neither clear nor certain about some of his claims in “Two Dogmas.” Obviously, there appears to be an inconsistency between the position Quine took in the triangle correspondence and what he said in “Two Dogmas.” But Quine did not show in “Two Dogmas” that analyticity was indefinable; he just criticized certain proposals for its definition – namely, that analytic sentences were interchangeable in an extensional language without change of truth value and Carnap’s definition by semantic rules. The first was easy; nobody thought analyticity could be defined in an extensional language. The second was part of his continuing argument with Carnap, but Quine’s objections had no effect on Carnap. Quine did not in fact reject the term “analytic”; rather, he found his behavioral definition in WO.

Quine’s statement of his answer to the second dogma is very condensed in “Two Dogmas,” and given the importance that he attached to it, it will be useful to spell it out. It is (and was) well known that in observations and experiments, the test always assumes a number of auxiliary hypotheses. Hence the test of an hypothesis “h” has the logical form

$$h \cdot a_1 \cdot a_2 \cdot \ldots \cdot a_n \supset e$$

where the “a_i” are the auxiliary hypotheses and “e” is the predicted experience. If the result is negative, then what follows is

$$-e \supset -h \vee -a_1 \vee -a_2 \vee \ldots \vee -a_n$$

Hence the negative result can be accounted for by rejecting any or several of the auxiliaries; the failure does not determine which adjustment of the theory is to be made. Therefore, experience underdetermines scientific theories. The fact that the auxiliary hypotheses involved in scientific experiments are at risk as well as the test hypothesis was not news; many other philosophers had remarked on this fact. Lewis had noted it;[132] so had Cohen and Nagel in 1934.[133] But before 1950, only Pierre Duhem had drawn the conclusion Quine drew – that a negative experimental result implied that the conjunction of the hypotheses involved in the experiment was false, and so the test hypothesis could be saved by rejecting one or more of the auxiliary hypotheses.[134] Quine did not know Duhem's work at the time he wrote "Two Dogmas." That may help to explain why he overstated his holism; he should not have said that all of science is at stake in every experiment, only that "chunks" are. That led him later to the idea of "semantic mass"; "a cluster of sentences had critical semantic mass if it implied an observation categorical." "In so far as mathematics gets applied in natural science, I see it as sharing empirical content.... as for inapplicable portions of mathematics, say higher set theory, I sympathize with the empiricist in questioning its meaning." Quine also remarked that he declared for pragmatism in "Two Dogmas" without thought of what it is to be one.[135] But as noted above Quine had come to the view that the justification of mathematics and of all science had to be pragmatic before he wrote "Two Dogmas." It is an interesting question whether one can be a pragmatist while subscribing to Tarski's definition of truth. In any case Quine's claim that he did not know what it meant to be a pragmatist was absurd. He had studied pragmatism under C. I. Lewis, who was the leading pragmatist of the generation after Peirce and James, and had written a paper on "Conceptual Pragmatism." He was trying to disguise how at sea he had been in 1950. The web metaphor needed unpacking, and this was largely his concern between "Two Dogmas" and WO. As noted, in "Two Dogmas," Quine brought in the second dogma to support the first, but he later concluded that it is the second dogma that is the really important one.[136]

But there are further points that need emphasis. As we have seen, Quine was in certain respects a phenomenalist; he considered phenomenalism more fundamental epistemologically than realism, but he was a realist regarding ontology. Hence his statement that physical objects are posits or myths, comparable in *epistemological* footing to the gods of Homer. This is a view that he continued to hold throughout his life, although his basis for it radically altered. Moreover, in "Two Dogmas" Quine does not address the questions of logical and mathematical truth. But in rejecting analyticity, he raised those issues and would have to deal with them.

Overall, "Two Dogmas" was for Quine more a report on a work in progress than a finished essay. He raised a series of major problems and issued a number of IOUs that he would have to pay off in the coming years. It is in a certain sense ironic that "Two Dogmas" propelled him to international fame as the man who destroyed positivism. Quine was much closer to Logical Positivism than has generally been recognized. The decade between "Two Dogmas" and WO would be devoted to trying to make good on his IOUs.

Finally, in describing science as a web or field that is anchored to experience only at the periphery, Quine left the relation between the center of the web, where logic and mathematics are supposed to reside, and the periphery, vague; he spoke of a relation of "germainness" but did not clarify what that meant. Yet he speaks of the connections within the web as being logical. "Reevaluation of some statements entails reevaluation of others, because of their logical interconnections – the logical laws being in turn simply certain further statements of the system, certain further elements of the field."[137] But there is a problem here. If the relations that link statements to form a system are logical, then the truth of logic is assumed in constructing the system. How then can logic and mathematics be central to the web? How can logical laws be reputable, as Quine says they are, if the system itself rests on the assumption of the truth of logic? Is "germainness" logical implication? If not, what is it? And if so, how can logic be central to the web? This was another problem that Quine would have to solve.

1950 had been an eventful year, and had brought Quine the celebrity he had long sought. It also brought something else; in December, Marjorie gave birth to his first son, whom they named Douglas. Like many twice married men, Quine began a second family. Elizabeth and Norma were now in their teens; a second family no doubt seemed perfectly natural. The day he brought Marjorie and Douglas home from the hospital, Quine left for Toronto where he read "Two Dogmas of Empiricism" before the American Philosophical Association.[138]

Quine was now fully engaged with the problems of epistemology and language, and particularly with semantics and ontology. In 1951, Quine delivered a lecture in Ann Arbor that was subsequently published as the third essay in FLPV. He entitled it "The Problem of Meaning in Linguistics,"[139] and it is in many respects a preview of WO. The "problem" Quine has in mind, although he does not quite put it this way, is how to do linguistics without using the concept of meaning. Quine takes his usual position: meaning, he claims, like Moliere's dormative virtue, "engenders an illusion of having explained something." The lexicographer deals with the problem of what expressions are "alike in meaning" – hence are synonyms. The grammarian deals with the problem of what expressions "have meaning" – i.e., are "significant." The problem then becomes one of how to cope with "synonymy" and "significance" without using "meaning."

Suppose a grammarian studying a hitherto unstudied language that he has come across in his field work. His problem, Quine says, "is to discover the bounds of the class K of significant sequences of the language."[140] Members of K may be of any length, but even to determine that the list is a list of phonemes – "the single sounds, distinguished as coarsely as possible for purposes of the language," requires analysis. Since the sequences of K are sequences of phonemes, Quine admits that the definition of "phoneme" requires the notion of "synonymy," but he supposes a case, which he freely admits is fanciful, in which an alternative behavioristic method of defining "phoneme" is available. This enables the grammarian to concentrate on finding a recursive definition for K. The problem here is that what the grammarian needs is a "prior notion of significant sequence, or possible normal utterance."

> Without this notion, or something to somewhat the same effect, we cannot say what the grammarian is trying to do – what he is trying to match in his formal reproduction of K – nor wherein the rightness or wrongness of his results might consist.[141]

The point is important. Many readers would later find Quine's doctrine of the indeterminacy of translation confusing. The point here is that unless the linguist has some criterion of what constitutes a significant sequence of sounds in the native language, he cannot match sequences of native noises to English expressions, nor would there be any way of telling whether he got it right.

Worse yet, what the grammarian needs is not just a list of significant sequences the native has used but also a way of specifying the significant sequences that *could* be uttered by the native speaker. What the grammarian actually does, Quine suggests, is to frame K so as to include all the actually observed cases, all observable cases that he can predict on the basis of whatever rules he can fit to his data, all the unobservable cases that his rules permit him to conjecture, and exclude all cases that the natives find bizarre or unacceptable. That, Quine holds, is a reasonable facsimile for "could." So much for significance.

The lexicographer's concern is synonymy. Suppose first that the dictionary he is making is an English to English dictionary. Clearly the expressions he pairs must be interchangeable, but only subject to two conditions,

> (1) In just what sorts of contextual position, if not in all, are the two forms to be interchangeable? (b) The forms are to be interchangeable *salvo quo*?[142]

Quine showed in "Two Dogmas" that "*veritate*" in an extensional language will not do for "*quo*." He specifies three reasons why he believes that the study of synonymy must involve fairly long expressions. First, synonymy must be attributed to segments of discourse sufficiently long to provide independence of their containing contexts. Second, they must be long enough to overcome the difficulties posed by homonymy. And third, they must be long enough to avoid the use of what Quine calls "stage directions" – e.g., "Addled" as "spoiled, said of an egg." But even with these, Quine finds no adequate definition of synonymy, and to illustrate the problem he turns to interlinguistic translation.

Presumably, the utterance of synonymous expressions ought to take place under similar conditions. But to say what conditions are similar for a Kalaba native and an English speaking linguist is an extraordinarily difficult problem. If we could assume that the native's world view was similar to our own, finding equivalent words for a common content would be easy. But, Quine says, citing Whorf and Cassirer, "it is not clear even in principle that it makes sense to think of words and syntax as varying from language to language while the content stays fixed; yet precisely this fiction is involved in speaking of synonymy, at least as between expressions of radically different languages." In other cultures, the line between animate and inanimate, or between human and animal, or between dream experience and waking experience, may be drawn in ways very different for our own. There are matters concerning which cultures overlap because of human necessities; all humans have certain needs in common, and in such areas a common content may be presumed. But beyond this narrow base the lexicographer comes "to depend increasingly on a projection of himself, with his Indo-European *Weltanschauung*, into the sandals of the Kalaba informant."[143]

> The finished lexicon is a case, evidently, of *ex pede Herculem*. But there is a difference. In projecting Hercules from the foot we risk error, but we may derive comfort from the fact that there is something to be wrong about. In the case of the lexicon, pending some definition of synonymy, we have no statement of the problem; we have nothing for the lexicographer to be right or wrong about.[144]

It is clear here that Quine had reached a number of conclusions about the problems of synonymy and significance as early as 1951, if not earlier. He had not yet found solutions. The period from approximately 1950 to 1960 was devoted to the effort to find them. Quine had talked in "Two Dogmas" about the "fabric" and the "edge"; he needed to figure out just how these were connected. He had raised problems about analyticity and synonymy; now he needed to solve them. He had already formulated the thesis that in the translation of a radically different language there might be no fact of the matter. Thus the ideas that would come together in WO were already being formulated.

In another article entitled "Logic and the Reification of Universals," Quine drew on a number of his prior works, but this article first appeared in complete form in FLPV.[145] We need, Quine says, a standard for judging what the ontological commitments of a theory are, and he reasserts his maxim noted above.

> An entity is assumed by a theory if and only if it must be counted among the values of the variables in order that the statements affirmed in the theory be true.[146]

This of course assumes that the theory is in standard quantificational form, but Quine shows that "every statement containing a variable can be translated, by known rules, into a statement in which the variable has only the quantificational use." This is true even of the statements of Schoenfinkel's system which contains no variables. Quine admits the possibility of an alien language so different from ours that we could find no translation that would indicate ontological commitment, but that is not something to be pursued in this article. He then repeats his arguments concerning schematic letters and expressions, such as the "F" of "Fx" and the "p" of the propositional calculus. These letters do not stand for values but stand in for expressions; they carry no ontological freight. But if, and when, these distinctions are not maintained and we permit quantification over predicate letters, we find ourselves committed to the existence of abstract objects such as classes. If the system in which we are working is extensional, nominalists may be able to interpret some such quantifications as mere figures of speech, but this evasion of commitment does not always work. Quine remarks.

> The maneuver of extending quantification to predicate letters, as a means of expanding quantification into class theory, can be represented as a provision merely to allow predicate letters all the privileges of the variables "x", "y", etc. ... Actually, however, it turns out to proclaim a a realm of classes *far wider* than the conditions that can be written in the language.[147]

Since classes are universals, we then find ourselves committed to a vast universe of universals. Here, Quine says, "we can no longer see what we are doing, nor where the flood is carrying us."[148] Our efforts to avoid contradiction are purely ad hoc, and are justified only by the fact that they seem to work.

There are however different approaches to the creation of classes. Quine outlines what he calls a "conceptualist" approach in which classes are constructed from concrete individuals as classes of such objects defined by conditions statable in quantification theory, which he calls here L_0. We can then create a next level of classes defined by conditions statable in L_0, and add new variables for classes to enrich our language to L_1, which includes L_0. We can so proceed to construct more classes by a series of such steps, further enriching our languages as we go. The process can be continued, level by level, up to the limit L. Paradox can be controlled by adopting rules amounting to stratification. The resulting theory Quine compares to Russell's ramified theory of types, though it is much simpler. "It represents . . . a position of conceptualism as opposed to Platonic realism; it treats classes as constructions rather than discoveries."[149]

"The conceptualist theory of classes requires no classes to exist beyond those corresponding to expressible conditions of membership." The Platonist theory admits all classes not prohibited by the ad hoc restrictions necessary to avoid contradiction. The "heroic or quixotic" position, Quine says, is that of the nominalist. He must renounce all genuine classes – those that cannot be explained away as mere figures of speech. This means the renunciation of much of mathematics. The nominalist could regain the ability to quantify over numbers if he could somehow correlate them to the particulars of the universe that he does recognize, but since the number of such particulars is not known to be infinite, he would have to do without the infinity of numbers required by classical mathematics. This allows Quine to specify the differences between Platonism, conceptualism, and nominalism in a very clear way. The nominalist is a finitist; the conceptualist admits but one degree of infinity, and the Platonist admits the full Cantorian infinite hierarchy of infinities.[150]

> Tactically, conceptualism is no doubt the strongest position of the three; for the tired nominalist can lapse into conceptualism and still allay his puritanic conscience with the reflection that he has not quite taken to eating lotus with the Platonists.[151]

Thus although Quine now recognized that nominalism was an impossible position for him in view of its cost in foregone mathematics, he was still looking for ways to avoid Platonism, and ways that would be consistent with his epistemological phenomenalism.

Quine continued his efforts to clarify reference in "Notes on a Theory of Reference" which was published as part of FLPV.[152] Although it drew on some of Quine's earlier work, this article was substantially new in 1953. Quine's purpose here is to contrast the state of the theory of reference with that of the theory of meaning. The chief concepts of the theory of reference are *naming*, *truth*, *denotation* (or truth of), and *extension*; those of the theory of meaning are *synonymy*, *significance, analyticity*, and *entailment*, (meaning analyticity of the conditional). Having adopted the terms "ontology" and "ideology," where the latter means "what ideas can be expressed in it [the theory]." Quine concludes that ideology is an imprecise word for "definability," which he assigns to the theory of reference.

The theory of reference suffers from certain paradoxes – the liar, Grelling's, Berry's, and others. These center around the problem of truth. But to deal with these problems in their generality, it is necessary to relativize them to language; "true-in-L," for example, and then take "L" as variable. But as Tarski showed, if "L" is formalized, and predicates such as "true-in-L," "true-of-in-L" etc. are moved from the object language into the metalanguage, then the famous definition of truth is forthcoming, viz.

"Snow is white" is true if and only if snow is white.

Tarski's methods can be applied to the other concepts of the theory of reference with similar success. But all of this lies within the theory of reference; for the theory of meaning, Quine says, we have nothing of comparable clarity or utility.[153]

In "Reference and Modality,"[154]Quine drew on his previous articles "Notes on Existence and Necessity" and "The Problem of Interpreting Modal Logic." The resulting fusion was published as a part of FLPV in 1953. Quine's purpose here is to develop the distinction between referential transparency and referential opacity. He begins with singular terms; identity of such terms presumably permits the substitution of one for the other in all contexts. But using the examples of

1. Giorgione = Barbarelli
2. Giorgione was so-called because of his size.
3. Cicero = Tully
4. "Cicero" contains six letters.

both of which he had employed in previous writings, he shows how the substitutivity fails. Here the occurrence of "Gioegione" in (2) and of "Cicero" in (4) are not referentially transparent, but opaque. A similar opacity affects belief contexts. Hence

> We may speak of the contexts "is unaware that . . .", and "believes that . . ." as *referentially opaque*. The same is true of the contexts "knows that . . .", "says that . . .", "doubts that . . .", "is surprised that . . ." etc.[155]

Modal contexts Quine holds to be similarly opaque, at least when strict modality is involved. Quine interprets "Necessarily p" as meaning "'p' is analytic" and "Possibly p" as false only if "–p" is analytic. He then repeats his arguments that although "the number of planets is 9," and "Necessarily 9 > 7," Substitution would yield "Necessarily the number of planets is > 7" which is false.

These examples involve singular terms. But similar problems arise in quantification theory. Existential generalization can lead us from "Cicero" contains six letters to (∃x)("x" contains six letters), which means that "The twenty-fourth letter of the alphabet contains six letters." Similar problems arise in belief contexts. Suppose that "Philip is unaware that Cicero denounced Catiline" is false. We have

(∃x)(Philip is unaware that x denounced Catiline)

But what is "x"? It cannot be Cicero, since the claim is false by supposition. But Quine makes the important distinction between

5. (∃x)(Philip is unaware that x denounced Catiline)

and

6. Philip is unaware that (∃x)(x denounced Catiline).

(5) involves quantifying into a belief context, (6) does not. Quine concludes "We cannot properly *quantify into* a referentially opaque context."[156] This holds for modal contexts as well.

Does this make quantified modal logic impossible? Quine thinks not, if we reject "all objects which, like 9 and the planet Venus, or Evening Star, are namable by names which fail of interchangeability in modal contexts." What would be left? Since Quine believes that strict modality involves analyticity, he holds that "putting one name of x for another in any analytic statement must yield an analytic statement. Equivalently: any two names of x must be synonymous." Quine further holds that "necessity does not properly apply to the fulfillment of conditions by *objects* ... apart from special ways of specifying them." But if any two conditions that specify "x" uniquely are analytically equivalent, then "the universe is to consist of things which are never contingently identical, which are necessarily identical if identical at all." But this would mean that concrete objects would be eliminated in favor of concepts or attributes: in short, unrestricted quantification into modal sentences has been bought at the price of adopting an "ontology of exclusively intensional or idealistic type."[157] This is an altered form of the argument he had used in "The Problem of Interpreting Modal Logic." There is, Quine notes, an escape from this; if extensional objects and intensional ones are segregated so that all quantification into modal contexts occurs only if the variables there quantified are limited to intensional ones. The same problems, Quine holds, arise if attributes are adopted in place of classes, and similar ones arise with the introduction of propositions as intensional entities. Further, Quine suggests that the same problems would probably arise with the introduction of the Aristotelian distinction of essence and accident.

To round out his argument, Quine turns to the question of statement composition.

> Let us suppose that we are dealing with a theory in which (a) *logically* equivalent formulae are interchangeable in all contexts *salva veritate* and (b) the logic of classes is at hand. For such a theory it can be shown that *any* mode of statement composition, other than the truth functions, is referentially opaque.[158]

Note that in this paper, Quine has no hesitation in using "analyticity."

"Meaning and Existential Inference"[159] was composed in 1953, although it drew on an earlier reviews of Quine's. He begins with the problem that existential statements are not true in an empty universe. That the problem is thus limited, Quine shows by the fact that it is provable that logical schemata that are true in a universe of size n are true in all universes of less than n members, except for the empty universe. Since all universal quantifications are true in the empty universe and all existential ones false, we have an easy test of whether a theorem holds in the empty universe. He then considers arguments put forward by Langford to the effect that "Fa" and "–Fa" imply "a exists." The argument, Quine says, assumes the "Fa" has a meaning which is a proposition, and since the proposition exists, so must "a." Hence, then "a exists." "The flaw in the reasoning is quickly picked out ... namely, existence of the meaning of 'a' was confused with the existence of 'a.'"

Quine notes that logic does not always follow the model of ordinary language, and quite properly deviates from it when the deviation simplifies the system and does not interfere with the process of deriving truths from truths. A well known example is the assignment of "true" to a material conditional whose antecedent is false. Problems of the sort raised by Langford usually arise with atomic sentences where the terms do not name or involve statements such as "a exists." We can assign truth values to these atomic statements arbitrarily; Quine suggests making them all false. He argues that we ought to eliminate singular terms by the same sort of Russellian maneuver that he used to eliminate "Pegasus." We thus escape from such troublesome forms as "a exists," and the rules of inference by existential generalization and universal instantiation become derivable rules and "are thus eliminated from the theoretical foundations of logic."[160]

In 1953, Quine published a collection of his articles under the title of *From a Logical Point of View*.[161] The title was suggested to him by his colleague Henry Aiken. Quine was a jazz buff who enjoyed night clubbing. He recounts that he and his wife Marjorie went with the Aikens to a night club in Greenwich Village where they heard Harry Belefonte sing a Trinidad calypso that ended with the lines

> And so, from a logical point of view
> Always many women uglier than you.

Aiken remarked that "from a logical point of view" would be a great title for the book of essays Quine was preparing, and so it came to pass.[162.] (Bruce Kuklick tells me that there is no other record of this song. Belefonte did however sing one in which the two lines quoted above appear but with "logical" replaced by "personal." Could Quine have misremembered? I do not know.) The book included "On What There Is," "Two Dogmas of Empiricism," "The Problem of Meaning in Linguistics," "Identity, Ostension, and Hypostasis," "New Foundations" together with some reflections on it, "Logic and the Reification of Universals," "Notes on the Theory of Reference," "Reference and Modality," and "Meaning and Existential Inference." Quine's previous writings had been confined largely to technical papers and books on logic, and even the semantic and ontological writings he had done since the war had appeared as articles in specialized journals. *From a Logical Point of View* introduced him to a far wider audience, not only of philosophers but lay readers as well. It is from this point that Quine's fame as a *philosopher* rather than purely a logician spread. "Two Dogmas" particularly was celebrated as marking the end of Logical Positivism (or Logical Empiricism, as it was then called). That "Two Dogmas" was less a definitive set of conclusions than a bundle of promissory notes that Quine had yet to redeem was generally overlooked. But Quine knew well that it was a progress report on a project that he had yet to complete.

The articles in *From a Logical Point of View* are chiefly on semantics and reflect where Quine stood circa 1953. But it was approximately at this time that Quine scored a breakthrough that altered his philosophical stance. The breakthrough was the development of his new notion of observation sentence. In a letter to Quine in 1965, Davidson suggested a possible way of eliminating observation sentences. On August 13th, Quine replied.

> I don't share your glee over disposing of observation sentences. For me, the glee has been on the other foot. I felt that everything the fumbling old positivists needed of their dim notion of observation sentence was suddenly right here on a behavioral basis and no mystery. My criterion of observationality, viz. social uniformity on the part of eliciting stimuli (a matter of degree, of course) nicely epitomized the scientist's proverbial recourse to evidence for purposes of intersubjective agreement. I thought it a triumph of the Neurath attitude, the embedding of epistemology in natural science in the spirit of the *Schiffsumbau* simile. Still I was disturbed, as you know, by the homology problem that rises in the intersubjective equating of stimulation. Dependence on this notion is indeed a drawback of the notion of observation sentence.[163]

As we have seen, Quine was seeking a way to combine his realistic ontology with his phenomenalist epistemology. What he did was to externalize his phenomenal observation sentences. Instead of trying to deal with the phenomenal result of impinging light rays, he took observation sentences as directly conditioned to the sensory stimulation caused by the light rays, sound waves, etc. There was no need to worry about how neural stimuli were transformed into mental representations, nor even to hold that people are aware of their neural stimulations; it suffices that the stimulations exist, of which science assures us, and that sentences can be learned by conditioning to those stimulations. And here his doctrine that sentences do not denote served him well. Observation sentences in his new form do not refer to stimulations, or to anything else; their terms can refer to the external world but the sentences do not.

This breakthrough brought a further significant change. Epistemology now became a part of science, for neural stimulation is a scientific fact and so is conditioning. Quine could use science to undergird his epistemology since he was not now seeking to *justify* science; epistemology became the science of science – that is, the science of how we have come to have knowledge, which for Quine meant how we have come to have science. Suddenly he saw a way to combine ontology and epistemology; ontology as the ontology of science, and epistemology as the science of science. I cannot give an exact date for when this reconfiguration of his philosophy occurred, but it was clearly after "Two Dogmas" and after he had written the "Introduction" to *Methods of Logic.* That "Introduction" is a nice summary of his position after "Two Dogmas" but before he discovered his theory of observation sentences.

Methods of Logic[164] was Quine's long delayed revision of his elementary textbook, his newly revised elementary logic text. In the "Introduction," he says that logic is a science.

> Furthermore, the crucial point of contact between description and reality is to be sought in the utterance of a statement on the occasion of an experience which that statement utterance directly reports. The seeing of a green patch, and the simultaneous utterance "Green patch now," constitute the sort of composite event which, in its rare occurrences, gladdens the heart of the epistemologist.[165]

This is clearly a phenomenalistic view, and one that did not appear in subsequent editions of the book. "Green patch now" reports a visual experience; it is not conditioned to stimulations. Quine remains a phenomenalist in holding that everything beyond our sensory experience is a posit; physical objects, he says repeatedly, are "theoretical." Physical object statements are part of a system of statements that touches experience only at the periphery. There is no reference here to stimulations. The statements at the periphery, Quine says, must be jealously guarded, but so must the statements of logic and mathematics that lie at the center of our scientific structure. It is between these two poles – center and periphery – that revision of the system is most likely. Quine here maintains the logistic thesis that logic is the foundation of mathematics. It is their central location in the system that accounts for their "necessity."

Before describing *Methods of Logic,* it is worth quoting a letter from Goodman addressed to an issue in Quine's holism.

> I have always been in agreement with your point (and Poincare's) that logical principles are like physical ones in that they are subject to revision if a disturbance in theory occasioned by a new observation is best accommodated by such a revision. But what constitutes a disturbance or lack of coherence in our theory will be determined by what logical principles we are applying. In other words we are using some principles of coherence when we decide that lack of coherence demands a change somewhere in our system; and the principles of coherence thus have a different status than the statements among which we demand coherence. It seems that we can no more put them all on one level than we can, analogously, reduce all the rules of logic to formulae.[166]

Goodman' point is that the transformation rules of a logical system are not formulae of the system; they are about the ways in which the formulae of the system can be manipulated. Furthermore, if logical relations create the system, they can hardly be said to be at its center. To say that the truth of the system confirms logic, as it does mathematics, is circular since the truth of logic is assumed in creating the system. To say that the falsification of the system would falsify logic contradicts the assumption of the truth of logic used to create the system. This problem did not go away; as we will see, it proved troubling.

Quine divided his new elementary text, *Methods of Logic*, into four parts: "Truth functions," "Uniform Quantification," "General Theory of Quantification," and "Glimpses Beyond." The major change in Part I has to do with proof. Quine never liked truth tables because he thought them cumbersome; a truth table for a schema involving four statement variables requires sixteen rows, for five variables thirty-two rows, for six variables sixty four rows and so on. He therefore introduces a variant of the truth table method that he considers easier and much quicker to carry out. "Truth-Value analysis"[167] as he calls it involves eight rules. Where "t" and "f" stand for true and false respectively; they are

1. Delete "t" as a component of a conjunction.
2. Delete "f" as a component of an alternation.
3. Reduce a conjunction with "f" as a component to "f".
4. Reduce an alternation with "t" as a component to "t".
5. If a conditional has "t" as an antecedent or consequent, drop the antecedent.
6. If a conditional has "f" as an antecedent or consequent, negate the antecedent and drop the consequent.
7. Drop "t" as a component of a biconditional.
8. Drop "f" as a component of a biconditional and negate the other side.

Thus the "resolution" of a schema proceeds by substituting "t" or "f" for the most frequently occurring letter, and resolving. If that does not suffice, choose a further letter and resolve and so on. Thus the resolution of "$pq \vee \overline{pr}$.⊃. $q \equiv r$" is

$tq \vee f\bar{r}$.⊃. $q \equiv r$		$fq \vee t\bar{r}$.⊃. $q \equiv r$	
q ⊃. $q \equiv r$		$\bar{r}$ ⊃. $q \equiv r$	
t ⊃. $t \equiv r$	f ⊃. $f \equiv r$	f ⊃. $q \equiv t$	t ⊃. $q \equiv f$
$t \equiv r$	t	t	$q \equiv f$
r			$\bar{q}$

The schema is consistent but neither valid nor contradictory. But even this method allows of improvement if the schema is simple enough so that one can see at once that only one assignment of truth values can make it true or false. So for example, it is clear that "$p\bar{q}$" is true only if "p" is true and "q" is false. To determine whether or not "$p\bar{q}$" implies another schema, say "$p \supset . q \supset r$", substitute into the second schema the values that make "$p\bar{q}$" true and resolve.

$$p \supset . q \supset r$$

$$t \supset . f \supset r$$

$$f \supset r$$

$$t$$

Quine called this method the *fell swoop* and comments that it is a method that "though not general, works for an important range of simple cases." "The general test of implication, applicable in every case, is truth-value analysis of the conditional: the *full sweep* as opposed to the fell swoop."

In Part II, Quine presents a method of proof for "uniform quantification," meaning quantified monadic schemata that have the same variable throughout. After introducing and explaining quantifiers, he distinguishes open schemata (those with free variables) from closed (those with no free variables). Any open sentence can be represented by "Fx". Closed sentences are either true or false, open sentences are not true or false, but are "true of" or "false of" something. "If open sentences are compounded by truth functions and quantification into a statement, the truth value of the compound will depend on no features of the component open sentences beyond their extensions." A true sentence must hold for every interpretation of its components, that is, for every class that can be its extension. But this in turn depends on what the universe of discourse is, and that can vary from problem to problem. As Quine puts it, "A uniform closed quantificational schema is *valid*, if and only if, no matter what non-empty universe 'U' is chosen, the schema comes out true under all interpretations of 'Fx', 'Gx', etc." Having pointed out why the empty universe must be excluded, Quine then defines validity for open schemata: "validity of an open schema is validity of its universal quantification." Open monadic schemata behave like truth functional schemata where "Fx", "Gx", etc. are treated like "p," "q," etc. and the truth value analysis of Part I can be applied to them in testing for consistency, implication, and equivalence. Quine notes that one schema implies another if and only if the conjunction of the first with the negation of the second is inconsistent.

Quine wants a general decision procedure for uniform quantificational schemata. This can be achieved, he says, by a series of transformations that result in either a "t" or an "f" or in what he calls a "canonical" schema, for which a separate test is necessary. He points out that the existential quantifier is distributive through alternation but not conjunction, and that the universal quantifier is distributive through conjunction but not alternation. He then lays down the rule that for uniform quantificational schemata: If S_1 and S_2 are equivalent and P_1 is like P_2 except for containing S_2 in place of one or more occurrences of S_1, then P_1 and P_2 are equivalent. He defines a "fundamental schema" as any literal (an expression "p", "q", "p", etc) or any conjunction of literals in which no letter appears twice, and a "normal schema" as any fundamental schema or alternation of fundamental schemata. Then for any closed quantificational schema S, (1) translate all universal quantifications into existential ones by the equivalence "(x) = (−∃−)". Then (2) transform the whole open schema into an equivalent which is either normal or equal to "Hx • − Hx". (3) Put "f" for any occurrence of "∃(H · −H)", and resolve. If resolution terminates in "t" then the schema is valid; if in "f" then the schema is invalid. If neither, then (4) distribute all existential quantifiers through alternations. The result is a schema that is a truth function of existential quantifications of fundamental open schemata. (5) If the resulting schema is neither "t" nor "f", convert it to normal form, and resolve. If no decision has been reached, then the resulting schema is canonical – that is, it is like a normal truth functional schema except that in place of each sentence letter, it has an existential quantification of a fundamental open schema. Quine then shows how the validity of such schemata can be tested.[168] He thus has a decision procedure for uniform quantificational schemata that can be carried out mechanically.

Part III covers general quantification theory. There are no great surprises here. Part IV is substantially changed. He treats singular terms and identity, but he also adds a discussion of classes and of number. He presents the Frege-Russell theory of number and introduces Russell's theory of types, together with a discussion of the paradox that made it necessary. He also discusses briefly the theories of Zermelo, von Neumann, and his own theories of NF and ML, along with Wang's correction. Proof theory is also discussed and Godel's incompletability proof. But while the limitations imposed by Godel's proof are made clear, Quine also remarks on the surprising fact that such a proof was possible. It has given birth to the new field of metamathematics, and Quine mentions some of the new developments in this field due to Skolem, Tarski and Godel himself. This section, which Quine entitled "Glimpses beyond," is designed as a teaser to draw students into further exploration of the field.

In the period from the end of World War II until 1960, Quine was primarily working on the ideas that were published in WO. At least by 1953, Quine had adopted the view that there is no first philosophy; there can be no justification of science beyond what science itself provides. The task of philosophy is not the justification of science or the analysis of the language of science, but the use of science to explain how we have come to have the science we do. From this perspective, there is no circularity in using science to account for science, since any thought of justification is abandoned; philosophy should be the scientific study of science. As is obvious, Quine was drawing heavily on psychology and linguistics.

Yet this period was the one in which the Cognitive Revolution swept psychology and the Chomskian Revolution swept linguistics. Behaviorism had dominated academic psychology from Watson's appearance on the scene in 1913 until the 1950s. In the 1940s the leading American behavioral psychologist was Clark Hull at Yale. Hull tried to create a hypothetical-deductive theory that included some internal constructs such as "drives," but he was clearly in the behaviorist movement. Hull inspired great confidence in many and there were high expectations for his theory. But in the late forties and the early fifties, it became clear that the theory did not work; experiments carried out to test its implications returned negative results. It was in the late forties and early fifties, with the failure of Hull's theory, that Skinner emerged as a leader among the behaviorists.[169]

> Appointed a Junior Fellow at Harvard, he developed all the basic concepts of what he called *operant conditioning* between 1930 and 1935. Operant conditioning was a method for carrying out the Watsonian program of behaviorism – to show that behavioral analysis could, in fact, account for all of human behavior, and to do so in a way that was far more effective and safe from criticism than Watson's approach had been.[170]

What made Skinner famous was his experimental work. The "Skinner Box" that he invented has been very widely used in experimental psychology – it is, as Boring describes it, "a simple box which originally was made to contain the rat, a lever, a device for delivering a pellet of food when the rat pressed the lever, and nothing else."[171] Outside of the box were recording instruments that kept a record of what the rat was doing. The box was used to study conditioning and extinction of responses. Skinner refused to postulate intervening variables between the stimulus and the response; all that mattered was the correlation between them. His concept of "operant behavior" avoided many of the problems that vexed other behaviorists. Operant behavior occurs without an observable external stimulus. When a rat that has been deprived of food for twenty four hours presses the lever to obtain a food pellet, there is no visible stimulus. Hull would have attributed the behavior to a hunger drive within the rat; Skinner refused to do so and made the rat's behavior simply a function of the time since last feeding – something that can be observed and measured objectively. Skinner's refusal to postulate internal states of the rat that intervened between the onset of deprivation and the response led some to say he dealt "with the empty organism." In his 1947 William James lectures at Harvard, Skinner extended his theory of operant behavior to language. This led to his publication of *Verbal Behavior* in 1957. Skinner regarded this work as the crowning achievement of his career.

It is not possible to specify an exact date for the Cognitive Revolution. The basic insight that underlay it was the recognition that people do not respond directly to the external stimulus but to mental representations of the stimulus. Some psychologists, like Jerome Bruner, were saying such things in the 1940s but the message did not catch on. It was first necessary for experiments to disconfirm Hull's theory. One of the areas in which the new ideas took hold was in the study of memory. It did seem clear that memories must be internal psychological states of some sort, but experiments showed not only the difference between short term memory and long term memory, but the fact that long term memory was structured and classified memories. This made psychologists aware that there was more going on there than they could account for in stimulus-response terms. Certainly one of the leaders of the revolution was George Miller, a Harvard psychologist who managed to ride the leading edge of the wave. In 1956, there was a meeting at M.I.T. where Miller presented his famous paper, "The Magical Number Seven, Plus or Minus Two" on the limits of short term memory. Also at that meeting was Noam Chomsky, who gave a paper outlining the work that subsequently appeared in *Syntactic Structures* in 1957. And also among those present were Herbert Simon and Allen Newell who presented their General Problem Solver.[172] Miller was one of the first psychologists to recognize the importance of Chomsky's work and to spread the word among psychologists. But many other movements led into the Revolution. Information theory was a new development that was only slowly absorbed in psychology. The computer itself became a major influence, as psychologists began to see the analogy between the program and machine on the one hand and the thought and neuropsychological structure on the other. Artificial Intelligence was a further parallel development but one that tended increasingly to involve psychology. All of these developments were happening more or less at the same time, together with new experiments on attention, imagery, memory, etc. There was little recognition at the time that a revolution was taking place, only great excitement over the sudden expansion of the field. It was not until Ulric Neisser published *Cognitive Psychology* in 1967 that the movement got its name or that in retrospect psychologists saw how radically the field had changed.[173]

While this was happening in psychology, there was a parallel development in linguistics, led by Chomsky. And there, no one was in any doubt that a revolution was taking place. Chomsky changed radically the conceptual basis of linguistics. The introduction of Transformational Grammar changed linguistics from a classificatory science to one employing highly sophisticated formal methods. The impact of Chomsky's ideas in psychology was made clear when his smashing review of Skinner's *Verbal Behavior* appeared in *Language* in 1957[174] – a review that left practically nothing of Skinner's book intact. But the importance of Chomsky's work went far deeper. The method that behaviorists had used to deal with the serial order of behavior was a chaining theory in which a stimulus evoked a response that had stimulus properties and so evoked a new response, and so on. This was Hull's method of dealing with the problem. Such a chain is a Markov process – the response at stage n depends only on the response at stage n – 1. Chomsky gave a formal proof that no Markov process could be adequate for learning such simple linguistic strings as mirror image strings (strings where the first half is the mirror image of the second half.). What this meant was that it was impossible in principle for the behaviorist theories to account for language learning. In due course it turned out that Chomsky's transformational grammar is formally equivalent to a Turing Machine, which meant that no theory less powerful than one that can compute any function whatever can account for language.[175]

George Miller and Jerome Bruner were Quine's colleagues at Harvard; Chomsky was at MIT. Quine knew Chomsky and had in fact recommended him for the position at MIT. Whether he knew Bruner personally I don't know, but he knew Miller, who was a fellow at the Stanford Center for the Behavioral Sciences when Quine was there in 1958–1959. He could hardly have avoided knowing about the revolutionary developments in psychology and the founding of the Center for Cognitive Studies at Harvard in 1960. Yet Quine was very slow to adopt any of the results of these movements, and only gradually incorporated some of them; he never wavered from his behaviorism.

In 1952, Quine received an offer from the University of California at Berkeley. He was also offered the George Eastman Visiting Professorship at Oxford. Quine used the offer from Berkeley to force the Harvard administration not only to raise his salary but to guarantee that after his year at Oxford, "I would receive the top Arts-and-Sciences salary, continuing with it as it rose."[176] Thus assured that Harvard treasured him, he happily accepted the Oxford invitation.

During the 1950s Quine continued to publish on logic. In 1952, he published "On ω-inconsistency and the So-Called Axiom of Infinity".[177] This article is really a reply to Rosser on the questions mentioned in the title. Rosser was particularly interested in NF which became the basis of his own system. Quine remarks that it is odd that the axiom of infinity has not been proven in NF; he thought it intuitively clear that the axiom ought to hold, but proof was lacking. Rosser had proposed "$\Lambda\ \bar{\varepsilon}$ Nn," but in 1952, no way of proving "$\Lambda\ \bar{\varepsilon}$ Nn" for NF was known. However, if "$\Lambda \in$ Nn" is added to NF, then the system becomes ω-inconsistent. But what this means is not that NF is inconsistent, but that there cannot be proven to be in NF a class that contains *only* the natural numbers; that is, any class of NF that contains the natural numbers also contains something else. Quine thinks this situation ought not to be called "ω-inconsistency" and proposed instead the new name "numerical insegrativity" defined as "Every sentence φ which is demonstrably true of all natural numbers (i.e., numerically general) is also demonstrability true of something else." Quine notes that "in such a system there is no proper translation of 'x is a natural number,' but only an infinite series of better and better approximations." This may be the case for NF, but Quine thinks the issue has yet to be settled. In fact, Specker did prove the axiom of infinity for NF and published his proof a year after this article was published.

Quine returned to his attack on modal logic in "Three Grades of Modal Involvement."[178] Why does Quine continue these attacks on modality? Partly it is because he sees modality as interfering with the functioning of quantification theory and the theory of reference. But it is also I think because he wants to block the implications of modality for ontology. Possible objects, Aristotelian essences – these are not in his view part of the ontology of science, and for him it is science that determines, and should determine, ontology.

Also in 1953, Quine published an article entitled "On Mental Entities."[179] This article contains some important indications of the development of Quine's thought. He begins with sensations, but immediately expands the argument to any form of sense data. He argues that we have on the one hand our language with its infinity of phrases and sentences; on the other we have our experience. The two connect at various points, but we do not immediately experience tables and sheep any more than we do molecules and atoms; there are no separate meanings in terms of direct experience. The statement that there is a table here, or that there is a planet somewhere in outer space, are not based on immediate experience of the table or the planet. What we have is rather a system of statements variously acquired, that, as Duhem pointed out, is "keyed" to experience as a whole.

Where does this linguistic system come from? Quine remarks "It would be irrational to suppose that those origins were rational." Surprisingly, Quine says.

> the prehistory of science was probably a composite of primitive unconscious symbolism of the Freudian kind, confusions of sign and object, word magic, wishful thinking, and a lazy acquiescence in forms whose motivation had been long forgotten,[180]

The path from these dim beginnings was guided by natural selection; "happy accidents" were favored over unpropitious ones.

How then, Quine asks, do we decide what things there are? His answer is a mixture of simplicity and pragmatic utility. "We posit molecules, and eventually electrons, even though these are not given to direct experience, merely because they contribute to an overall system which is simpler on the whole than its known alternatives." And Quine holds that tables and sheep are similarly posits that we make to simplify our theory of the world. "The crucial insight of empiricism is that any evidence for science has its end points in the senses."[181]

> Epistemologist have wanted to posit a realm of sense data, situated somehow just me-ward of the physical stimulus, for fear of circularity: to view the physical stimulation rather than the sense datum as the end point of scientific evidence would be to make physical science rest for its evidence on physical science.[182]

But Quine rejects this argument: "the science of science is a science." What this involves is the rejection of the whole traditional concept of epistemology as somehow grounding physical science. There is no circularity in using science to study science, so long as we abandon the notion of justifying science by doing so. This recognition destroys the motive for the postulated sense data to begin with. We are all in Neurath's ship, repairing it as we go.

> To repudiate mental entities is not to deny that we sense or even that we are conscious; it is merely to report and try to describe these facts without assuming entities of a mental kind.[183]

Repudiation of mental entities, Quine says, converts the iron curtain between the private and the public into a smoke screen, "a matter of varying degree of privacy of events in the physical world." This is not to deny consciousness, if consciousness is taken "as a faculty of responding to one's own responses."[184] Mental entities, Quine holds, contribute nothing to our understanding of the world.

One should note here how easily Quine conflates language and science. On his view, science is continuous with common sense knowledge, and that knowledge is inseparable from the language in which it is formulated. One should also note how smoothly Quine shifts from a repudiation of sense data to a repudiation of all mental entities. His behaviorism is everywhere apparent. Further, it is clear that sensory stimulations are being substituted for sense data; such stimulations are as it were sense data externalized. What lies behind this is his new theory of observation sentences. He no longer needs such things as sense data; stimulations are now what tie us to the world. And perhaps most important, one should note that Quine has here redefined what he is doing as science rather than philosophy. Or, perhaps it is more accurate to say that Quine has moved from the Positivist view that philosophy is the study of the language of science to the view that it is the scientific study of the acquisition of science. Quine now sees himself as a scientist, and he rejects the view that philosophy is somehow prior to science. Philosophy, as Quine now understands it, is the science of science.

Also in 1954, Quine delivered a lecture at Columbia that was subsequently published in 1955 with the title, "The Scope and Language of Science".[185] The article shows that Quine had already formulated much of the position he would set forth in WO. Quine argues that all we know or can know comes from the stimulation of our sensory surfaces, together with what is innate in us. We cannot question the reality of the external world for to do so is to surrender the terms "reality" and "evidence." Science he holds is an extension of common sense knowledge. If we want to know how we came by our science, we must use science to discover the answer. This, Quine holds, means discovering how we came to have language. "This is a question for the natural science of the external world: in particular, for the psychology of human animals." The learning process is well known: association, conditioning, training, habit formation, reinforcement and extinction and induction. But these alone are not enough; there must be some type of similarity scale that has an innate basis. Otherwise, the child could never learn such statements as "It is red." Further, similarity judgments have to be intersubjective since they are acquired from the child's caregiver. Indeed, Quine holds "mother is the source of our idea of externality." As the child's knowledge of language increases, his knowledge of the world increases, since most of what he will know about it is acquired through language. "Thought, if of any considerable complexity is inseparable from language – in practice surely and in principle quite probably." Thus the child's common sense knowledge expands with his growing mastery of language. Since science is a continuation of common sense, the concepts of "reality" and "evidence" are taken over in science. What enables us to go beyond common sense is system. "System moreover dictates the scientist's hypotheses themselves: those are most welcome which are seen to conduce most to simplicity in the overall theory."[186] Science departs from ordinary language in important respects. Indicator words such as "I", "he," "here", etc. are eliminated. Time, which also provides indicator words through tense, is replaced by four dimensional bodies. Basic sentences, Quine says, are of the form "Fx." Then, by use of truth functions and quantification, we build the system. We can then eliminate singular terms in favor of general terms. Sentences give us the notions of truth and falsity. Of course there are problems: propositional attitudes, indirect discourse, counterfactual conditionals, and sentences in the causal idiom all create trouble, but the latter is being eliminated from science. Mental objects are not needed, but abstract ones are if we are to have mathematics, which is itself a part of our science. All science, Quine says, is tentative and growing. Physical objects are becoming questionable in quantum mechanics, and even the status of the law of excluded middle is being debated.

This paper contains a number of assertions about how we come to know certain things. It is not clear whether Quine thinks this is a scientific description of how language acquisition takes place or simply a conjectural history of what might be the process. But the implicit assumption is that to explain how we came to have the science we do, what is required is a genetic history of how language is learned.

Why is the acquisition of science identified with the learning of language? Why is epistemology identified with language learning? Quine's behaviorism required that there cannot be unobservable "ideas"; we can only think in words, the utterance of which is observable. As we have just seen, Quine denies the existence of mental entities. This is not only his behaviorism at work, but also his physicalism. Quine interprets talk of mental entities as antiphysicalistic – as spiritualism or idealism. Here the influence of Watson is clear. Watson defined thought as silent speech. Thinking involves incipient movements of the musculature involved in speech. Watson called it "laryngeal thought." Hence for Quine there is no thought without language. Accordingly, Quine denies that prelinguistic children think. To trace the development of thought is therefore to trace the learning of language.

This is an odd view. It identifies our having science with how the current individual learns science, rather than with how the conceptual structure of science grew over time from its beginnings in antiquity to the present. But Quine believes in recapitulation theory; he believes that ontogeny recapitulates phylogeny. Hence he often equates the steps in language learning to the steps by which our prehistoric ancestors created language. It is extraordinary to find Quine in the 1960s using recapitulation theory which by that time was thoroughly discredited among scientists. In fact, we know nothing whatever about the origin and early development of language; we do not know when it first appeared or in what form.

There is clear evidence that the relation between thought and language is not one to one. There are societies in which it is demonstrable that the members have concepts for which there are no words in their language. It is also obvious that animals think, and that prelinguistic children think, though how they do so is now a field of intensive research, and was in Quine's time. One would have thought that if Quine believed that epistemology was a matter of psychology and linguistics, as he said, he would have devoted considerable study to these subjects, yet the evidence is clear that he did not do so in the 1950s. Later, as we will see, he grudgingly tried to incorporate some of the results of cognitive psychology into his system, but the adjustments were minimal. It is strange that having defined his project as the science of science, he based his system on outmoded doctrines and resisted making changes that progress in the science the acquisition of which he was studying required.

Quine was writing at a time when the history of science was a blooming field of research and scholarship. He knew something of the work of Kuhn and presumably others as well – Harvard had one of the foremost departments of the history of science in the country, and published *Isis* – the leading journal in the field. Yet Quine has very little to say about the actual history of science, nor about the theories of scientific development of which Kuhn's was the most popular. Quine was close to Hempel, who was thoroughly versed in these developments, yet they apparently did not interest him. His project remained one of showing how a present day child could learn language, meaning by "language" the logic of truth functions, quantification, and identity which he identified with the language of science. With what success he carried out this program, we shall see.

The Quines started off 1954 with the birth of a daughter that they named Margaret. She was born in Oxford at the Radcliffe Infirmary. Since Quine was teaching at Oxford, he could help out with the new baby and keep Douglas amused[187] (and perhaps he did). But such things did not diminish his relentless productivity.

In 1954, Quine published several articles on logic that can be briefly noted. "Church's Theorem on the Decision Problem"[188] is a simplification of Church's original proof. "Interpretations of Sets of Conditions"[189] presents a simplification of the Skolem-Lowenheim theorem following the general line of Kleene's treatment. "Reduction to a Dyadic Predicate"[190] shows that any interpreted theory formulated in quantification theory can be translated into a theory, also in the notation of quantification theory, in which there is only one predicate letter – a dyadic one. (One should recall that this was Carnap's claim in the *Aufbau*.) In "Quantification and the Empty Domain"[191] Quine shows how, by using Hailperin's proof of how a quantificational system can be modified to include the empty domain, he can modify ML to do so also by minor changes. Also in 1954, Quine wrote an encyclopedia piece on symbolic logic that was published in 1957. The coverage is very broad for such a short piece and ranges from the Stoics to Godel, Kleene, and Quine himself. It is I think remarkable that Quine wrote this piece without a single mention of Aristotle, particularly since Lukasiewicz's book on Aristotle's syllogistic had appeared four years before.[192]

Schilpp was soliciting papers for his Library of Living Philosophers volume on Carnap, and of course Quine was a contributor; his piece was entitled "Carnap and Logical Truth."[193] The problem, Quine says, is "how is logical certainty possible?" It was, Quine says, largely this problem that precipitated Logical Positivism – a movement that Quine holds began with Wittgenstein and reached maturity with Carnap. As Quine puts it, "the linguistic doctrine of logical truth, which is an epistemological doctrine, goes on to say that logical truths are true by virtue purely of the intended meanings, or intended usages, of the logical words." Quine then asserts that a first order calculus is "obvious," which seems to mean that it is obviously true. Just what this means is not clear, for Quine subsequently remarks that the logical truth "(x)(x = x)" depends on how the world is as well as language, on self-identity being a property things have in our world. Set theory is not obvious; the higher ranges of transfinite numbers are matters on which agreement is lacking. Quine then takes up the question of whether logic is true by convention,– a claim he had rejected in 1936, and still rejects. Nor will definition solve the problem of logical truth. Legislative definitions (stipulative definitions) do establish the truth of the defiens by convention. But what Quine calls discursive definitions do not; rather they rely on pre-existing equivalences. Thus definition does not solve the problem of logical truth.

These problems were recognized by Carnap when he divided the rules of his system into P-rules, meaning those true in the physical world, and L-rules, meaning logical truths – that is, analytic statements. Carnap switched from syntax to semantics when he found the constraints on his syntactic system too restrictive. But Quine has problems with the concept of analyticity, as we have already seen. Quine resumes his attack here. A statement such as "All bachelors are unmarried men" depends for its analyticity, according to Quine, on the fact that "unmarried man" is a synonym for "bachelor," so that substitution of "bachelor" for "unmarried man" turns the statement into the logical truth "All As are As." But, Quine argues, "synonymy" is just as obscure a notion as "meaning." This being so, the claim for the linguistic truth of "all bachelors are unmarried men" cannot be sustained. It is Quine's position that since the claim that sentences are true by L-rules must be rejected, all statements of logic and mathematics, like those of physical science, are synthetic.

Quine argues that Carnap's errors stem from at least four causes. The first is the failure to recognize that the attempt to define truth by convention leads to an infinite regress; this was the argument he had used in "Truth by Convention" in 1935. The second is "failures to appreciate that postulates, although they are postulates always by fiat, are not *therefore* true by fiat." The third is an "overestimation of the distinctive nature of postulates, and of definitions," because of their use in formal systems. The fourth has two parts: failure to recognize that the legislative character of postulates, and of definitions, is true also of physical science, and failure to recognize that the legislative character of postulation applies to particular acts of postulation or definition that introduce an expression at a given time, but once introduced the expression is as liable to revision as any other; the legislative character pertains to the act that introduces the expression, not to the expression itself.

The article was published in the Schilpp volume on Carnap in 1963, and the volume also contains Carnap's reply which should be noted here.[194] Carnap objects to the characterization of his position as "truth by convention"; he says "the logical truth of the sentence 'all black dogs are dogs' is not a matter of convention even in the looser sense. Once the meanings of the individual words in a sentence of this form are given ... the truth of such a sentence is determined by the logical relations holding between the given meanings."[195] Quine has also described logical truth as truth based on meanings. Carnap agrees with this: "to ascertain the truth of a given sentence, it is necessary, first, to know the meaning of each part and thereby that of the sentence as a whole; in other words, it is necessary to understand the sentence."[196] Carnap has some fun with Quine's assertion that the linguistic doctrine of logical truth "seems to imply nothing not already implied by the fact that elementary logic is obvious." Carnap agrees, and says that this statement shows that Quine apparently supports the linguistic doctrine, which of course he does not. He then turns to Quine's more serious criticisms.

Referring both to Quine's article in the Schilpp volume and to "Two Dogmas," Carnap points out that his own use of analyticity has always referred to a formalized language – "a language for which explicit semantic rules are specified that lead to the concept of truth." Quine argued that such rules are marked only by the label "semantic rules." Of course that is how they are marked, Carnap replies, just as the axioms of a formal system are marked out by being called "axioms." Why should this be an objection to meaning postulates?

Quine's second argument, Carnap says, is that there is no behavioral criterion for analyticity. Why, Carnap asks, should there be a behavioral criterion for "analyticity" but not for "truth", "naming," and the other semantic terms? But Carnap interprets Quine's view as saying that there is no clear explicandum. Carnap accepts the idea that "a pragmatical concept, based on an empirical criterion," could be used as an explicandum. Indeed, Carnap holds, referring to an earlier paper that he had written in reply to Quine, it is "plausible" that "there should be an empirical criterion for the concept of the meaning of a word or a phrase." Since lexicographers do determine meanings empirically, Quine's argument that lexicographers have no such criterion, Carnap dismisses as not convincing. So Carnap proceeds to give a way of determining empirically whether a sentence is analytic or not.

> Let us suppose that two linguists study the natural language L as used by the person X. Let us suppose that L consists of some English words and English sentences, among them the following sentence (S_1) "All ravens are black." We assume that the two linguists agree on the basis of previous experience that X uses the words "all" and "are" in the ordinary sense, and that X has repeatedly affirmed the sentence (S_1) and hence presumably regards it as true. Now the first linguist states the following hypothesis:
>
> [1] The sentence S_1 is analytic in the language L for the person X. The other linguist denies this hypothesis. In order to obtain evidence relevant for the hypothesis [1], the linguist says to X: "Mr. Smith told us that he found a raven which is not black but white, and that he will show it to you tomorrow. Will you then revoke your assertion of [1]?" Let us consider the following two of many possible responses by X:
>
> [2] "I would never have believed that there are white ravens; and I still do not believe it until I see one myself. In that case I shall, of course, have to revoke my assertion."
>
> [3] "There cannot be white ravens. If a bird is not black, then I just would not call it a raven. If Mr. Smith says that his raven is not black, then (assuming he is not lying or joking) his use either of the word 'raven' or of the word 'black' must be different from my use."
>
> It seems obvious to me that a response like [2] will be disconfirming evidence for the hypothesis [1], while as response like [3] would be confirming evidence for it.[197]

Carnap then takes up Quine's claim that any statement in a scientific theory can be altered to accommodate a negative outcome of an experiment. Carnap accepts that, but he rejects Quine's conclusion that there is no division between analytic and synthetic sentences. The concept of an analytic statement which I take as an explicandum is not adequately characterized as "held true come what may."[198] Carnap distinguishes between a change of the language and a change of a truth value of a statement whose truth value was not fixed by the rules of the language. The first kind signifies a scientific revolution and a change to a different language; the second is trivial. Carnap says his concept of analyticity refers to the status of a statement within a given language; if the language is changed, the statement is still analytic for the original language. To say that S is analytic in L means that the truth of S in L is based on the meanings of the terms occurring in S.

The Quine-Carnap exchange shows very clearly Quine's commitment to behaviorism. The basic issue between the two men is clearly the meaning of "meaning." Had the experiment Carnap suggested ever been carried out, I doubt very much that Quine would have accepted outcome (3) as evidence of analyticity. Indeed, I doubt if any experimental outcome would have led Quine to change his theory.

Quine's attack on analyticity was directed primarily at Carnap. Throughout his career Quine continued his debate with Carnap. He deplored Carnap's use of intensional and modal notions and framed his own writings to counter them. But Quine remained engrossed by Carnap's semantics. Carnap moved on to work on probability and confirmation; Quine did not follow. It is not that these issues did not interest Quine, but he did not work or write on them. It was the early Carnap that he debated, and interestingly not just Carnap's work on syntax and semantics but also the *Aufbau*. There is, as we will see, an interesting relation between Quine's later epistemology and Carnap's position in the *Aufbau*.

In 1955, Quine published an article entitled "On Frege's Way Out."[199] The article was occasioned by a remark by Geach and Black in their translation of Frege's work that Frege's way out of the paradox discovered by Russell was similar to Quine's. This led Quine to a brief review of the history of the paradox. Russell had completed his *Principles of Mathematics* before he discovered Frege's work. When he did discover it, he added the section of the *Principles* dealing with Frege and wrote the famous letter to Frege informing him of the paradox. Frege's book was already being printed when he received Russell's letter, and all he could do was to add an appendix dealing with the paradox. That means that Frege had very little time to formulate his reply. His proposal was a modification of the principle of abstraction

$$(\exists y)(x)(x \in y \equiv \varphi x)$$

to $$(y)(y \neq \hat{x}(\varphi x)\,.\supset :\, y \in \hat{x}(\varphi x) \equiv .\, \varphi y)$$ [200]

This, Quine points out, is not adequate and is circular.

Russell took a different approach in his theory of types, but Frege's approach of restricting the membership of "$\hat{x}(\varphi x)$" has proven viable. As Quine notes, all the well known methods of dealing with the paradoxes, Russell's type theory excepted, involve making exceptions to the principle of abstraction. Zermelo was the first (in 1908) to understand the situation clearly; his set theory is the ancestor of von Neumann's system and of Quine's NF, and so of ML.

> A striking feature of von Neumann's theory is that it, unlike its predecessors, provides for their being some classes which are not members of any classes at all. Such classes I called *non-elements* when I gratefully carried his idea over into my *Mathematical Logic* ... The

ancestry of *Mathematical Logic* is an open book: the parents are von Neumann's system and "New foundations".[201]

Quine goes on to state "all modern logic owes an incalculable debt to Frege. If anyone can be singled out as the founder of mathematical logic, it is by all odds he." This thesis about the history of logic became gospel at Harvard, but not elsewhere. There are many, including no less a figure than Russell, who believe that the title of "founder of mathematical logic" belongs to Boole.

In 1955, Quine also published "A Way to Simplify Truth Functions"[202] which continues his earlier paper ("The Problem of Simplifying Truth Functions" in 1952). Claude Shannon had shown that truth functional logic could be applied to problems in electrical engineering. Quine, who had long hoped for scientific applications of logic, developed ways of simplifying truth functions that were important for the engineering problem of electrical circuits. Using alternational normal form, Quine gives rules for simplification.

But by 1955, Quine's chief interest was no longer logic, although for some years he continued to publish logical papers; it was semantics and ontology. A piece that he wrote in 1955 but did not publish until 1960 was originally intended for the beginning of WO, but was superseded. Nevertheless, Quine published it under the title "Posits and Reality."[203] He begins by noting that solid and substantial objects are, if current physical theories are true, really swarms of vibrating molecules. Having thus compromised the apparent reality of everyday objects, he poses a similar question about the reality of molecules. We believe in them because by doing so we achieve a simpler, more successful scientific theory. But all that our scientific theories rest upon is our sense data. Sense data, however, are not known in any direct way either. What we really have are stimulations of our senses. It is from these that we posit sense data as we do physical objects. These sensory stimulations are all the evidence that we have or can ever have about the world.

> Sense data are *evidentially* fundamental: every man is beholden to his senses for every hint of bodies. The physical particles are *naturally* fundamental, in this kind of way: laws of behavior of those particles afford, so far as we know, the simplest formulation of a general theory of what happens. Common-sense bodies, finally, are *conceptually* fundamental: it is by reference to them that the very notions of reality and evidence are acquired, and that concepts that have to do with physical particles or even with sense data tend to be framed and phrased.[204]

These three types of fundamental things are linked together in "our one serious conceptual scheme" – that of science. And since science is our one serious conceptual scheme, we must work within it to find answers to our questions.

> Epistemology, on this view, is not logically prior somehow to common sense or to the refined common sense which is science; it is part rather of the overall scientific enterprise, an enterprise which Neurath has likened to that of rebuilding a ship while staying afloat in it.[205]

This is a further statement of Quine's view that the only way we can understand the process of how our science came to be is by the scientific study of that process.

At the beginning of 1956, salary issues arose once again. Quine wrote "Someone's chance remark in January of 1956 led me to the discovery that the top salary in Arts and Sciences at Harvard had risen to seventeen thousand dollars." Quine's was fifteen thousand dollars. He confronted the dean, McGeorge Bundy, with the letter he had been given four years earlier promising him the top salary. Bundy haggled; Quine then threatened to go to Berkeley, and Bundy gave in. Quine was not a modest man; he had a high opinion of his own worth, and with the celebrity that had come to him after "Two Dogmas" and the Berkeley offer, he had the leverage to get his way. Not only did he get the top salary, but in July he became the Edgar Pierce Professor of Philosophy. This was the chair that C. I. Lewis had held and that had not been filled since his retirement.[206] Quine, I am told, gloated that the chair was now his.

In 1956, Quine retired as president of the Association for Symbolic Logic. His address on this occasion was then published, in an expanded and revised form, as "Unification of Universes in Set Theory."[207] Quine argues that standard quantification theory is that in which all set theories ought to be formulated. "This means representing the truths of the special theory as a class of quantificational schemata with interpreted predicate letters and a chosen universe of discourse." Since a number of well known set theories are formulated with multiple universes of discourse, he proposes to show how they can all be brought into his standard form. The first theory that he discusses is the von Neumann-Bernays theory, then Russell's theory of types, Zermelo's theory, and Frege's. Quine shows how each of these in turn can be standardized, and suggests modification that would increase the elegance of some of them. He also deals with his own system NF in a long footnote where he points out that Specker has proven both that the axiom of choice is inconsistent in NF and that the axiom of infinity is provable in NF.[208]

Chapter 3
From *Word and Object* to *Roots of Reference*

In 1960, Quine published *Word and Object*[1] (hereafter WO) – a book that has generally been regarded as his major work. It will therefore require some careful examination, for it is here that he began to pay off the IOUs he had issued in "Two Dogmas."

Chapter one of WO is a brief overview of what is to follow in the book. Quine takes it as a fact established by science that our only contact with the world is through the stimulations of our senses that we receive. The problem, then, is how, from these stimulations, do we reach the science that we have. The process is that of acquiring language, for Quine identifies thinking with talking, whether to others or to ourselves, and learning language is the precondition of that. Quine then lays out his behavioristic view of learning. All learning, he holds, is by stimulus and response conditioning and habit formation. Sentences are therefore learned by conditioning to stimulations, even if they are one word sentences such as "Ouch!" or "red." The former can be taken as "That hurts," and the latter as "that's red." For Quine, sentences are prior to words; only after sentences have been learned can the learner analyze them into words. The child hears certain noises used in different sentences and so comes to separate them out as distinct units. He can then start to formulate sentences of his own, given suitable rewards for doing so. Sentences are linked together through stimulus-response chains where a sentence that is a response to a prior sentence serves as a stimulus to a subsequent one. Quine never deals, here or elsewhere, with Chomsky's argument that such stimulus-response processes cannot account for the child's acquisition of language.

Since children learn languages from their parents, each child's learning will be different, but social pressure will insure uniformity in the result.

> Different persons growing up in the same language are like different bushes trimmed and trained to take the shape of identical elephants. The anatomical details of twigs and branches will fulfill the elephantine form differently from bush to bush, but the overall outward results are alike.[2]

Given the way the child has acquired words by learning them in context, he learns what other combinations are rewarded and what punished. He thus builds up a system of interconnected sentences and words that constitute a theory, and

M.G. Murphey, *The Development of Quine's Philosophy*, Boston Studies in the Philosophy of Science 291, DOI 10.1007/978-94-007-2424-2_3,

through which a given stimulation may be transmitted by chaining to issue in a new response.[3]

Immediate experiences do not cohere in any sensible way until we posit the existence of objects that serve to account for them. Such learning about objects is intersubjective, since the sentences are taught to the child by another person. If objects did not serve as common coin such intersubjectivity would be impossible. But objects are not directly known; we do not experience objects but only their stimulatory effects upon us. Hence, objects are "posits" – something we postulate to account for the various stimulations we receive. What the objects posited in the child's theory are is what the conceptual scheme of his culture says they are. Generally, the child's theory is governed both by the non-verbal stimulations he receives and by his desire for the simplest account of what is going on that he can form consistent with the numerous stimulations he receives. Changes therefore in his growing theory will be required by new stimulations, but he makes as conservative changes as he can. Quine emphasizes that stimulations always underdetermine the theory, so there is considerable slack that may be exploited. Indeed, Quine says:

> We have no reason to suppose that man's surface irritations even unto eternity admit of any one systematization that is scientifically better or simpler than all possible others. It seems likelier, if only on account of symmetries or dualities, that countless alternative theories would be tied for first place. Scientific method is the way to truth, but it affords even in principle no unique definition of truth.[4]

This passage foreshadows Quine's latter piece on empirically equivalent systems of the world. Meanwhile, we must work from within our own scientific conceptual scheme and pursue truth as that scheme defines it.

In chapter two, Quine tries to prove one of his major theses. He states it in the third paragraph of the chapter.

> The infinite totality of sentences of any given speaker's language can be so permuted, or mapped onto itself, that (a) the totality of the speaker's dispositions to verbal behavior remains invariant, and yet (b) the mapping is no mere correlation of sentences with *equivalent* sentences, in any plausible sense of equivalence however loose. Sentences without number can diverge drastically from their respective correlates, yet the divergences can systematically so offset one another that the overall pattern of association of sentences with one another and with non-verbal stimulations is preserved. The firmer the direct links of a sentence with non-verbal stimulations, of course, the less that sentence can diverge from its correlate under any such mapping.[5]

This thesis, as stated, is applied to the home language – that is, to English. As stated, it is not a thesis about translation at all, unless one takes it as applying to the translation of English into English. One should note that here, as throughout the chapter, Quine takes a person's language to consist in his dispositions to verbal behavior.

Something needs to be said here about "dispositions" as Quine employs it. As Quine uses the term, "disposition" has the same meaning that it has in physics. Thus one says that sugar is soluble in water, meaning that if a lump of sugar is immersed in water it will dissolve. Quine accepts this usage on the grounds that, from our knowledge of the microstructure of sugar, it is predictable on the basis of physical laws that if sugar is placed in water it will dissolve. A disposition may therefore be formulated as

$$(x)(\exists y)(\exists z)(x \in y(\text{sugar}) \text{ and } x \text{ has microstructure } z \ .\supset.\ x \text{ is immersed in water} \supset x \text{ dissolves})$$

As applied to verbal behavior, the disposition becomes

$$(x)(\exists y)(\exists z)(\exists w)(x \in y(\text{human}) \text{ and } Px \ .\supset\ x \text{ receives stimulus } z \supset x \text{ makes verbal response } w)$$

The question is, what is "P"? All Quine has to say about that is that it is some unknown neurological factor which (perhaps) will someday be determined. The presently vacuous character of "P" renders "dispositions to verbal behavior" useless for any scientific explanatory purpose; Quine sees them rather as descriptive terms, denoting an unknown physiological state by one of its characteristic effects.

To make his point about language, Quine adopts the strategy of supposing that a field linguist comes into contact with a hitherto unknown tribe whose language is completely alien to the linguist. How is the linguist to establish communication with members of this tribe? Quine supposes the following series of events. As the linguist and the native confront each other, a rabbit runs by. The native thereupon says "gavagai." Is the linguist justified in concluding that "gavagai" means "rabbit"? No.

Quine's answer needs some unpacking. First, consider the case in which the stimulus that led the linguist to think "rabbit" prompts the native to say "gavagai". The linguist may think that the rabbit stimulus evoked the native's response, but of course he cannot be sure. To find out, he tries to find other rabbit appearances, and queries the native, "gavagai"? If the native assents each time to the linguists utterance, and dissents whenever he is prompted in a situation where no rabbit is present, then the linguist may conclude that "gavagai" means "rabbit." But what this really gives the linguist is not the synonymy of "gavagai" and "rabbit" but a positive stimulus meaning for "Gavagai." Quine defines "the *affirmative stimulus meaning* of a sentence such as 'Gavagai,' for a given speaker, as the class of all the stimulations ... that would prompt his assent."[6] The negative stimulus meaning is then the class of all stimulations that would prompt dissent. Thus the stimulus meaning is the ordered pair consisting of the class of stimulations that would prompt assent and the class of stimulations that would prompt dissent. But it is also necessary to determine how long an exposure to the stimulus is allowed. Quine's answer is all stimulations up to a certain time, which he calls the "modulus." Then:

> Fully ticketed, therefore, a stimulus meaning is the stimulus meaning *modulo n* seconds of sentence S for speaker *a* at a time *t*. (See Note 6)

A stimulus meaning so defined is a universal; we need to be able to say that the native makes the same response to the same stimulations on different occasions.

But now what of "assent" and "dissent"? Quine notes that there are people whose gestures of "assent" and "dissent" are nearly the opposite of ours – the Turks, for example. The linguist must guess what native verbalizations indicate assent and dissent, and then confirm his guess by trying them out on the native. But this is hardly an adequate answer. The linguist must establish reliably what locutions of the native mean "assent" and "dissent" *before* he can begin to investigate the meanings of any of the native's other noises. At this point, the linguist does not know how to segment the stream of native noise into words, if indeed it is composed of distinct words, and it is not only circular to try to give stimulus meanings for "assent" and "dissent" but it is not clear that the terms should have stimulus meanings. Quine is here trying to levitate by his own bootstraps.

The gavagai case is meant to illustrate what Quine calls "radical translation" – translations between languages that have no common etymological link. It is important here to note that stimulus meaning as Quine defines it involves stimulations only, not objects. The goal here is the relating of sentences in the native language to sentences in ours. Quine holds that sentences do not refer. The component terms of sentences refer, but not sentences themselves. Further, at this beginning stage of radical translation we do not even know if the native perceives the world as containing objects. It is also important to note that the native's responses are not simply to sensory stimulation but also to the linguists questions; the linguist must "prompt" the native to respond, rather than simply waiting for the native to remark upon what from the native's point of view may be events too humdrum to require note.

But there are problems in taking "gavagai" and "rabbit" as synonymous. The native's responses are inevitably conditioned by more than the pure stimulations. One disturbing factor is the collateral knowledge that the native has and the linguist has not. For example, there may be a particular fly that for whatever reason usually accompanies rabbits. Knowing this, the native may proclaim "gavagai" on sight of the fly even though the rabbit itself is not visible. The linguist, of course, not knowing about the fly, will be totally confused by this event. Again, there may be different words in the native's language that refer to rabbits, just as there are in English.

"Gavagai" is an *occasion sentence* – a sentence prompted by stimulations at a particular time. Even "bachelor," taken as a one word sentence, may be an occasion sentence for one who knows that the man currently recognized is indeed a bachelor. Quine therefore defines what he calls "observation sentences" which are defined thus: "occasion sentences whose stimulus meanings vary none under the influence of collateral information may naturally be called *observation sentences*."[7] The border lines between occasion sentences, observation sentences, and standing sentences – those relatively independent of immediate stimulation – are never clear. Observationality grades off from one to another.

Quine believes that *sentences* do not refer; they are true or false but there is no reference. As I have noted above, this doctrine is a consequence of Quine's adherence to Tarski's theory of truth. But *terms* do refer on Quine's view. Quine of course takes terms as extensional, so they refer to those things of which they are true. But in the case of radical translation there is a problem in determining what the term refers to. Consider "gavagai" as a term. If we translate "gavagai" as "rabbit" on the basis of the sensory stimulations we have, those stimulations would equally be caused by a series of temporal rabbit slices, or by undivided rabbit parts, or by an occasion of rabbit stuff, so that "gavagai" would mean "lo, it rabbiteth" on analogy to "lo, it raineth." In any of these cases the stimulations received by a witness would be identical. One could of course settle this question if one had available the standard English apparatus of individuation – articles, pronouns, plural and singular indicators, the copula, and the identity predicate. But in the case of radical translation, we have none of these. Suppose that we could identify in the natives stream of noise something we take to be an identity predicate; then we could ask questions such as, "Is this gavagai the same as yesterday's?" But if the native assents, what is he assenting to? What we take to mean numerical identity he may understand as "is this rabbit slice part of the same series as yesterday's?" and we would be none the wiser. We cannot therefore conclude from our translation of terms such as "gavagai" as "rabbit," that the native refers to a physical object at all. It thus appears that co-extensiveness of terms is no clearer in radical translation than is sameness of meaning.

We cannot therefore establish co-referentially between the native's terms and our own. We can also follow the same device as with stimulus synonymy by doing the same for analyticity "calling socially stimulus-analytic just the sentences that are stimulus-analytic for almost everybody." But this will not be adequate, for it would make analytic such sentences as "there have been black dogs."

> Let us face it: our socialized stimulus synonymy and stimulus analyticity are still not behavioristic reconstructions of intuitive semantics, but only a behavioristic ersatz.[8]

What can be satisfactorily translated by radical translation? Quine's answer is truth functions. If we have translations of assent and dissent, we can translate negation in that it turns any native sentence to which the native assents into one to which the native dissents. Conjunction comes over as that which produces compounds to which the native assents only if he assents to the conjuncts. And similarly for alternation. But this Quine holds is as far as we can go. We cannot achieve satisfactory translation of categoricals since they depend on the reference of their terms, and the reference of terms is, as noted, indefinite under radical translation.[9]

But curiously enough we can define more adequately a form of stimulus synonymy among sentences as long as we stick to the *intralinguistic* case. If, for native speaker *z*, the two occasion sentences "bachelor" and "unmarried man" always yield a response of assent whenever *z* is questioned under the same stimulus conditions, we may conclude that for *z* they are stimulus synonymous. Further, we can test this sort of synonymy across the whole tribe. In fact, we do not need to know what either sentence means; an alien from the planet Zinoozi, ignorant of all earth languages, could recognize such stimulus synonymy. And since for the *intralinguistic* case we can establish stimulus synonymy, we can also define *stimulus-analyticity*. "I call a sentence *stimulus-analy*tic for a subject if he would assent to it, or nothing, after every stimulus (within the modulus)."[10] For if "F" and "G" are for our native stimulus synonymous, then he will assent to "all Fs are Gs" and vice versa. But it must be emphasized that this form of synonymy and analyticity obtain only for the *intralinguistic* case.

How in the face of these obstacles does the linguist manage to come up with a translation or a dictionary? The answer, Quine holds, is *analytical hypotheses*. The constraints upon these hypotheses are multiple. First, observation sentences can be translated, even though some uncertainty remains. Second, truth functions can be translated. Third, stimulus analytic (and contradictory) sentences can be recognized if not translated. Fourth, intralinguistic stimulus synonymy can be recognized though not translated. Given these constraints, the linguist partitions the native's stream of noise into segments that he assumes correspond to words, and he equates these to English words and phrases so as to meet the constraints and at the same time to produce coherent discourse, as judged by the linguist. Where his hypotheses lead to absurd results, he imputes that to the failure of his translation, and revises his analytical hypotheses accordingly. In proceeding thus, the linguist may take certain native words as equivalent to English ones, but in doing so he is going beyond any data offered by stimulations. The yield will be a Jungle to English dictionary that provides "an *infinite semantic correlation* of sentences: the implicit specification of an English sentence, or various roughly interchangeable English sentences, for every one of the infinitely many possible Jungle sentences."[11]

So far as radical translation being the correlation of stimulations goes, the translation has an empirical base despite all the slack; the matching of sound to stimulation has an empirical content. But it should be clear that the situation is very different for analytical hypotheses. There are indeed some empirical constraints on such hypotheses, but chiefly the standard is coherence. As long as such coherence is achieved, there is for Quine no other standard that can be applied. As one would expect, given this procedure, one result will be that several English sentences which are very different from each other may turn out to be translations of the same Jungle sentence. But where this happens, one cannot say that one sentence is correct and the other incorrect. As Quine points out, there is no fact of the matter, nothing to be right or wrong about, for the only control on translation by analytical hypotheses is coherence, and given the freedom from empirical constraint that analytical hypotheses enjoy, coherent translation can take many forms.

Quine's statement that in radical translation there is no fact of the matter as to whether a given translation is right or wrong has puzzled many readers. But here one must refer back to Quine's "The Problem of Meaning in Linguistics" which we have discussed above. The linguist has no empirical basis for deciding what constitutes a significant utterance for the native, nor has he any empirical basis for assuming that the native's world view is similar to his own. But without some notion of what the native takes to be a significant utterance, the linguist's semantic correlation of English sentences to native ones makes no sense; one of the two series being correlated may be so different from the linguist's that no correlation exists. To try to match our ontological discourse to that of a people who regard dream experience as real experience and who believe in cannibal giants and supernatural thunderbirds is very likely to produce nonsense. Lacking a knowledge of what constitutes a significant utterance for the native, even the statement of the linguist's problem become impossible and there is nothing to be right or wrong about.

But suppose the linguist were to learn the native language so that he became a bilingual; would he not be able to match words and sentences between languages accurately? Would there not be a fact of the matter for him? Quine's answer is no, for in learning the native's language, the linguist would all along have been using analytical hypotheses; an adult can never learn a new language in the same way a child does because he inevitably uses his own language as a way of getting into the new language. This claim is simply asserted without proof, but Quine has a further reason. Since Quine denies the existence of mental entities such as meanings or concepts, all the data the bilingual linguist has to go on are the same sort of correlations of dispositions to verbal behavior that the monolingual linguist has. The bilingual linguist cannot compare mental representations because there are none in Quine's philosophy.[12]

Finally, Quine brings the lesson back to the thesis he proclaimed at the beginning of the chapter. The same problems of the linguist in radical translation recur with the home language. Homophonic translation does not guarantee sameness of content. Each of us has his own idiolect, and translating the idiolect of another, even within one English language community, one faces the same sorts of indeterminacies that the linguist and the native face.[13]

Quine published *Word and Object* in 1960. It was not many years later that the work of Elizabeth Spelke and Renee Baillargeon began to appear in the psychological journals, but Quine was still very much alive and working when it did. Spelke, Baillargeon and others have presented very compelling evidence that human beings are hardwired to perceive the world as a world of objects. The experimental literature contains reports of experiments on prelinguistic infants showing that at as early an age as four months they perceive physical objects as objects. The implications of this for Quine's doctrine of term reference are obvious.[14]

Further, suppose for the moment that we step outside Quine's behavioristic world and accept the notion, now standard in cognitive psychology, that the meanings of words and sentences are mental representations. Then even in radical translation, there is a fact of the matter –namely, how are the meanings in the native's mind related to those in the linguist's mind? Verbal behavior on that hypotheses comes to constitute evidence concerning that relation. This was a psychological theory available to Quine in the 1950s, and one that he knew something about, to judge from his references to Chomsky and George Miller, but although many people date the end of the reign of behaviorism in American psychology from Chomsky's 1959 review of Skinner's *Verbal Behavior,*[15] Quine remained a committed – one is inclined to say a dogmatic – behaviorist to the end of his days.

But it will not do to say that Quine's adherence to behaviorism was dogmatic. Though Quine was certainly influenced by Skinner, his behaviorism is fundamentally the behaviorism of Watson, and for Watson, all non-behavioral psychology was involved in religious metaphysics. "Ideas," "consciousness," "mental states" were holdovers from the psychology of the soul. Watson identified behaviorism in psychology as empiricism, as the only scientific approach to psychology, the only approach free of religious obscurantism. Accordingly, Quine saw behaviorism, not as just one psychological theory among many, but as the only *scientific* theory of psychology. Rightly or wrongly, Quine held this view throughout his career, though he was later forced to make some modifications in it.

Quine entitles Chapter three "The Ontogenesis of Reference." Unlike the earlier chapters, this one is specifically limited to English, and so to the English apparatus of reference. "We shall ponder the accruing of those devices to the speech habits of the child of our culture."

> The phylogenetic aspect will be neglected, except in a few speculative remarks toward the end of the chapter; and in what I shall have to say even of the ontogenetic aspect I shall venture no psychological details as to the actual order of acquisition.

The fourth sentence after the last quote reads "The babbling is a case of what Skinner calls '*operant behavior,*' emitted rather than elicited." In fact, Quine's treatment of his subject is explicitly behavioristic in Skinner's sense and based on conditioning. On the next page, Quine remarks:

> Skinner, whose ideas the foregoing sketch is meant to follow in essential respects is not without his critics.

The footnote is to Chomsky. But that is the only reference to Chomsky in the book, and Quine proceeds as if there were no question regarding the truth of behaviorism.[16]

Quine describes the child's acquisition of his first words as the result of parental reinforcement when the child's babbling generates a sound close to the word "mama" or some other English word. In due course the child reaches the stage where he begins to mimic heard words. From there on, Quine says, "he proceeds to amass language hand over fist." Word acquisition is due to the reinforcement of "society." But if the child is to learn in this fashion words such as "red" and "blue," he has to find some qualitative experiences more similar to some than to others. This requires that the child be innately endowed with a "prelinguistic quality space" in which some resemblances count for more than others. There must be such quality spaces for each sense; one cannot extrapolate from visual similarity to auditory similarity. Furthermore, as the child learns words, they tend to cluster around norms. This, Quine remarks, is "a means of reconciling continuity and discreteness." Quine does not speculate on the origin of such norms, but stresses their utility. They are essential in allowing us to segregate the continuously varying stimulations we receive, and to group different stimulations together. And such norms apply equally on the response side, allowing moderately deviant responses to be properly associated with the appropriate norms. The saving of labor so involved is due, Quine says, to the law "the norms of segments of an utterance are segments of the norm of the utterance." This "law" is not strictly true, Quine says, but it is true enough to facilitate the child's learning.[17]

Quine then moves from phonetics to phonemics. "The phonemes of a language are to speech in that language what letters are to writing." Quine describes the uses linguists make of phonemes, and describes them in terms of norms instead of classes of approximations. But he has little to say about them. The surprise is that he introduced them at all since phonemes are usually defined as meaningful linguistic units.

Quine then turns to mass terms. He takes the infant's learning of "mama" and such other words as "red" and "water" to be the learning of mass terms. "Water" is a mass term referring to a stuff that is only individuated by the use of some auxiliary such as "glass of" or "lake of"; "red" can be taken in the same way if we conceive of red as a stuff that crops up here and there. For in fact, Quine holds, "mama" is similarly conceived. Mama comes and goes, and her reappearances are greeted by the child as "Hello! Mama again."[18] For the child to attain the notion of divided reference – of singular and general terms – he must acquire our apparatus of articles, plurals, etc.

> How can we ever tell, then, whether the child has really got the trick of general terms? Only by engaging him in sophisticated discourse of "that apple", "not that apple", "an apple", "same apple", "another apple," "these apples."

The child, Quine argues, first picks these up in context, and then gradually sorts them out and learns how to use them by being rewarded for doing so. These particles are acquired simultaneously as "the child scrambles up an intellectual chimney, supporting himself against each side by pressure against the others." Each child's history of acquisition is doubtless different, but social pressure assures that the outcome will be the same. But once he has caught on to divided reference, he has the basis for singular and general terms and a world of individual objects.[19]

It is through grammatical role that singular and general terms are distinguished, and that means through predication. It is predication that unites a singular and a general term to form a sentence capable of being true or false. As terms take the predicative position in sentences such as "this is a bit of sugar," Quine holds that they can be treated as general terms true of each portion of the stuff, at least above the molecular level. Adjectives, which like "red" have a similarity to general terms, verbs, and substantives, Quine lumps together. And some general terms serve also as mass terms. Quine's example is "Mary had a little lamb," where "lamb" cam be read either way.[20]

The demonstratives "this" and "that," when prefixed to general terms, give us demonstrative singular terms – "this river", "that woman." These versatile particles are like the indicator words "I", "you", "now", "here", "today" etc. in their variability in securing reference. They also facilitate the device of ostension, which serves as one of the major aids in learning further words of divided reference. Then comes attributive position as a further way of compounding terms, usually by the application of adjectives to substantives, but substantives also play this role, as in "student prince," and "lady cop." Adjectives can also produce queer expressions, as in "mere child" which does not refer to something that is both "mere" and a "child." Such modifiers Quine terms "syncategorimatic."[21]

Quine then discusses relative terms, thus introducing a new category of expressions. These produce composite general terms. They can also yield absolute terms of the sort Quine has already discussed, as "brother of" gives rise to "brother." But their use goes far beyond that. As Quine says, "the peculiar genius of the relative clause is that it creates from a sentence '...x...' a complex adjective summing up what that particular sentence says about 'x.'" Such relative clauses may be singular terms or general terms. "Which", "who", "whom", "that" etc. may serve to form relative clauses and where such a clause is joined attributively to a substantive and has a "the" prefixed, we have a singular description such as "the car which I bought from you." Even more versatility is introduced by what Quine calls the "unlyrical 'such that.'" The construction, Quine says, demands none of the tricks of word order demanded by "which," because it divides the two responsibilities of "which": the responsibility of standing in a singular-term position within the clause is delegated to "it," and the responsibility of signaling the beginning of the clause is discharged by "such that."

This of course leads Quine to the discussion of indefinite singular terms, such as "a lion," or "a ball." When "a lion" is substituted for "the lion," the reference to a particular lion is lost, and the truth of such terms raises the question of whether there are lions to be seen. Of course, "the lion" is also true of the particular lion, and if there is none such the usage becomes questionable. But with the indefinite singular, any lion can suffice. Pronouns are definite singular terms, and their ability to substitute for other definite singular terms greatly facilitates repetition. And in sentences such as "I saw a lion and you saw a lion" where there is no indication that the same lion is referred to by both "lion" terms, the substitution of "it" for the last occurrence of "a lion" fixes the identity. Used with "such that," as in "the car such that I bought it from you," "it" serves to fulfill the responsibility of the singular terms in the construction.[22]

"Identity" is a relative term that holds between terms that designate the same thing. It is essential for divided reference because "the dividing of reference consists in settling conditions of identity." Where do you have one apple, and where do you have two? Quine notes that it is identity that solves the problem of Heraclitus and the river: you can step twice into the same river, but not into the same river stage. Quine comments that in the "earliest stage of reference" the identity sign may be flanked by two names of substantives, but as this would require both substantive terms to be conditioned to the same stimulations, he thinks it unlikely.

"At length," Quine says, there comes a phase at which a drastically new kind of posit sets in. This phase is marked by the advent of terms like "roundness": abstract singular terms, purported names of qualities or attributes. Abstract singular terms must appear with abstract general terms if they are to fit the form of predication. This is only one use of abstract terms, as Quine notes, but it requires assuming the existence of abstract objects.

> For I deplore that facile line of thought according to which we may freely use abstract terms, in all the ways terms are used, without thereby acknowledging the existence of any abstract objects.

The introduction of such terms brings with it plural endings, articles, etc., and Quine then "speculates on the development." Quine thinks that mass terms may have been the entering wedge. Instead of taking water to be a stuff that occurs in various places, the child may take it as an attribute shared by puddles, lakes, bathwater, etc. With even more plausibility, he may even construe "red" as an attribute variously instantiated. Attributes may also serve as abbreviations; instead of listing many different things that are food, one may adopt the term "edible" for the lot, and "once we start admitting abstract objects, there is no end." Classes come in as easily as attributes. Quine sums up the chapter by referring to Neurath.

> The interlocked conceptual scheme of physical objects, identity, and divided reference is part of the ship which, in Neurath's figure, we cannot remodel save as we stay afloat in it. The ontology of abstract objects is part of the ship too, if only a less fundamental part. The ship may owe its structure partly to blundering predecessors who missed scuttling it only by fool's luck. But we are not in a position to jettison any part of it, except as we have substitute devices ready to hand that will serve the same essential purposes.[23]

Quine said at the beginning of the chapter that "I shall venture no psychological details as to the actual order of acquisition." But in fact Quine repeatedly refers to how the child acquires his language, and how one development leads to another. This is especially evident in the pages 83 through 95 dealing with early acquisition, but it runs all through the chapter. And in the opening paragraph of the next chapter, Quine writes: "In the preceding chapter we have imagined the progressive acquisition of terms and auxiliary particles by the child of our culture." The purpose is clearly to describe the process by which the child has acquired the language step by step.[24]

In chapter four, Quine examines what he calls "vagaries of reference." The first case dealt with is vagueness itself; for many terms, the exact boundaries of their extensions are unclear. Colors are a notorious example; when is something orange rather than red? What precisely is included in a mountain such as Mt. Rainier; the peak of course but just how much more? These sorts of problems having to do with boundaries and borderline members of classes are well known. Quine notes that some aid is available by going comparative: "redder than", "bigger than" etc., but even so some vagueness remains. Similar questions arise with respect to the "longest river": does the Missouri River count as part of the Mississippi or not? And how is such a length measured? Does it include all bends and twists – the distance a boat would have to cover from beginning to end, or the linear distance from source to mouth?

Similar problems arise with ambiguity. Quine's example is "our mothers bore us." Is this a reference to one's mother's pregnancy, or to her lack of wit? Is a "poor violinist" one who is a violinist and in poverty, or one whose performance is bad? As Quine notes, lexicographers often deal with these problems by declaring the different meaning homonyms. Where the etymologies of the two are different, as with "bore," the homologous solution is common, or when the grammatical status is different, as with "bear." But this does not solve all problems. The process-product ambiguity ("assignment" for example) is not so resolved. Similarly, proper names such as "Paul" are ambiguous as among the various Pauls of the world. More importantly, Quine notes the words "truth" and "exist" are ambiguous in their use by many philosophers. Is "true" used univocally in "mathematical truth" and "physical truth" or are there different kinds of truth? Do numbers exist in the same sense in which trees exist? A further type of ambiguity arises with composite terms where the adjective can be used attributively or syncategorimatically. One of Quine's favorite examples of this is "intellectual dwarf": is an intellectual dwarf a dwarf and an intellectual, or is he a person with a stunted intellect? The former use is attributive, the latter syncategorimatic. And there is the time-honored case of "nothing." If one says "nothing happened," is one saying that there was no event, or that an event called "nothing" occurred?

There are also what Quine calls syntactic ambiguities. Plural subjects or objects form one type: "Lions like red meat" really means "all lions like red meat." "I hear lions" stands in for "I hear some lion(s)." "Tabby eats mice" does not mean that she is eating one just now but that she has a disposition to eat mice. But the ambiguities that interest Quine the most are those of pronominal cross reference and scope. This leads Quine to introduce the notion of *bondage* as a way of solving the cross reference problem. Problems of grouping he illustrates by the five possible groupings for "pretty little girls' camp." This leads him at once to the ambiguities of scope and the scopes of "any" and "every." "Every" Quine says, "by a simple and irreducible trait of English usage, always calls for the shortest possible scope," whereas "any," "by a simple and irreducible trait of English usage, always calls for the longer of two possible scopes." Quine illustrates this by the pair of sentences

(1) I do not know any poem.
(2) I do not know every poem.

(1) of course means that there is no poem that I know, whereas (2) only says that there is some poem(s) that I do not know.[25]

The second half of the chapter is devoted to the problem of referential opacity. Quine repeats here much that he has already published in articles, but without the explicit use of quantification, which has yet to appear in the book. The basic point however is that one should not quantify into opaque contexts. Quine is particularly occupied with propositional attitude constructions, and focuses on "believes." Some uses of "believes" he holds to be referentially opaque and some referentially transparent; the problem is to determine which is which.

> A way of doing that is to agree to localize the failure of transparency regularly in the "that" of "believes that" and the "to" of "believes to," and not in the "believes." Thus we may continue to write "Tom believes that Cicero denounced Catiline" when we are content to leave the occurrences of "Cicero" and "Catiline" non-referential, but write rather
>
> Tom believes Cicero to have denounced Catiline
>
> if we want to bring "Cicero" into referential position.

What holds for belief, Quine says holds for all cases of propositional attitude.[26]

Chapter five, Quine entitled "Regimentation." As has already become clear in the earlier chapters, departures from ordinary language are often required to clarify and simplify what is being said. Quine now argues for the advantages of a canonical notation that can avoid the ambiguities and confusions of daily talk. "Paraphrasing into logical symbols is after all not unlike what we all do every day in paraphrasing sentences to avoid ambiguity." The aim, as always for Quine, is the simplification and clarification of science.

> The simplification and clarification of logical theory to which a canonical logical notation contributes is not only algorithmic; it is also conceptual. Each reduction that we make in the variety of constituent constructions needed in building the sentences of science is a simplification in the structure of the inclusive conceptual scheme of science.

So Quine now undertakes to show how ordinary language can be paraphrased into the canonical notation. He emphasizes several times that the paraphrase is not synonymous with the original. This is not only because he considers synonymy a dim notion but because the point of the paraphrase is to remove the ambiguities of the original; if they were synonymous no clarification would be gained. But Quine does add a remarkable passage about the relation of logic to reality.

> The quest of a simplest, clearest overall pattern of canonical notation is not to be distinguished from a quest of ultimate categories, a limning of the most general traits of reality. Nor let it be retorted that such constructions are conventional affairs not dictated by reality; for may not the same be said of a physical theory? True, such is the nature of reality that one physical theory will get us around better than another; but similarly for canonical notations.

This statement enunciates an important doctrine of Quine's. Science, as we have seen, determines what there is, but this can also be put as the language of science determines what there is, and the language of science is first order logic. It is clear therefore why Quine repeatedly attacks modality, attributes, propositions, and mental states; they do not belong in the austere palace of science.[27]

The first step in regimentation is the introduction of quantifiers that replace indefinite singular terms. And here Quine takes a moment to pay tribute to Frege whose *Begriffsschrift* he calls "a thin book that may be said to mark the start of mathematical logic." Since the concepts of variables, of "such that," and of bondage have already been introduced, Quine can bring them together under the concept of the quantifiers. The singular description and class abstraction are also variable binding and yield singular terms. Quine does not here quibble about intensions or propositions, being content to postpone their elimination to a later point in the book. The addition of quantification enables him to clarify the problem of referential opacity by the rule: "no variable inside an opaque construction is bound by an operator outside." But since singular terms other than variables are to be eliminated, they need not detain us. Quine argues strongly for construing objects as four dimensional. Relativity theory indicates that this conception fits the way the world is, and it gets rid of tenses; verbs can be taken tenselessly and philosophical problems such as those of Heraclitus and Zeno and personal identity solved.[28]

The constant singular terms also pose a problem. This is obvious with terms such as "Pegasus," but even when the term has a designatum, there is a problem of the proper construction. Quine solves this by construing such singular terms as general terms in predicative position, though true of one thing or none. Thus "Socrates exists" becomes "(∃x)(x is Socrates)" which is true, and "Pegasus exists" becomes "(∃x)(x is Pegasus)" which is false. Quine limits this device to singular terms that have no internal structure and proposes to call them "names." Singular descriptions also require elimination, and Quine does so as follows: Let "$y = (\iota x)(\ldots x \ldots)$"

be taken as equivalent to "... y ... and y only", "so (ιx) (... x ...) exists" becomes "(∃y)(... y ... and y only)," following in Russell's footsteps.

But there is another class of singular terms such as "x + y", "x + y^z", etc. that are vital for mathematics. Quine proposes to eliminate "+" by adopting a triadic relative term "S" such that "Swxy" is true if and only if "w = x + y." But although "w = x + y" can be eliminated in favor of "Swxy," Quine's dictum that to define a term is to show how to avoid it operates in reverse. Taking "Swxy" as fundamental, we can then reintroduce "w = x + y" as an abbreviation for it. The classic example of this is the definition of the truth functional operators in terms of the Sheffer stroke. Having done so, we restore the standard operators as abbreviations and proceed merrily on our way.

> Attribute abstraction can be treated analogously [to class abstraction], by reparsing "a = x [... x ...]" as formed by an irreducible two-place variable-binding operator in the fashion "aO_x (... x ...) ;" then the old "x [... x ...]" comes out as "(ιw) (wO_x (... x ...))". Similarly for the abstraction of relations. Thus evidently nothing stands in the way of our making a clean sweep of singular terms altogether, with the sole exception of the variables themselves.

As Quine remarks, "It is one of the consolations of philosophy that the benefit of showing how to dispense with a concept does not hinge on dispensing with it."[29]

In chapter six, Quine lays out his own position. His first move is to define "eternal sentences" – "a sentence whose truth value stays fixed through time and from speaker to speaker." For this to be the case, if the sentence describes an event, the date, time, and place, must be objectively specified to rule out any variation in truth value. To avoid the difficulty that an eternal sentence may never actually be uttered, Quine proposes to take them as sequences.

> We can take each linguistic form as the *sequence*, in a mathematical sense, of its successive characters or phonemes. A sequence $a_1, a_2, \ldots a_n$ can be explained as the class of the n pairs $a_1, 1 \quad a_2, 2 \quad \ldots \quad a_n, n \quad \ldots$ We can still take each component character a_i as a class of utterance events, there being here no risk of non-utterance.

Quine notes that many philosophers have considered sentences to be the names of propositions and have held that truth and falsity apply to the proposition rather than to the sentence. But Quine holds that sentences do not name; their terms may refer but they do not. Accordingly, he dismissed propositions as useless metaphysical baggage. By doing so he rejects the notion of translation as equating sentences of different languages because they refer to the same proposition and dismissed the idea that propositions are meanings of sentences.[30]

Quine then turns to modality. Here he rehearses the arguments that he had previously made in his articles, especially in "Three Grades of Modal Involvement," which has already been discussed. He also rejects intensions and attributes for reasons that he has previously stated in print. There is no criterion of identity for attributes, unlike classes, and classes can do the work of attributes. Further, one cannot quantify into intensional abstraction, whereas one can quantify into class abstraction. But having rid himself of propositions, meanings, intensions, and the like, one might wonder how Quine can proceed to talk in these terms. Quine was well aware of this objection, and he answered it.

> What is involved here is simply a grading of austerity. I can object to using a certain dubious term at crucial points in a theory, on the grounds that to use it would deprive the theory of its desired explanatory force, but I can still use and condone the term in more casual or heuristic connections, where less profundity of theoretical explanation is professed. Such grading of austerity is a natural adjunct of the scientific enterprise, if we see that enterprise in Neurath's way.[31]

Quine can go slumming in the ghetto of intensionality without moving from his pristine lodgings in the halls of science. Or so he says.

But Quine has still to deal with the problem of propositional attitudes. He focuses on indirect quotation as being the most tractable since we have the direct quotation to use as a standard, and discusses a number of possible ways of dealing with the problem. Indirect quotation is similar to translation, and indeed becomes translation if the direct quotation is in a language different from our own. But none of the devices Quine discusses seem adequate to him. Finally he refers to Brentano's and Chisholm's claim that intensional terms such as "meaning" and "synonymy", etc. cannot be translated into a non-intensional terms. Quine concludes that "relativity to non-unique systems of analytical hypotheses invests not only translational synonymy but intensional notions generally." This poses for Quine a choice:

> One may accept the Brentano thesis either as showing the indispensability of intensional idioms and the importance of an autonomous science of intension, or as showing the baselessness of intensional idioms and the emptiness of a science of intension. My attitude, unlike Brentano's, is the second.

Quine holds that "if we are limning the true and ultimate structure of reality," we should adopt "the austere scheme that knows no quotation but direct quotation and no propositional attitudes but only the physical constitution and behavior of organisms." But if our aim is "only to dissolve verbal perplexities or facilitate logical deductions we are often well advised to tolerate the idioms of propositional attitudes."[32]

But the Gordian knot is not yet completely severed. Quine has still to deal with the problem of subjunctive conditionals and dispositions. With regard to the latter, Quine repeats the analysis he has given elsewhere. Disposition terms such as "soluble" are admissible because they are based upon the microstructure of the substance involved that makes the dissolving in the liquid named predictable from physical laws. Where the microstructure is unknown, use of dispositions is often accompanied by "ceteris paribus." The utility of the disposition terms so qualified rests on the scope of the "ceteris paribus" or other clues from the context. Dispositions to verbal behavior rest upon the presumption of an underlying microstructure, to be provided in due course by neurology. The subjunctive conditional however leads Quine to remark:

> The subjunctive conditional depends, like indirect quotation and more so, on a dramatic projection; we feign belief in the antecedent and see how convincing we then find the consequent.

This is not far from the possible worlds analysis that would later emerge with Lewis and Stalnaker, but Quine goes no further. And Quine also rejects the causal idiom as lying outside the austere bounds of science.[33]

What is left within those bounds? Truth functions, quantification, and predication. "The doctrine is that all traits of reality worthy of the name can be set down in an idiom of this austere form if in any idiom." This is Quine's "philosophical doctrine of the categories." The doctrine, Quine says, "is philosophical in its breadth, however continuous with science in its motivation." This austere science is to be applied to the universe, for Quine accepts only one. But is it not possible to discover some set of categories in the more traditional sense – some set of general terms "on the basis of which all traits and states of everything could in principle be formulated?" Quine's answer is no, for "we can prove that openness is unavoidable, as long anyway as the sentences of a theory are included as objects in the universe of that theory." Indeed, Quine notes, all terms of the theory are reducible to a single dyadic term.[34]

Quine entitles his final chapter "Ontic Decision" and adds a subtitle "Nominalism and Realism." But he does not turn at once to a defense of classes as one might expect. Rather, he starts by seeking to justify our confidence in the existence of physical objects rather than sense data. One might think that sense data have a better claim to empirical certainty than objects have, if objects are known only through the senses. But Quine argues that on the contrary the direct conditioning of observation sentences to sensory stimulations counters that claim. But his chief argument is that physical objects are posits that have greater utility for theory than sense data. "In a contest for sheer systematic utility to science, the notion of physical object still leads the field." Further, the posit of physical objects works better as a focus of successful communication, and terms for such objects are basic to our language.[35]

Quine also defends the priority of physical objects as against attributes. His argument is that those who believe that the use of a general term to refer to an object carries with it an attribute of that object are in error. The attribute then comes to be considered as an object referable to by a singular term, and so is taken as a suitable object for quantification. There are also, Quine says, those who, failing to distinguish between abstract singular terms and concrete general terms, have concluded that concrete general terms carry no commitment to attributes, and the same holds for abstract singular terms. The result is a confusion that then leads to distinguishing different meanings of exist. This mare's nest can be cleaned out, Quine holds, by holding fast to quantification theory as a way of settling ontological commitment.

Quine inveighs against what he considers as defective nouns such as "sake" and "behalves", "miles", "minutes", and the like. If one asks for the identity conditions of "miles," and they are said to be alike, then how can they be many? "'Length in miles' is said to be understood as true of this or that number relative to this or that body or region." Similarly, possible objects, terms such as "Pegasus" that refer to no objects, "facts" – a term which Quine thinks serves no purpose, infinitesimals, and ideal objects are all defective in one way or another. But not all such terms are useless. "Infinitesimal" served a purpose for Newton and Liebniz, but that purpose is better served by the theory of limits. And there are myths that similarly are useful.

Relativity theory has replaced Newtonian theory as our view of reality, but the simplicity of Newtonian theory compared to Relativity theory and the fact that for low velocities (compared with that of light) its equations yield results that are an excellent approximation to those of Relativity theory, give us every reason to continue to use it.[36]

Not all defective nouns should be banned. Quine uses as his case example that of the ordered pair. This provides Quine with an opportunity to discuss the process of explication. The concept of the ordered pair goes back to Peirce and Frege, but a satisfactory explication of it was first given by Wiener who defined it as $\{\{x\},\{y,\Lambda\}\}$ and then by Kurotowsky who gave it as $\{\{x\},\{x,y\}\}$. Which is right? Both are, according to Quine. The point of explication, he says, is not to find an expression synonymous with the original. "We fix on the particular functions of the unclear expression that made it worth troubling about, and then devise a substitute, clear and couched in terms of our liking, that fills those functions. Beyond those conditions of partial agreement, dictated by our interests and purposes, any traits of the explicans come under the head of 'don't cares.'" Other examples of this process are Russell's theory of descriptions, and the explications of number by Frege, von Neumann, and Zermelo.

> The condition upon all acceptable explications of number (that is, of the natural numbers 0, 1, 2, ...) can be put ... succinctly ...any *progression* – i.e., any infinite series each of whose members has only finitely many precursors – will do nicely.

The choice among these alternatives is, at most, a question of convenience; they all do equally well. Quine also applies this doctrine of explication to the mind-body problem. Here, as usual, Quine takes the physicalist line.

> If there is a case for mental events and mental states, it must be just that the positing of them, like the positing of molecules, has some indirect systematic efficacy in the development of theory. But if a certain organization of theory is achieved by thus positing distinctive mental states and events behind physical behavior, surely as much organization could be achieved by positing certain correlative physiological states and events instead.

"Introspection may be seen as a witnessing to one's own bodily conditions, as in introspecting an acid stomach, even though the introspector be vague on the medical details."

Explication, Quine portrays as a method of replacing an unclear term by a clear one that performs the same functions as well or better. But explication does not serve to eliminate classes. The paradoxes certainly provide a reason for wishing it so, but no alternative is available. And the theory of classes "confers a power that is not known to be available through less objectionable channels." Hence the problem is how, while keeping the theory of classes, to avoid the paradoxes. Russell's theory of types, Zermelo's set theory, von Neumann's theory (and of course Quine's own) are all attempts to do this, all equally effective. So like it or not, classes must be kept if mathematics is to be kept. Nominalist proposals to do without classes require too high a price in foregone mathematics. And in a lengthy footnote on page 243, Quine tries again to make the point that he is not a nominalist and never has been.[37]

Quine then turns to semantic ascent. The idea is not new; as he notes this is Carnap's distinction between the material mode and the formal mode, though Quine has some differences with Carnap. Nevertheless, the doctrine is essentially the same.

> The strategy of semantic ascent is that it carries the discussion into a domain where both parties are better agreed on the objects (viz., words) and on the main terms concerning them.

Quine illustrates it by supposing a dispute about the existence of miles. The argument he thinks will accomplish nothing until we "ascend" to talking about the word "mile" and the contexts in which it can be usefully employed. Semantic ascent has been much used in mathematics; the axiomatizing of a theory was seen as a way of gaining rigor, but too often the derivations made implicit use of knowledge not given in the axioms. One response was disinterpretation – i.e., taking the system as purely formal. But Quine holds that with the advent of logic, which as usual he credits to Frege, "alternative and more refined precautions against question-begging became available to axiomatic studies; and it is a case, precisely, of what I am calling semantic ascent."

In closing, Quine reiterates his doctrine that logic and mathematics owe their seeming certainty to their central position within the overall scheme. "The question what there is is a shared concern of philosophy and most other non-fiction genres." All sciences are concerned with that question. So too, at some more abstract level, is mathematics. "What distinguishes between the ontological philosopher's concern and all this [science] is only breadth of categories."

> It is scrutiny of this uncritical acceptance of the realm of physical objects itself, or of classes, etc. that devolves upon ontology. Here is the task of making explicit what had been tacit, and precise what had been vague, of exposing and resolving paradoxes, smoothing kinks, lopping off vestigial growths, clearing ontological slums.[38]

This the philosopher must do from within some conceptual scheme, for there is no working without one. He must work on and improve the scheme from within, but he, like all scientists, must preserve its links to sensory stimulation.

In WO, Quine has defined what he considers to be the language of science; it is first order logic. His philosophic project is the reconstruction of this language by showing how it is acquired, hence the emphasis on language learning. But it is also vital for Quine's project that this language not be encumbered with extraneous matters such as modalities, intensions, and mental entities. For Quine, what there is is what there can be said to be in the language of science – i.e., in first order logic; it is in this sense that it limns the structure of reality. Hence Quine's repeated efforts to banish intensions, modalities, and mentalism. The preservation of the austerity of the language of science is for him a fundamental commitment.

Quine finished writing WO by 1958 and evidently sent a copy of some parts of it to Goodman. On December 15, 1958,[39] Goodman wrote questioning Quine's use of "surface stimulations." The retina is not on the surface; few nerve endings are. Goodman questions just what Quine means by "surface" since the relevant nerve endings appear to be somewhere inside the body. Goodman also added that he was trying to help Chomsky get a job. Quine wrote back on December 24, 1958,[40] defending his use of "surface" and says he does not use nerve endings although at some points he had. He also says that WO is finished. Goodman replied on the 30th,[41] again questioning Quine's use of "surface." Why not say nerves? or the brain? Nothing about the surface is critical, Goodman says, and internal sensations are not mentioned.

> (1) Much of what hits my surface plays no role in knowledge. (2) Some things that originate beneath the surface give rise to perceptions, etc. (3) What you say about the surface is equally true about many things other than the surface.[42]

Then on May 10, 1960, having seen the published work, Goodman wrote

> I am relieved, bemused, and almost frightened by the way the big bad behaviorism so forthrightly declared is immediately thrown out the window with the adoption of a shameless theory of innate ideas.... involving of course prelinguistic organization into elements, prelinguistic natural kinds, and prelinguistic ordering. – Wheeee!![43]

One should note here how Quine resolved the problem of the epistemologically real and the ontologically real. Having decided to redefine epistemology as the science of science, Quine can base his epistemology on sensory stimulations to the existence of which science attests and whose status as real is not different from objects. But while Quine continues to talk of objects as posits or even myths, he does not apply these terms to sensory stimulations.

WO has been called Quine's major book. It is therefore appropriate to note some of the problems with it here. First, Quine holds that a young child learning the word "ball" learns to say "ball" when he receives certain stimulations. "Ball," Quine takes as acquired by conditioning. But by Quine's account, "ball" is a sentence; it is really something like "that's a ball." Quine further holds not only that sentences do not refer, but that terms of the child's sentences cannot refer to anything in the external world. So what does the child really learn? He apparently cannot refer to the ball; the light reflected from the surface of the ball may lead him to say "ball," but the word (sentence) has for him no reference. Quine is clear that such one word sentences are not about the stimulations that prompt the child's response. In fact, he appears to believe that they are not about anything. So what does the child learn? Learning a language is not just about learning when to say "ball." The word (sentence) must have meaning and reference. According to Quine, it has neither.

Second, Quine gives no indication of how the linguist learns the Jungle words for "assent" and "dissent." According to Quine, the linguist guesses what jungle noises mean "assent" and "dissent." But if the linguist can do that, why can he not learn other jungle words the same way? Quine thinks he cannot; the most he can do is to learn the stimulus meaning for words like "gavagai." But he can't learn the words "assent" and "dissent" that way; he can only learn a stimulus meaning if he already has the jungle words for "assent" and "dissent." It is not even clear how the linguist can learn to partition the stream of jungle noise into words, if indeed the natives use words.

Third, Chomsky showed that a single state Markov process is inadequate for the child's learning of language. But Quine's causal chains, where one word serves as the stimulus for the next, is just such a process. We need an explanation here of how Quine deals with Chomsky's argument.

Fourth, Quine's dispositions to verbal behavior involve reference to neurophysiological states of which we know nothing. That of course makes them useless for explanatory purposes. Quine says they are purely descriptive, but when he employs them in S-R chains, they are being used for explanatory purposes.

Fifth, Quine claims to be carrying out a scientific investigation of how people learn language, starting from the very first words (sentences) uttered. But he repeatedly claims his account is "speculative." If that were true, it is difficult to see what he hopes to accomplish. But it is clear by the end of the book that he thinks his speculative history of language learning is a true account, although in skeletal form. Quine's conjectural history is not a plausible substitute for an investigation of the actual process.

Sixth, Quine says that the child first learns sentences, then learns words by analyzing the sentences into constituents. How does one analyze a one word sentence? Or a multiword sentence, first learned by conditioning?

Seventh, Quine uses metaphors to cover ignorance; "hand over fist," "scrambling up the chimney", etc. The metaphors may be well chosen, but they are not explanations.

Eighth, he previously agreed with Brentano and Chisholm that intensional terms form a closed circle that is not translatable into non-intensional terms. But he cannot avoid propositional attitudes, and although he says they are intensional, he claims to deal with them in non-intensional ways.

But one should also note that Quine's claim that observation sentences are holophrastically conditioned to stimulations allows the child to acquire language without already having linguistic skills. This, if true, solves the problem of learning first sentences (or words).

In the decade following the publication of WO, Quine had a number of problems that he needed to solve. First, he needed to defend, clarify, and expand the position he had advanced in WO. Many of Quine's philosopher colleagues, as well as the popular audience he had acquired since the publication of "Two Dogmas of Empiricism," were flummoxed by WO. A book on how children learn language was hardly what they had expected from a man regarded as a distinguished logician. Some thought he had abandoned philosophy for psychology, others that he had abandoned logic for linguistics, but most were simply puzzled and confused. Quine needed to explain and clarify his doctrines.

But Quine wanted to do more than that. He was well aware of the status he had acquired since the publication of "Two Dogmas" and he wanted to use that standing to spread his doctrines not only among philosophers but also among his lay readers. If this meant doing some popular writing, Quine was more than willing to oblige. It should be recalled that rather than use introductory logic texts such as Tarski's, Quine had insisted on having his own introductory text, which he revised and republished repeatedly over the years. He now felt himself in a position to put his stamp on the field of philosophy.

But Quine also had to deal with other problems stemming from his earlier work. Abandoning analyticity as he had in "Two Dogmas of Empiricism" left him with no adequate foundation for logic and mathematics, and he needed to find one. This would not prove to be an easy task and would not be completed before RR was published, thirteen years after WO. And there were also other papers that he wrote, some for special occasions, some on the history of logic, some on logic, and some on subjects of general interest, particularly on science and scientific method. Quine was always working on several projects at once, and his publications reflected that fact. He had therefore much to do if he was to consolidate his position as the leading American philosopher.

When he finished the manuscript of WO, Quine turned to another project – one he had postponed. During his year at Oxford in 1953, he had lectured on the comparison of set theories formulated by different writers. He had decided then to do a short book on comparative set theory. But when he returned to Harvard, the completion of WO took precedence and the book on set theory was postponed. With the completion of the manuscript of WO in 1957, he returned to the set theory project.

But, as was his custom, while he worked on set theory, he also published articles on other matters. In WO, he had emphasized simplicity as an important characteristic of scientific theories. Why was this so? In "On Simple Theories in a Complex World,"[44] he suggested some answers. Simple theories, Quine remarks, are usually regarded as superior in "beauty and convenience" to empirically equivalent ones that are more complex.[45] But what is remarkable is that they are also regarded as more probable. Why should this be so? Quine proposes four reasons. One is wishful thinking. A second is "a perceptual bias that slants the data in favor of simple patterns."[46] Third, "there is a bias in the experimental criteria of concepts, whereby the simpler of two hypotheses is sometimes opened to confirmation while its alternative is left inaccessible." And fourth, "there is a preferential system of score keeping, which tolerates wider deviations the simpler the hypothesis."[47] Are these reasons enough to account for the role that simplicity plays in science? Quine leaves that question open; at this point he had no answer to it.

Also in the year he published WO, he published "On the Application of Modern Logic."[48] This was a long standing interest of Quine's, who had long hoped that applications of logic would prove as valuable in science as those of mathematics. That applications of mathematics are vital in science everyone knows. But the applications of logic are less in evidence and Quine points to several areas in which these applications are important. He deals first with the applications to mathematics. While there are spectacular examples of this, such as Godel's incompletability proof, Quine also stresses the importance of the process of translating ordinary language into logical notation. This "paraphrasing" requires a clarification of what is being paraphrased that brings out difficulties and presuppositions otherwise obscured. But there are applications also in engineering. Claude Shannon discovered that the simplest type of logic – truth functional logic – can be applied to the designing of electrical circuits. It is remarkable, Quine comments, that so simple a part of logic should prove so valuable, for in truth functional logic "every specific question of logical implication or equivalence ... can be settled by a routine calculation."[49] Quantificational logic admits of no such decision procedure, as Church proved, but there is nevertheless a general routine of computation whereby any specific question of logical implication or equivalence can be settled affirmatively if the right answer happens to be affirmative, though not if the right answer is negative.[50] Quine points out the similarities between programming a computer and paraphrasing ordinary language into logical notation. But the relation goes beyond that, into machine design and the theory of computing, and was developed in the 1930s by Church, Turing, Kleene, and Post. And computers are now being employed to find ways of simplifying truth functions. As Quine remarks:

> The utterly pure theory of mathematical proof and the utterly technological theory of machine computation are thus at bottom one, and the basic insights of each are henceforth insights of the other.[51]

From 1960 to 1963, Quine was working chiefly on his book on set theory. But he had other problems to confront. In WO, he had been unable to find an adequate way of dealing with propositional attitudes. The problem did not go away. On May 8, 1962, he wrote to Hempel, "I despair of reducing the idioms of propositional attitude to scientifically admissible constructions ... I think the gulf is important: that it represents essentially the difference between final and efficient cause, and between mind and matter."[52] Propositional attitudes were to remain a thorn in Quine's side for years to come. He could neither abandon them nor find an adequate way to deal with them.

One of the objections to WO raised by critics was that it was psychology rather than philosophy. Quine's reply was

> I'm unmoved by the objection that this ceases to be philosophy and becomes psychology. Let it be psychology; still there are aspects having what has been called philosophic interest – those bearing on the nature of evidence, of scientific method, of scientific knowledge, or of knowledge, verifiable or otherwise.[53]

The next year, Quine published a revised edition of *Methods of Logic.* The chief change from the earlier editions occurs in Part III where Quine gives revised rules for EI and UG. There is also an appendix that gives Godel's proof of the consistency of quantification theory, and Lowenheim's theorem that "Any consistent quantificational schema comes out true under some interpretation in the universe of positive integers." The other changes are minimal.[54]

In 1962, Quine was working on set theory. Appropriately, he published an article entitled "The Ways of Paradox" in *Scientific American.*[55] This was a nontechnical article on a technical subject – paradoxes. He classified paradoxes into the veridical, the falsidicus, and antinomies. He takes the paradox of the barber as an example of the first type, since it leads to the true conclusion that there can be no barber in any village who shaves all and only the village men who do not shave themselves. A better example is Grelling's paradox. Some adjectives apply to themselves. Thus "short" is a short word and "English" is a word in English. Some words do not apply to themselves: "long" is not a long word, and "German" is not a word in German. Let those adjectives that do not apply to themselves be called "heterological." Then is "heterological" heterological? If it is, then it isn't. If it is not, then it is. So if it does apply to itself, then it does not apply to itself, and if it does not apply to itself, then it does apply to itself. As an example of the second type he uses DeMorgan's proof that $2 = 1$. Let $x = 1$. Then $x^2 = x$. Hence $x^2 - 1 = x - 1$. Dividing through by x – 1, we have x + 1 = 1 or 2 = 1. The fallacy is of course that $x - 1 = 0$. But the antinomies are the interesting paradoxes. Quine discusses Grelling's paradox, the liar paradox, Zeno's Achilles and the tortoise, Berry's paradox, Russell's paradox, and Cantor's paradox. He gives Tarski's solution to the liar and suggests stratification as a way of dealing with Russell's paradox. Finally, he discusses Godel's proof of the incompletability of arithmetic; this is not an antinomy but a veridical paradox. "That there can be no sound and complete deductive systematization of elementary number theory, much less pure mathematics generally, is true."[56]

In 1963, Quine published *Set Theory and its Logic*.[57] Sets, it should be recalled, are classes that are members of something (other than themselves). There are, as Quine points out, various set theories in the literature; hence the importance of a work in which the various versions are compared. This was the original idea of the book. But as he began to write the book, his drive for simplicity and elegance, and his desire for minimal ontological commitment, led him to expand the early chapters. As noted previously, Quine distrusted set theory because of the paradoxes. As he remarked in the Preface, "Intuition here *is* bankrupt"[58] since it led us into contradiction. Therefore, one of the chief objectives of this book is to explore how much of set theory can be constructed on the basis of very weak assumptions. He begins the book with what he calls a "virtual theory of classes," – "virtual" because in this section of the book he does not use real sets at all but only imitation sets constructed from first order logic with " ϵ" imbedded into unanalysable constructs. Beyond that point, he has to invoke real sets but only finite ones, and is able to construct a good deal more of mathematics on this basis than one might have thought could be done. As Quine comments, "I am merely concerned to get the foundations of number theory before positively assuming that there are infinite classes."[59] Thus instead of specifying numbers by an initial number and a successor function, which would yield an infinite set, he takes numbers to be the predecessors of some finite number, and yet is able to define addition, multiplication, and exponentiation on this basis. Eventually, of course, he has to introduce infinite sets, and for this he uses von Neumann's theory. But Quine remarks:

> Any objects will serve as numbers so long as the arithmetical operations are defined for them and the laws of arithmetic are preserved. It has sometimes been urged that more is wanted; it is not enough that we account for pure arithmetic; we must also account for the application of numbers in the measurement of multiplicity. But this position, insofar as it is thought of as contrary to the other, is wrong. We have seen how to define not only arithmetical operations but also the *Anzahlbegriff*, "α has x members," without having yet decided what numbers are.[60]

In fact, Quine holds, it is indifferent whether we use the Frege-Whitehead-Russell numbers, where "n" is the class of all n-membered classes, or the von Neumann numbers, where "n" is the number of predecessors of "n." Any progression will do, so long as we use it consistently.

Although Quine gives an account of transfinite set theory, he had little interest in it. Unlike Godel, Hilbert, Sierpinski, and others, he gave little attention to the continuum hypotheses, which are barely mentioned here. He was of course concerned about the status of the axiom of choice, but Godel had shown in 1940 that the axiom is consistent with the axioms of set theory, and can therefore be added without danger of contradiction. But Quine's preference was for the finite; it is only in the infinite that the paradoxes of set theory bloom. Despite his ingenious constructions, Quine found that he could not do without infinite sets if he was to capture the full power of mathematics. Even after completing the book, he remained skeptical about set theory, and ended the volume by saying "Let me seize this opportunity to leave the reader with a sense of how open the problem of the best foundation for set theory remains."[61]

In 1964, Quine published "Ontological Reduction and the World of Numbers."[62] This is an important paper, for it introduces – I believe for the first time – the notion of *proxy functions*. Quine builds up to it by noting that Frege's reduction of numbers to set theory and von Neumann's reduction are equally successful despite the fact that they define number differently; what they preserve and share is structure.

> Since, according to the Lowenheim-Skolem theorem, any theory that admits of a true interpretation at all admits of a model in the natural numbers, G.D.W. Berry concluded that only common sense stands in the way of adopting an all purpose Pythagorean ontology: natural numbers exclusively.[63]

But Quine finds this sort of ontological reduction too loose. It is true that by the Lowenheim-Skolem theorem any true theory has a model in the natural numbers, but what we want is "to accommodate all the *truths* of θ [the original theory] – all the sentences, regardless of axiomatizability, that were true under the original interpretation of the predicates of θ," and the Lowenheim-Skolem theorem does not guarantee that this can be done.[64] What is required for a reduction of a theory θ to a different theory θ' is a *proxy function.*

> We specify a function, not necessarily in the notation of θ or θ', which admits as arguments all objects in the universe of θ and takes values in the universe θ'. This is a proxy function.[65]

Quine then shows that the reductions that ought to be regarded as satisfactory, like those of Frege and von Neumann, meet this standard and those not satisfactory do not. Thus was born one of Quine's most important notions – the proxy function. The consequences of this notion for his theory of reference would soon become apparent.

The publication of *Set Theory and Its Logic* marked the completion of a significant change in Quine's work. The change had been underway since the war, and was already clear in *WO*. Referring to a meeting at MIT that involved Wang, Dreben, and some mathematicians, Quine remarked that he participated little, "my preoccupations having receded further from the mathematical pole with the completion of *Set Theory and Its Logic*".[66] In fact, after *Set Theory and Its Logic*, Quine did little work *in* logic; instead he turned to writing *about* logic in an effort to establish his views of logic, and mathematics, as the dominant perspective in the field.

Quine decided in 1964 that he wanted a reduction in his teaching load. Harvard President Nathan Pusey did not agree. Quine thereupon began soliciting offers from rival universities – Penn, NYU, CCNY, and Case. He then went to Europe on a sabbatical, where, he happily recalled, "we were lionized."[67] While in England he received a major offer from the University of Chicago. Pusey gave in, and Quine's load was reduced to one course per term.[68]

In "Two Dogmas" Quine had called himself a pragmatist. Was that any different from being an empiricist? In his course, Philosophy 148, on the philosophy of language, he was led to think about Peirce's famous maxim. As Quine paraphrased it, "whatever sense it may make to talk of a presumed object is to be [sought] in the difference which the truth or falsity of the sentence would make in our experience." The paraphrase, he said, can be accepted as the essence of empiricism. But he rejected Peirce's definition of truth as the view to which scientific inquiry will converge in the long run. We lack a clear notion of scientific method, and there is no clear way to apply the concept of limit to such a process as inquiry.

In 1964, Quine published a popular piece on the "Foundations of Mathematics" in the *Scientific American.*[69] One must take Quine's popular writings seriously because he often used them to present new ideas or refine old ones, as he does here. He treats his reader gently, starting off with the need to clarify and eliminate such notions as the "infinitesimal" and "imaginary numbers" in terms of limits and ordered pairs. He then takes up number, and indicates how it can be explained by either Frege's definition or von Neumann's. What these examples show is the "process, we see, of reducing some notions to others, and so diminishing the inventory of basic mathematical concepts."[70]. He carries this to the point where the only terms required are conjunction, negation, universal quantification, and membership. Indeed, he says that all mathematical problems can be transformed into problems of set theory. "Either this augers well for the outstanding problems of mathematics, or else set theory is itself as deep in problems as any part of mathematics." "The latter is the case,"[71] Quine says. Noting the paradoxes of set theory, he remarks that set theory is far less secure than the mathematics founded on it. "Foundation ceases to be the apt metaphor; it is as if a frail foundation were supported by suspension from a sturdy superstructure."[72] So the great hope of Frege, Whitehead, and Russell that the problem of mathematical certainty could be solved by reducing mathematics to logic has been abandoned, and Quine takes mathematics as better founded than set theory, on which it was supposed to rest. Note that this leaves Quine with the problem of finding a new basis for mathematical truth.

Quine was also struggling with the homology problem that had plagued WO, where he had assumed that the linguist and the native have the same stimulations. In September of 1965, he was playing with the idea of defining synonymy by taking pairs of statements that the community members thought synonymous and seeing what agreement there might be. He was also pondering whether bilinguals could be used to define synonymy for different languages and between them. Is there, he wondered, a place for a theory of mechanisms? "Observation sentences are those closest to the untrained perceptual mechanism ... certainly that was the idea originally, dimly." But it would be some time before Quine found what he thought was an adequate answer to these problems.

In 1966, two collections of Quine's articles were published. *Selected Logical Papers,*[73] is a collection of twenty-three of his papers on mathematical logic, twelve originally published in the *Journal of Symbolic Logic*, nine in other publications, and two not previously published. Quine did the actual editing while a Fellow at the Center for Advanced Studies at Wesleyan in the spring of 1965. Also published that year was *The Ways of Paradox*[74] – a collection of twenty-one essays, three of them from the 1930s, nine from the period 1951 to 1955, and nine from 1960 to 1964. While less technical than those in *Selected Logical Papers*, they were hardly popular writings and included such topics as "Ontological Remarks on the Propositional Calculus" from 1934 and the "Foundations of Mathematics" from 1964.

Quine had an interest in the history of logic, and although he wrote relatively little on the subject aside from the historical notes in his other works, his views were well known and were very influential at Harvard, though much less so elsewhere. He had little interest in the algebraic tradition, and held that mathematical logic began with Frege. In 1967, Jean van Heijenoort edited a volume of readings entitled *From Frege to Godel*,[75] which was put together largely by Quine and his students. The selections were made chiefly by Quine, Dreben, Wang and other of his students. Quine himself wrote the introductions for the readings from Russell, Whitehead and Russell, and Schoenfinkel. The preface begins with the statement "The second half of the nineteenth century saw a rebirth of logic." The algebraic tradition is dismissed in the first paragraph with mentions of Boole, De Morgan and Jevons (none of Peirce who appears in the volume only in footnotes and an occasional reference.) The "great epoch" of logic is said to begin with Frege's *Begriffsschrift.*[76] Second in his pantheon of logical greats stood Russell, and in 1966 he published an article entitled "Russell's Ontological Development" in the *Journal of* Phillosophy.[77] Quine pays eloquent tribute to Russell's enormous impact on philosophy in the first half of the twentieth century, and even beyond. "Russell's books have run to forty, and his philosophical influence, direct and indirect, over this long period has been unequalled."[78] "Russell's name is inseparable from mathematical logic, which owes him much, and it was above all Russell who made that subject an inspiration to philosophers."[79] It was Russell's work that inspired Quine to become a logician and a philosopher. In 1962, he had written to Russell that "*Principia Mathematica* was what, of all books, has influenced me the most."[80] It was Russell's work that inspired the logical positivists, Carnap among them, and it was also what inspired Wittgenstein's *Tractatus* which so profoundly influenced the Vienna Circle and English philosophy.

> I think many of us were drawn to our profession by Russell's books. . . We were beguiled by the wit and a sense of new found clarity with respect to central traits of reality. We got memorable first lessons in relativity, elementary particles, infinite numbers, and the foundations of arithmetic. At the same time we were inducted into traditional philosophical problems such as that of the reality of matter, and that of the reality of minds other than our own. For all this emergence of problems the overriding sense of new found clarity was more than a match. In sophisticated retrospect, we have had at points to reassess that clarity, but this was a sophistication that we acquired only after we were hooked.[81]

Quine's reassessment here focuses on Russell's ontology. In his early work, Russell was prodigal in giving existence and reality to objects, attributes, numbers, and Homer's gods. But in "On Denoting" in 1905, Russell criticized Meinong's impossible objects – the golden mountain, for example – and showed how to eliminate them through his "theory of singular descriptions, that paradigm, as Ramsay has said, of philosophical analysis."[82] Quine credits this theory with having shown that the unit of communication is the sentence rather than the word. Russell discovered his paradox in 1901; his theory of types did not appear until 1908. But Russell saw problems with classes – the problem of the one and the many. In his 1908 paper, he used his theory of incomplete symbols to explain classes away.[83] But the paradox held also for class concepts or propositional functions, and so the theory of types deals with propositional functions and was transmitted to classes through contextual definition.[84] Russell, Quine says, was careless about the distinction between use and mention, and so propositional functions were sometimes attributes and relations and sometimes open sentences or predicates. Russell accorded existence to things, and what he called "subsistence" to universals. Quine says that he confused meaning and reference, and hence took terms that had no reference as meaningless.[85] Subsistence was dropped by 1914, but qualities and attributes then became existents. While Quine applauds the death of subsistence, he does not approve Russell's adoption of "facts" as what a proposition corresponds to. Quine thinks Russell's fact ontology is the result of his confusing meaning with reference. But in *Our Knowledge of the External World*, he took sense data as "the logical atoms for the construction of the rest of the world,"[86] and external objects as classes of sense data. Something of the excitement generated by that book creeps into Quine's exposition.

> It was a great idea. If executed with all conceivable success, it would afford translation of all discourse about the external world into terms of sense data, set theory, and logic. It would not settle induction, for we should still be in the position of predicting sense data from sense data. But it would settle the existence of external things. It would show that assumption superfluous or prove it true: we could read the result either way.[87]

Russell did not claim that this reconstruction could actually be done.

> But the illustrations gave a vivid sense that the concepts of *Principia Mathematica* could be helpful here and the many ingenious turns and strategies of construction that went into *Principia* could be imitated to advantage.[88]

This of course is what Carnap attempted to do in *Der logische Aufbau der Welt*. Despite that heroic effort Quine wrote that "One must in the end despair of the full definitional reduction dreamed of in recent paragraphs."[89]

Meanwhile, Russell turned away from phenomenalism to what he called "neutral monism," which, although it did not reach physicalism, became increasingly naturalistic. And Quine quotes from Russell a passage that might well serve as a description of Quine's own later philosophy:

> There is not any superfine brand of knowledge, obtainable by the philosopher, which can give us a standpoint from which to criticize the whole of the knowledge of daily life. The most that can be done is to examine and purify our common knowledge by an internal scrutiny, assuming the canons by which it has been obtained.[90]

During the 1960s, Quine was approached by David Duskin of Random House to write a small book meant for freshman English.

> When I protested that I had lost touch with the freshman mind, he suggested that I think of a former student who might still have the touch: hence Ullian.[91]

Quine's former student Joseph Ullian was teaching philosophy at Washington University, and agreed to join Quine in the project. The result was Quine's only co-authored book, *The Web of Belief.*[92] Why did Quine agree to do such a book? As noted above, Quine often used popular pieces to present ideas that he did not develop in his technical writings. This book offered him a chance to deal with questions of general scientific method. And Quine knew that such a book bearing his name would reach a far wider audience than freshman English. The manuscript was completed in 1967 and appeared in print in 1970. Quine subsequently described it as a book about rational belief, and interestingly the web metaphor used in the title was barely mentioned in the book.

Since they focus on belief, a definition is necessary, and Quine and Ullian define belief as "a disposition to respond in certain ways when the appropriate issue arises."[93] Rather than taking "belief" as applying to some abstraction such as a proposition, they take it as meaning "believes true" applied to sentences. "Disbelief" or belief that the sentence is false, is distinguished from nonbelief or the absence of belief in a sentence. Similarly, they distinguish belief from knowing, which requires not only believing a sentence to be true but its being true. Beliefs are often changeable; sometimes because they contradict each other, sometimes because new beliefs conflict with old ones and one or the other must give way. To illustrate, they make use of a murder mystery. "Let Abbot, Babbitt, and Cabot be suspects in a murder case." The problem then is how to reconcile conflicting beliefs about who the murderer is as new evidence is acquired, and the example is well used to show how beliefs change.

Chapter two is entitled "observation." Here the authors restate Quine's position that "beliefs face the tribunal of observation not singly but in a body." But they make an exception for observation sentences which can be (almost) confirmed by the phenomena they report. Even so, the claim is not that they are infallible, but that speakers of the language will ordinarily agree on the spot as to their truth or falsity. They admit that an observation may be so deviant from the others that confirm a hypothesis that it is set aside, but they hold that the deviant observation still has to be accounted for. A theory will not be rejected despite deviant observations until there is a better one to take its place. Here they are following Kuhn's theory of scientific revolutions. Nevertheless, observations are "the boundary conditions of our body of beliefs."[94]

Observation sentences do not rest on other beliefs, they say. But they also claim that there is another body of beliefs that "look for support neither to other beliefs nor to observation. To understand them is to believe them." These are the truths of logic, which are either immediately self-evident, or can be derived from self-evident truths by steps, each of which is self-evident. And all statements that are instances of logical truths are true. They even say that "if observation is relevant to our acceptance of a claim then the claim fails to qualify as self-evident." The authors do not consider mathematics to be entirely self-evident but only partially so.

> Logic and mathematics seem to be the only domains where self-evidence manages to rises above triviality; and this it does, in those domains, by a linking of self-evidence on to self-evidence in the chain reaction known as proof. And even mathematics lends itself only partially to such treatment; this was brought home to us by Russell's paradox, Euclid's postulate of parallels, and Godel's incompleteness theorem.

Indeed, they say that "mathematics generally (including geometry and number theory as well as set theory) is from an evidential point of view more like physics and less like logic than was once supposed."[95]

Self-evidence is the author's substitute for analyticity, but it is a dubious one. Given the controversy that has surrounded the law of excluded middle, it can hardly be called self-evident. Nor will "if grass is red then $2 + 2 = 7$" be taken as self-evident by anyone not already indoctrinated with logic. But our authors are writing for freshman English and so do not want to argue the issue of analyticity for such a group; nevertheless, self-evidence is not an adequate substitute.

The authors also include a chapter on testimony. Humans have, they suggest, an innate tendency to tell the truth; natural selection has seen to that. But many also lie. Expecting truth, people are credulous, and so testimony cannot be taken at face value. One can appeal to others in the hope that they will correct our errors, but the whole community can be wrong; the belief in the geocentric universe was held by almost everyone (in Europe) until the sixteenth century. The authors' remedy is to appraise what others say in terms of their motives and morals, and their opportunity to have witnessed what they claim to have seen.[96]

The next chapter is devoted to hypotheses. "A man adopts or entertains a hypothesis because it would explain, if it were true, some things that he already believes." But the possibilities for false hypotheses are so vast that the authors offer guidance in choosing good ones. These they call "virtues." Virtue one is conservatism. This embodies Quine's maxim of minimum mutilation: "preserve as much of past belief as possible – other things being equal." Virtue two is generality. "The more general the hypothesis is by which we account for our present observation, the less of a coincidence it is that our present observation should fall under it." Virtue three is simplicity. Thus in fitting a line to a set of data points on a graph, we want the simplest line we can get. Their rule is that when two hypotheses are equally supported on other grounds, we should choose the simpler. Simplicity is subjective, yet it works in practice; why? One reason, the authors say, is that we are less likely to count observations as deviant the simpler our hypothesis. But the chief reliance is natural selection. Standards of simplicity that have worked in the past have enabled their bearers to survive where others perished. It is not clear whether they think innate simplicity standards led to survival, or that whatever the standards that led to survival, they seem to us the simplest. But this is Quine's solution to the problem he raised in "Simple Theories in a Complex World." Virtue four is refutability. A hypothesis that is irrefutable explains nothing. The authors say "the degree to which a hypothesis partakes of virtue IV is measured by the cost of retaining the hypothesis in the face of imaginable events." Astrology is their example of an irrefutable hypothesis. Virtue five is modesty. Given a choice between a modest hypothesis and an extravagant one, the modest one should prevail. These virtues are meant as guides to hypotheses that "explain the past and predict the future." But they emphasize again that predictions and explanations depend on combinations of beliefs, not single ones alone. Obviously, these "virtues" are normative; thus the authors do invoke normative considerations.[97]

The authors then turn to "induction, analogy, and intuition." In induction, "the central factor is the expectation that future cases will work out like past ones." But Goodman's paradox of green and grue emeralds shows how little that claim tells us. Goodman's paradox is this: consider the predicate "grue," defined as meaning "green before tomorrow, blue thereafter." Then are all emeralds grue or green? The problem is that all past experience equally confirms the application of both predicates to emeralds. Not every trait of past events carries on into the future; the problem is to specify which ones do. Goodman's answer is that the traits that do carry on into the future are those that are "projectible." What that means, the authors say, is that things already seen to be appreciably similar will prove similar in further ways. The question, what traits are projectible? can as well be put as simply this: what counts as similarity? And why do certain similarities catch our notice? The author's answer is again natural selection. In a crude form induction is the process of learning from past experience, something all other animals do as well as humans. And if the mode of learning is universal among animals, it has to be due to evolution and natural selection. Analogy also provides a method of learning – one that is often not recognized as such. This is another case of projecting similarities from the past into the future, but a less reliable one than induction. "We pictured analogy as by-passing inductive generalizations." "We often frame a new generalization on analogy to an established one." But intuition, the authors hold, is based on unconscious analogies at work.

> Where intuition has anything at all to be said for it, it has something making no mention of intuition to be said for it: sensory clues that may not have registered as such, long forgotten beliefs, analogies more or less vague.

So much for intuition.[98]

The next topic the authors tackle is "confirmation and refutation." "The obvious way to test a hypothesis is by testing its consequences." But this alone is too vague, since an observation can confirm many hypotheses. To narrow this vagueness, they call for precision, which they add to the list of virtues as virtue VI. "The more precise a hypothesis is, the more strongly it is confirmed by each successful prediction that it generates." Precision is mainly increased by measurement, which allows testing for functional dependence. Similarly, redefinition of terms, especially through explication, will usually sharpen a hypothesis. If the hypothesis was reached by induction, predictions are just instances of the generalization. But Goodman's example of grue shows that not every generalization is confirmed by its instances. To be so, the generalization must be "lawlike," which the authors explicate as "couched in projectible predicates." But this runs afoul of Hempel's paradox of the ravens. "All ravens are black" is logically equivalent to "All non-black things are non-ravens," which is confirmed by a yellow ribbon. To meet this problem the authors widen the definition of "lawlikeness" to include not just the sentences confirmed by their instances but all their logical equivalents. This throws the burden back on projectibility, and the authors claim that "in practice we seem to be able to recognize projectibility."

Any prediction derived from a generalization that turns out to be false refutes that generalization. But since a hypothesis alone rarely implies testable consequences, the predictions are actually derived from multiple sentences, and the false result can be accommodated by dropping any one of the sentences that led to it. The authors note that some philosophers have tried to apply probability to hypotheses. Quine and Ullian find this an unsatisfactory procedure since they see no general way of assigning such probabilities, although they admit that there are cases where this can be done. That is, they hold to the frequency theory of probability rather than the subjective theory. But they are careful to note that they are not here discussing statistical hypotheses.[99]

General hypotheses or laws explain the observations they imply. The authors take explanation and prediction as roughly correlative.

> Since an explanation implies what it explains, it would have sufficed for prediction if it had been what we started with. Conversely, most knowledge that sanctions a prediction serves also as an explanation of what it predicts if what it predicts comes true.

This claim is too strong. A doctor, looking at a patient's current symptoms, can often predict what his future symptoms will be, but symptoms do not explain each other. This is of course a common cause phenomenon, but it shows that the authors' claim needs modification. What explanations explain is the truth of the predicted sentences. But as noted, it is groups of sentences that imply the truth of the predicted sentence.

> The explanation must imply more than what was to be explained, but each part of the explanation, short of the whole, must fall short of implying what was to be explained. Note incidentally that together the two stipulations prevent an explanation from containing the sentence explained.

The authors also consider statistical explanations, but they do not consider them to be true explanations. They believe statistical regularities ought to admit of non-statistical explanation. It is surprising that they seem unaware of the probability models being used to explain learning by philosophers like Patrick Suppes. But they make an odd error in presenting their argument: they present their example in terms of "inoculation" when it is clear from the context that they mean "vaccination".

There is, the authors hold, a "mutual reinforcement" between an explanation and what it explains. The explaining hypotheses gain support if they explain something taken as true, and the truth of the statements implied gives support for the truth of the hypotheses. And a hypothesis gains credibility "from the mere circumstance that it would, if true, explain something for which we have no other explanation." But a hypothesis that always works, such as "whatever God wills happens," is not testable and so tells us nothing. The authors are also skeptical of explanation in terms of motives and character traits. In line with Quine's behaviorism, they hold that such hypotheses are virtually irrefutable. They also reject teleological hypotheses such as "we have eyes so that we can see," rather than "we can see because we have eyes." Biological explanations in terms of fortuitous variation and natural selection are sound, but teleological ones are not.[100]

The authors note "two basic purposes of language: getting others to do what we want them to do, and learning from others what we want to know." The former purpose particularly shows "the utility of lying." But this, the authors claim, is offset by our inherited tendency to tell the truth, (surely an extraordinary claim). But there is a further purpose: teaching others to believe what we believe. This makes the question of what supports our beliefs crucial.

> We convince someone of something by appealing to beliefs he already holds and by combining these to induce further beliefs in him, step by step, until the belief we wanted finally to inculcate in him is inculcated.

This is the method of convincing someone of something. But if our aim is just persuasion, the authors recommend what they call "shallow analysis": "appeal to a common ground of beliefs which are no more particular and detailed than necessary for agreement." Where we support our efforts by appeal to our own observations, we need a reputation for truthfulness and nice judgment if we expect our testimony to be accepted. If we are met by contrary beliefs, then the authors recommend "overwhelming" our adversary with evidence and "undermining" his contrary belief by directly challenging it. We may prevail; on the other hand, we may find ourselves persuaded, in which case we have learned something. This leads the authors to the cautionary note that we should distinguish between the desire to be right and the desire to have been right, which "is the pride that goeth before a fall."

Finally, our authors deal with evaluative beliefs. This they note is a matter of criteria, and "the criteria we resort to tend to be less evaluative than the original evaluative belief itself." Presumably what they mean is that the criteria may contain factual components that can be settled empirically, as Stevenson had argued in *Ethics and Language.*[101] Aesthetic evaluation may be a matter of taste where no agreement is possible. In the case of moral belief "we appraise the action's consequences in the light of rules and principles, and we appraise the rules and principles in the light of the expected consequences of their acceptance." "Happily," our authors say, "disagreements on principles are not common since men have common basic sentiments."[102]

The Web of Belief is more than a popularization of Quine's views on rational thought. It contains important material that does not appear elsewhere in his writings in such detail. The most important of these is the introduction of the six virtues for guiding hypothesis formation. One of the glaring omissions in most books on scientific method is that while they deal with the testing of hypotheses, they rarely have anything to say about the formation of hypotheses. Quine's virtues are very general, but they mark the point at which he was willing to introduce normative considerations into scientific method. Lacking any viable inductive logic, he thought this the most that could be done.

In 1968, Quine published a brief paper entitled: "Linguistics and Philosophy,"[103] a paper he had given at NYU earlier in that year. The paper is a defense of behaviorism against Chomsky's rationalism. Given Chomsky's status in linguistics, and his rationalistic doctrines, he was an opponent that Quine could not ignore. The defense however is not an attack on Chomsky's work but an attempt to co-opt at least part of what Chomsky claimed. Quine had previously admitted that humans have an innate quality space that is necessary for learning language. Thus Quine:

> The behaviorist is knowingly and cheerfully up to his neck in innate mechanisms of learning-readiness. The very reinforcement and extinction of responses, so central to behaviorism, depends on prior inequalities in the subject's qualitative spacing, so to speak, of stimulations.[104]

Quine appears to concedes more when he says "unquestionably much additional innate structure is needed." Since qualitative spacing of stimuli is observable in other animals, there must be something more that accounts for human ability to learn. Quine hopes that such further innate "mechanisms" will be found, but he claims "this would be no refutation of behaviorism, in a philosophically significant sense of the term; for I see no interest in restricting the term 'behaviorism' to a specific psychological schematism of conditioned response." Ostensive learning, Quine holds, is still a matter of conditioning. But he suggests that getting the child beyond that point probably requires further innate mechanisms "needed specifically to get the child over this great hump that lies beyond ostension." Chomsky's claim that behaviorism cannot account for language learning, Quine says, is "of a piece with my doctrine of the indeterminacy of translation." This claim is certainly not obvious and requires an argument that is not forthcoming here.

Quine goes on to argue that "behaviorism broadly so called" has no problem with Chomsky's innatism.

> The old empiricist looked inward upon his ideas; the new empiricist looks outward upon the social institution of language. Ideas dwindle to meanings, seen as adjuncts of words. The old inner-directed empiricists – Hobbes, Gassendi, Locke and their followers – had perforce to formulate their empiricist standard by reference to ideas; and they did so by exalting sense impressions and scouting innate ideas. When empiricism is externalized, on the other hand, the idea itself passes under a cloud; talk of ideas comes to count as unsatisfactory except insofar as it can be paraphrased into terms of dispositions to observable behavior. Externalized empiricism or behaviorism sees nothing uncongenial in the appeal to innate dispositions to overt behavior, innate readiness for language-learning.[105]

This is not as far from Watson's behaviorism as one might think, for Watson had no problem with instincts, and Quine concedes much less than he seems to. Having proclaimed that epistemology was now a matter of psychology and linguistics, Quine could hardly ignore Chomsky's ideas that were revolutionizing linguistics. Quine actually concedes very little; what he does do is to claim that "new empiricism" is his brand of behaviorism. But this is no longer Skinner's empty organism. Quine is postulating innate psychological mechanisms to account for overt behavior. He would, and will, claim that these "mechanisms" are physiological, but he has no idea what they are.

On February 13, 1969, Quine wrote to Davidson regarding a proposed meeting at the Center for Behavioral Sciences at Stanford:

> I would urge inviting Burt Dreben as a further participant. I didn't mention him before because he seldom goes to meetings. But he lately, on hearing of the plan from me,

> expressed interest in participating. He is very keen in both logic and philosophy, though publishing little. For years now he has read most of my writings and discussed them with me before I've sent them off; and has regularly caused improvements. He is a great clarifier of my philosophy to our students as well as to me. He is penetrating, articulate, and stimulating, and wholly free of philosophical nonsense and modality and such. He would be a boon to our discussions, with or without a paper of his own.[106]

When Dreben was appointed at Harvard and Wang was not, there was considerable criticism of the appointment among logicians. The letter may suggest why the appointment was made as it was. Dreben did his best work in discussions; he was a keen and constructive critic and he substantially aided many of his colleagues with their work by his critiques. Quine apparently found him a very useful aid.

The Vietnam war had sparked growing protests among American college students that mounted steadily through the 1960s. Quine's political views were conservative and he was outraged by the student movement, which inevitably reached Harvard. "Academic life," Quine wrote, "was abruptly deteriorating. Student disorders had been blandly indulged at Berkeley and were spreading." In due course there was a sit-in in University Hall at Harvard – the main administrative building. What Quine called the ugly parade of hippies in Cambridge and Boston led him to consider leaving Harvard. He arranged for a leave that he spent at Rockefeller University as a possible escape from Harvard if it blew up. It did not, and Quine consoled himself by going to Europe.[107]

On February 1, 1968, Quine gave a paper at the International Philosophy Year at Rockport, N.Y. entitled "Philosophical Progress in Language Theory."[108] The most interesting parts of this paper deal with truth functions and linguistic universals. Language universals, he claims, represent our imposition of our categories on foreign speakers.[109] "The more our translations rely on these self-imposed guides, the less can justly be said for linguistic universals." Again Quine asserts that the field linguist can determine which of the native's utterances signify assent and which dissent, but as before he does not explain how the linguist is to do this. But when he comes to conjunction and alternation, Quine modifies the position he had taken in *Word and Object*. He had thought conjunction was fully translatable, but he now says it is only partly so. "Lacking information, you may not be prepared to dissent from the statement that I was born in Pittsburgh, nor to dissent from the statement that I was born in Detroit; but still you will dissent from the conjunction of the two."[110] For alternation, one might not be willing to dissent from the alternatives, but could dissent from the alternation itself. Quine sees the same kind of quasi-determination for substitutional quantification, since the native may not give the same response to every substitution instance of a quantification. Objectual quantification is more difficult for translation since we would have to know what the native's language counts as names and singular terms and at least to some extent what can count as their referents.[111] The classical categoricals face similar problems, and so depend on analytical hypotheses. Quine does not hold that an alien culture is inscrutable. Much can be determined about it apart from language by observations as well as the translation of observation sentences. Quine neither adopts not denies the Whorf hypothesis. But he does make the point that for his doctrine, the more difficult the translation of an alien tongue the more unlike ours its world view may be. Just when we conclude that the alien's worldview is really alien instead of blaming our translation, Quine does not say. A similar problem is posed for the linguist when he finds native testimony that contradicts his theory of the native's language; when one should change the theory or throw out the informant's testimony will depend on one's reading of the native's psychology. We have the same problem in our home language when for example we find ourselves in debate on religious issues.

> If we master the religious vocabulary sufficiently to join substantive issue with religious speakers, we do so not just by taking their statements as true, but by reconstructing in some measure the psychology of their belief.[112]

But Quine makes it clear that by "psychology" here he means behavioral psychology. Even so, the introduction of psychology marks a significant change in his theory of translation.[113]

The article shows several things. Quine has retreated from his claim that all truth functions are translatable in the face of the examples given. He has however in no way softened his claims for the indeterminacy of translation or for behaviorism. And he has now gone beyond the Vienna Circle position that philosophy is the philosophic study of the language of science; he now sees *his* philosophy as part *of* science. He has yet to claim that he himself is a scientist, but he is not far from it.

In 1969 Quine published a second and revised edition of *Set Theory and Its Logic.*[114] The revisions are extensive, involving some fifty-five pages. Quine remarks that in section 23, he has added an axiom schema which provides in general that a class will exist if its members can be mapped onto the ordinals less than some ordinal. This is 23.12 on page 164,

$$\mathrm{Func}\,\alpha \cdot x \in \mathrm{NO} \,.\supset.\, \alpha\text{“}x \in \Omega$$

Quine comments, "13.1 assures of all classes up to any finite size. 23.12 assures us of all classes up to the size of any ordinal."[115] The section of transfinite induction has been extensively rewritten in the light of Charles Parson's "A Note on Quine's treatment of Infinite Recursion." Quine writes that Parsons:

> observed that some of my theorem schemata on the subject carried existence premises that would rule out some of the clearly desirable interpretations of the schematic letters. He showed how to remedy the situation, and I have adopted his central idea.[116]

Parsons pointed out that the "difficulty is that the theorems[,] that show that his [Quine's] device for defining functions by transfinite recursion (Section 25) accomplishes its purpose[,] do not apply to functions defined by recursion over all the ordinals."[117] But Parsons also shows that Quine's definitions are in fact adequate when combined with a weaker comprehension premise, and proves this by deriving the necessary theorems. These parts of the book have been extensively rewritten. There is also a significant revision of the section on transfinite cardinals that Quine credits to suggestions by Burt Dreben. There are other relatively minor changes as well. But this is still the same book that Quine had published in1963, although somewhat improved.

In 1969, Quine published a book of essays entitled *Ontological Relativity and Other Essays*.[118] This is the most important work that Quine had published since WO. It combines essays on semantics, epistemology, science, and logic, and mark a significant development of his position. He begins with the essay "Speaking of Objects."[119] This was Quine's presidential address to the eastern division of the American Philosophical Association in 1957, and was published in the Association's proceedings in 1958. Quine says he was persuaded to include the essay here by Burt Dreben, who thought it would serve as a good introduction to the volume, and so it is. Quine opens the essay by remarking that we speak constantly in terms of objects.

> It is hard to say how else there is to talk, not because our objectifying pattern is an invariable trait of human nature, but because we are bound to adapt any alien pattern to our own in the very process of understanding or translating the alien sentence.[120]

To support this claim, Quine uses again his example of the field linguist encountering a hitherto unknown tribe whose language has no affinity to our own. This is the "Gavagai" argument from *Word and Object*, and although Quine does not use the term "Gavagai," the argument is the same. And as before Quine simply assumes without justification that the linguist can begin his task with a knowledge of the native terms for assent and dissent. The linguist, Quine says, will build his translation by imposing our categories on the native, and so we will never be able to detect differences in ontology, should they exist.

> For the arbitrariness of our reading our objectifications into the heathen speech reflects not so much the inscrutability of the heathen mind, as that there is nothing to scrute.[121]

If Quine had read anthropologists such as A. I. Hallowell or Paul Radin, to name only two of many, he would have found evidence that this claim is false.

Having rehearsed this argument from *Word and Object*, Quine turns to our own children and asks how they learn English. Initially, Quine says, the terms "mama", "red", and "water" are alike for the child, since the case of the child resembles that of the heathen.[122] (It would be more accurate to say that the child is in the position of the linguist trying to master an unknown tongue). It is, Quine holds, "only when the child has got on to the full and proper use of *individuative* terms" that he can be said to be speaking of objects. It is when the child understands "that apple", "not that apple", "an apple", "same apple", "another apple," "these apples" that he can be said to individuate. And how does he do this? "In one's earliest phase of word learning" words such as "mama" and "water" are learned "by a process of reinforcement and extinction" as what we see in retrospect as "observed spatiotemporal objects." "In the second phase" the child acquires the individuative terms, and what Quine calls "a proper notion of object" emerges. In the third phase "demonstrative singular terms" appear, such as "this apple." "A fourth phase comes with the joining of one general term to another in attributive position," as e.g., "blue apple." "It is in a fifth phase" Quine says, "where the child learns to apply relative terms to singular terms," such as "smaller than a speck." Here it is that reference to things unobservable appears. Then "there comes yet a sixth phase" where abstract entities appear. "Let us speculate on the mechanism of this new move," Quine says.[123] The mass term Quine thinks is the "wedge." The child has to take the mass term "red" and reconceive it as a concrete general term such as "round"; since the child will still not clearly conceive of "red" as suddenly two words, "we have him somehow infusing singularity into the concrete general; and such is the recipe, however unappetizing, for the abstract singular." Hence the child can speak of attributes. We have, in these reflections, Quine remarks:

> Some material for speculation regarding the early beginnings of an ontology of attributes in the childhood of the race.

From the role of sensory stimulation in evoking assent, we "settle, to some degree, what is to count as empirical evidence for or against the existence of the objects in question." But Quine argues that statements are part of a "fabric" of interconnected statements and observations, and stimulations reverberate across the fabric. Statements of the existence of abstract objects "are scarcely to be judged otherwise than by coherence," or simplicity of the overall theory. But our ontology of abstract objects leads to the paradoxes of set theory and equally of attributes. We have therefore to restrict our ontology to avoid contradiction. Quine then touches briefly on the propositional attitudes and the questions about reference that they raise. He then argues for the replacement of attributes by classes since the identity of the latter is clear whereas that of the former is not. This opens the way for an attack on sameness of meaning. "I see no hope of making reasonable sense of sameness of meaning even for English."[124] Quine holds that attributes and propositions (taken as abstract objects) are impediments to clear understanding and that we should be better off without them. In his conclusion Quine writes:

> It seemed in our reflections on the child that the category of bulk terms was a survival of a pre-individuative phase. We were thinking ontogenetically; but the phylogenetic parallel is plausible too.[125]

Some day, Quine suggests, our individuative talk "may in turn end up, half vestigial and half adapted, within a new and as yet unimagined pattern beyond individuation." And Quine concludes that we will find that future language just as opaque as the linguist found the heathen's, since there will be multiple correlations between that future language and ours, and "there is nothing for such a correspondence to be uniquely right or wrong about."

Quine writes here as if his phases of linguistic development describe the actual process of language acquisition. But nowhere does he supply any evidence to support this conjectural history of the process of language learning. By 1968 there were empirical studies of language acquisition, but Quine does not cite them, or any other evidence to support his claims. His conjectured extension of ontogeny to phylogeny has no apparent empirical basis, and by 1969 recapitulation theories were no longer acceptable in science. Quine believed himself to be a scientist, but having said that epistemology was a matter of linguistics and psychology, he ignored the work in those fields that did not fit his preconceptions.

The second essay – the one that gave the volume its name – is "Ontological Relativity"[126] – a forty-two page treatise in itself. This was Quine's John Dewey Lecture at Columbia, the first of the Dewey lecture series, and Quine happily claimed Dewey as a precursor of the doctrine he advances. Quine first attacks the notion that meanings are entities of some sort; this he calls "the myth of the museum in which the exhibits are the meanings and the words are the labels."[127] As one would expect, Quine holds that language consists of dispositions to verbal behavior. But he also takes aim at ostensive definition. What is required in learning a word by ostension is another person to act as the teacher who must direct the subject's attention to the object and name it by a word. It is necessary here that both the teacher and the subject are looking at the same object and that each can see that the other is looking at it. Quine holds that even this is not enough, and to make his case he invokes the Gavagai argument once again. Both native and linguist focus on the same thing, but is that thing a rabbit, a temporal rabbit slice, undivided rabbit parts, or a piece of rabbit stuff? Quine's point is that there is no behavioral clue that will enable the linguist to determine what the native's reference is, and even learning the native's language will not help because in the process of learning it the linguist will so translate native expressions as to make the native's ontology the same as his own. Hence even if the native really is referring to temporal rabbit slices, the linguist will never find this out.

Quine then repeats his arguments for the indeterminacy of translation. For a given native sentence, there can be two equally plausible but incompatible English translations – that is, both translations fit all the dispositions to verbal behavior equally well, and there is therefore no way to decide between them because there is nothing but dispositions to verbal behavior to appeal to. Quine illustrates this by citing the different ways in which "ne ... rein" can be translated into English, and similarly for the "classifiers" of Japanese. But the same inscrutability thus illustrated, Quine holds, applies in our home language. And Quine argues for this from "deferred ostention."[128] Thus Quine cites the case of numbers. "Numbers ... are known only by their laws, the laws of arithmetic, so that any constructs obeying those laws – certain sets for instance – are eligible in turn as explications of number." Quine adds "any progression will serve as a version of number so long and only so long as we stick to one and the same progression."[129] Homophonic translation is of course the way we translate each other in the home language. But the inscrutability of reference is still there as the number example shows and we need a way around it.

The way, Quine holds, is relativity. We can fix the reference of our words only by using another language – Quine calls it a background language – in which we can specify the reference. Or, to put it differently, our language is the object language, and its references can only be fixed in a metatheory. Theories are fully interpreted systems of sentences only relative to a metatheory, and of course the same holds for the metatheory, the references of whose terms are fixed only by a metametatheory, and so on. Quine here invokes the "*proxy function*: a function mapping the one universe into a part or all of the other."[130] Godel numbering can serve as such a proxy function, carrying expressions in one theory into numbers. "One ontology is *always* reducible to another when we are given a proxy function that is one-to-one." It need of course not be one-to-one: Quine gives the example of an economic theory in which persons are replaced by their incomes and all persons with the same income are identified. Could everything be reduced to numbers so that we have a Pythagorean universe? Quine holds that "Pythagoreanism itself is meaningless. For there is no absolute sense in saying that all the objects of a theory are numbers, or that they are sets, or bodies, or something else; this makes no sense unless relative to a background theory."[131]

This leads, as one might expect, to the Lowenheim-Skolem theorem. According to this theorem, any theory, no matter how large its universe, has a model in the natural numbers. As Quine puts it, "the theorem is that all but a denumerable part of an ontology can be dropped and not be missed." If the theory to be reduced has an indenumerable universe, then of course there is no one-to-one mapping into the natural numbers possible unless all but a denumerable part of the theory can be dropped. Quine also makes the point that objectual quantification can deal with an indenumerable universe, but substitutional quantification cannot. Indeed, Quine says, "ontology is thus meaningless for a theory whose only quantification is substitutionally construed: meaningless, that is, insofar as the theory is considered in and of itself." The question of ontology, Quine holds, "makes sense only relative to some translation of a theory into a background theory in which we use referential quantification."[132]

The reader doubtless has noticed that half-way through my account, I switched from "languages" to "theories." This reflects accurately what Quine does: in the first twenty-four pages of the paper, Quine talks about language; from there on he talks about theories. In the last eighteen pages of the article, the word "language" occurs only four times: once on page 52 in Carnap's title, *The Logical Syntax of Language*, once on page 61, once on page 67 in the expression "background language," and once in the "Note Added in Proof." Quine simply assumes that languages are theories and that what is true of one is true of the other. Now it is well known that if a formal theory has a model, it has many models. It is also quite standard doctrine that the references of a theory's terms cannot be fixed (at least entirely) within the theory itself but must be fixed in a metatheory. But are natural languages theories in the sense that Quine requires? The answer is no. In a reply to Chomsky's charge that he confused theory with language, Quine said that theory and language are distinct. But he says that in WO and "other related writings," he is using "Theory" in a non-technical sense. "For these purposes a man's theory on a given subject may be conceived, nearly enough, as the class of those sentences, within some limited vocabulary appropriate to the desired subject matter, that he believes to be true." "Clearly," Quine says, "they are not interchangeable in all contexts, and they are pretty sure to be interchangeable in some." One must assume, therefore, that the context of OR is one where Quine does consider them to be interchangeable.[133]

Quine imputes to language powers of an extraordinary sort. He assumes that the individuation of the world into individual objects is done by language. But there is compelling evidence that pre-linguistic children perceive the world as largely composed of discrete objects. Since this is true before the child has *any* language, it is clearly an ability of human infants generally and therefore the linguist and the native can start with the recognition that each is pointing to a rabbit as a physical object. Further, when a child is taught by ostension that a word applies to a perceived object, the child takes it as referring to *that* object, not to something else. But the heart of Quine's argument for deferred reference is the proxy function argument. Quine's claim is that, for any natural language, proxy functions can convert its ontology into an alternative one. He had yet to recognize how powerful this argument was. Looking back on "Ontological Relativity" some years later, Quine commented:

> But a more startling point can be made: the ontology of any theory can be changed at will, by a proxy function, without falsifying any of its sentences or disturbing any of the empirical evidence that supports it. Its sentences can even be preserved verbatim, subject merely to reinterpretation. The upshot is that the ontology of a theory is empirically inscrutable, and indeed that the very ascription of any particular ontology makes nontrivial sense only relative to a proposed translation from [one] language to another... It was only in ensuing years that I came clearly to perceive the self-sufficiency and stark simplicity of the argument from proxy functions.[134]

Quine presented the third essay – "Epistemology Naturalized"[135] – in Vienna in 1968 at the fourteenth International Congress of Philosophy. It represents a further development of the philosophic position he had set forth in *Word and Object* and had been expanding ever since. "Epistemology is concerned with the foundations of science," Quine says. Note that he does not say the foundations of knowledge; for Quine "science" and "knowledge" appear to be co-extensive, for there is no sharp line between common sense knowledge and science and he seems to regard any knowledge worthy of the name as science. First he takes up the foundations of mathematics. "Studies in the foundations of mathematics divide symmetrically into two sorts, conceptual and doctrinal." The conceptual are concerned with meaning; the doctrinal with truth. As has been made clear in his earlier writings, Quine believes that the foundations of mathematics are logic and set theory, and that given the state of set theory, the foundation is less secure than the superstructure of mathematics itself. This leaves him with the problem of how mathematics can be "true," if it can be.

Quine remarks that one of the greatest advances on the conceptual side was Bentham's notion of contextual definition:

> He recognized that to explain a term we do not need to specify an object for it to refer to; nor even specify a synonymous word or phrase; we need only show, by whatever means, how to translate all the whole sentences in which the term is to be used.

It was this, he says, that led to the "recognition of the sentence as the primary vehicle of meaning."[136] Bentham is only one of the philosophers to whom Quine gives credit for establishing the priority of the sentence. But what of the doctrinal side – the question of truth? Quine thinks that the attempt to ground science in immediate experience has failed. The most brilliant attempt to do this he takes to be Carnap's

Aufbau, and that was an attempt that failed, as Carnap saw in 1936 when he published "Testability and Meaning." Efforts at rational reconstruction that would allow the translation of science into observations and set theory have failed. "We must despair," Quine says. "of any such reduction." Quine then sums up his argument so far as follows:

> The crucial consideration behind my argument for the indeterminacy of translation was that a statement about the world does not always or usually have a separable fund of empirical consequences that it can call its own. This consideration served also to account for the impossibility of an epistemological reduction of the sort where every sentence is equated to a sentence in observational and logico-mathematical terms. And the impossibility of that sort of epistemological reduction dissipated the last advantage that rational reconstruction seemed to have over psychology.[137]

So Quine claims "epistemology in its new setting . . . is contained in natural science, as a chapter in psychology."[138] There is no circularity here, Quine says. Once the idea of providing a foundation for science is scrapped, we are free to use science to study how we have come to have the science we do. It is clear from psychology that our contact with the world consists in stimulations to our sensory receptors. But our knowledge must be formulated in language, and so linguistics must be used along with psychology to explain how we got to where we are. The sentences closest to our stimulations are observation sentences. "An observation sentence is one on which all speakers of the language give the same verdict when given the same concurrent stimulation." And what defines the community of speakers? "Simply general fluency of dialogue."

> This criterion allows of degrees, and indeed we may usefully take the community more narrowly for some studies than for others. What count as observation sentences for a community of specialists would not always so count for a larger community.[139]

The observation sentence is fundamental, Quine holds, to both truth and meaning. It is located at the periphery of the web, where the web touches experience. The observation sentence is critical for both the problems Quine raised at the beginning.

> Its relation to doctrine, to our knowledge of what is true, is very much the traditional one: observation sentences are the repository of evidence for scientific hypotheses. Its relation to meaning is fundamental too, since observation sentences are the ones we are in a position to learn to understand first, both as children and as field linguists.[140]

The observation sentence, Quine says, "situated at the periphery of the body scientific, is the minimal verifiable aggregate; it has an empirical content all its own and wears it on its sleeve." Hence the observation sentences are largely immune to the underdetermination of language. As for meaning, it "ceases in general to have any clear applicability to single sentences." Furthermore, "epistemology merges with psychology as well as linguistics." So for Quine there is no foundational role for epistemology; there is no first philosophy. There can be no guarantee of the truth of science beyond what science itself provides. The real problem is to discover how we come to have the science we do, and this is a job for psychology and linguistics. The upshot is "reciprocal containment, though containment in different senses: epistemology in natural science and natural science in epistemology."

The fourth essay of the volume, "Existence and Quantification," was written for a colloquium at the University of Western Ontario in 1968. But bad weather prevented Quine from reaching the university, and Joe Margolis read the paper in his place.[141] Quine opens the paper with the question of how to distinguish metaphysical questions of existence from meaningful ones. Not surprisingly, Quine's criterion for existence is existential quantification. We have to assume the existence of something when a theory we hold requires that assumption in order to be true, and this is shown by the existential quantifications that it employs. But we must also specify what sort of universe the theory requires. Quine holds for single sorted quantification, and holds that the many sorted types are reducible to the single sorted. As for the term "exists" itself, Quine takes it as primitive, but there can be evidence for existential statements. One sort of evidence is sensory, but more generally the question is one of having evidence for the theory in which existence is postulated. On these grounds, if a scientific theory is well supported by evidence, the existence statements in the theory are well supported, whereas a metaphysical theory will, for want of evidence, carry off its existential statements when it is rejected on the basis of evidence.

The immaterialist will disagree on whether or not objects exist, but Quine notes that he will agree with the objectivist on some issues because of "a known holophrastic relation of stimulus synonymy between our sentence and some sentence geared to his different universe." For the immaterialist "we may say, I grant, that there are for him two senses of existence; but there is no confusion, and the theoretical use is rather to be respected as literal and basic than deplored as a philosophical disorder." The nominalist will also have problems with such things as classes, and will seek a way of eliminating them as "mere manners of speaking." Russell tried to avoid the assumption of classes by giving them contextual definitions, but Quine notes that this requires assuming attributes as values of bound variables. Quine mentions his own device of virtual classes that he put forth in his book on set theory, but he points out that "the virtual theory of classes affords no adequate foundation for classical mathematics even of the positive integers."

Quine then turns to substitutional quantification – quantification where the values of the variables are singular terms. The problem here, as Quine has argued on a number of occasions, is that the universe of singular terms is at most denumerable and so the theory fails when the universe must include the real numbers. Quine further objects that singular terms must have referents, so substitutional quantification does not escape ontological commitment. But Quine says:

> Substitutional quantification has its points. If I could see my way to getting by with an all-purpose universe whose objects were denumerable and indeed enumerated, I would name each object numerically and settle for substitutional quantification.[142]

"Where substitutional quantification serves, ontology loses point," since quantifiers can be turned into conjunctions and disjunctions. But Quine needs a more than denumerable universe since he wants to keep classical mathematics.

Quine then takes up a proposal of Wang's that suggests that classical quantification theory represents an arbitrary choice among the varieties of quantification theories available. Quine points out that Henkin has proposed an alternative quantificational theory. Consider, Quine says, the sentence

(1) Each thing bears P to something y and each thing bears Q to something w such that Ryw.

Quine says that (1) can go into classical quantification either as

$$(2) \quad (x)(\exists y)(Pxy \cdot (z)(\exists w)(Qzw \cdot Ryw)$$

or

$$(3) \quad (z)(\exists w)(Qzw \cdot (x)(\exists y)(Pxy \cdot Ryw).$$

Yet (2) and (3) are not equivalent. Quine comments

> There are interpretations of P, Q, and R in (1) that make both dependencies gratuitous; for instance, interpretation of P as "is part of"; Q as "contains" and R as "is bigger than."... One may suspect that the notation of quantification theory is at fault in forcing a choice between (2) and (3) in a case like this.

Henkin's proposed solution is to admit functions as values of bound variables, which gives

$$(4) \quad (\exists f)(\exists g)(x)(z)(Pxf_x \cdot Qzg_z \cdot Rf_x\, g_z)$$

or, staying with classical quantification theory, to allow branching quantifiers;

$$(5) \quad \begin{matrix} (x)(\exists y) & \\ & (Pxy \cdot Qzw \cdot Ryw) \\ (z)(\exists w) & \end{matrix}$$

Quine argues for the advantages of classical quantification theory on the grounds that it admits of a complete proof procedure for validity and a complete proof procedure for inconsistency, whereas the alternative proposed by Henkin has a complete proof procedure for the functionally existential version for inconsistency, and for the functionally universal version a complete procedure for validity. Henkin however has shown that formulae such as (4) and (5) admit of no general procedure "at any rate when identity is included," But Quine still holds for classical quantificational theory.

> Classical quantification theory enjoys an extraordinary combination of depth and simplicity, beauty and utility. It is bright within and bold in its boundaries. Deviations from it are likely, in contrast, to look rather arbitrary. But insofar as they exist it seems clearest and simplest to say that deviant concepts of existence exist along with them.[143]

The fifth essay, labeled "Natural Kinds,"[144] was written for a festschrift for Carl Hempel. It is appropriate therefore that Quine starts with the question of induction and with Hempel's paradox of the black ravens and Goodman's paradox of green and grue. Hempel's paradox is that the statement "all ravens are black" is logically equivalent to "All non-black things are non-ravens," yet it is queer, to say the least, that a green leaf confirms the statement "All ravens are black." Goodman's paradox is that the statements "All emeralds are green" and "All emeralds are grue" (where "grue" means green before tomorrow and blue thereafter) are equally well confirmed by the examinations of emeralds up to now, but we believe the former and not the latter. Goodman's answer to his paradox was "projectibility" – that is, we expect things to be the same in the future as in the past. And we account "green" as projectible and regard "grue" as not projectible. Quine suggests that this is because "green emeralds" are a natural kind. But this only leads us to the attempt to define a "kind," which Quine thinks must be done in terms of similarity. Indeed, he holds that the notion of kind and similarity are one notion. Yet Quine thinks neither can be defined in terms of logic or set theory. Since even the learning of words by ostension is based on similarity, Quine concludes that "there is an innate standard of similarity" that can be interpreted in behavioral terms, and so applied to animal learning as well as human. "Every reasonable expectation depends on similarity," and so does induction. "'Induction' itself is essentially more of the same." But to claim this is to claim "to trust induction as a way to access the truths of nature . . . is to suppose, more nearly, that our quality space matches that of the cosmos. The brute irrationality of our sense of similarity, its irrelevance to anything in logic or mathematics, offers little reason to expect that this sense is somehow in tune with the world."[145] Yet this is what we have to say if we are to hold induction trustworthy.

Part of the answer, Quine holds, is given by Darwin. Natural selection provides a reason to believe that our similarity standards do match nature's, since if they did not we would not be here. Given the prominent role of color in the natural world and the fact that it helps us to identity food, while at the same time science has shown that color is rarely a fundamental property of things in nature, Quine says natural selection has doubly endowed us "with both a color-slanted quality space and the ingenuity to rise above it."[146] We rise above it by developing scientific theories that permit the development of new classifications that have explanatory value. Thus, Quine says, consider dispositions. Using solubility as his example, he notes how we then acquire the class of water-soluble things. And closely connected with dispositions are subjunctive conditionals – e.g., if x were to be immersed in water, it would dissolve. It is, Quine says, the development of science that solves the problem. "The progress of similarity standards, in the course of each individual's maturing years, is a sort of recapitulation in the individual of the race's progress from muddy savagery."[147] Science now is able to explain solubility in terms of the microstructure of soluble objects, rendering the very notion of natural kinds obsolete. This state has not been achieved in all branches of science. In zoology for example, taxonomy is still vital. But Quine says:

> In general we can take it as a very special mark of the maturity of a branch of science that it no longer needs an irreducible notion of similarity and kind. It is that final phase where the animal vestige is totally absorbed into the theory. In the career of the similarity notion, starting in its innate phase, developing over the years in the light of accumulated experience, passing then from the intuitive phase into theoretical similarity, and finally disappearing altogether, we have a paradigm of the evolution of unreason into science.[148]

This use of natural selection, with suggestions of recapitulation theory, is one that will become more prominent in Quine's later work.

The final essay in the volume is entitled "Propositional Objects"; it was a lecture Quine gave at several universities and was soon to be published in *Critica.*[149] This is not the first time Quine has dealt with propositions, and his views have not changed substantially. He first discusses propositions as meanings of sentences and bearers of truth and falsity. He rejects them on both counts, arguing that eternal sentences can do the job better. A second use which propositions are supposed to have is as objects of propositional attitudes. But Quine raises the problem of the individuation of propositions. Is the belief that the faces of the great pyramid are equilateral triangles the same as the belief that the faces are equiangular? Here again we have the individuation problem. Or consider the case where "we like to say . . . that the cat wants to get on to the roof or is afraid the dog will hurt him." Can we accommodate such cases by taking states of affairs as the objects? Quine supposes that a state of affairs can be viewed as "the class of all possible worlds in which, intuitively speaking, the state of affairs can be realized." Quine then carries his reader through the process of defining a possible world – how it is to be located in time and space, which space, what coordinates, what is the origin of the coordinate system, and so on. Having done so, Quine says "we could as well limit our attention to the surface of our self-centered animal and take account merely of the possibilities of activation or inactivation of its several nerve endings. The possibilities in short of sensory input." Each pattern of such activation is a possible world, or "as we may better entitle it, a stimulation pattern." This is what in *Word and Object* Quine called an "affirmative stimulus meaning." And even after reaching this familiar terrain, that is still a problem.

> It seems vital that in correlating one subject's verbal behavior with another's, for instance as a basis for translating one language into another, we be able to equate one subject's stimulations to another's. Yet how are we to do so? If we construe stimulation patterns my way, we cannot equate them without supposing homology of receptors; and this is absurd, not only because full homology is implausible, but because it surely ought not to matter.[150]

And there, Quine leaves the reader. It was a problem with which Quine would struggle for some years, until Darwin offered him a way out.

One of the responses Quine received when *Ontological Relativity and Other Essays* was published was a letter from Hempel on November 21, 1969.

> On reading "Epistemology Naturalized," I suddenly thought that in view of what you say on p. 79-p.80 top, you might be taken to hold that a typical statement about bodies – or

> any single statement in a scientific theory cannot be significantly said to be either-true-or-false. Your discussion might seem to imply this by virtue of the remark that no empirical assertion at all is made by an isolated statement of the kind mentioned. But surely this reasoning would confound the possession of a truth value with observational testability, or with empirical meaning in Peirce's sense. Also, if at least a block of theory as a whole is held to be true or false, then its component sentences should have the same status.[151]

Quine replied that in his view the sentences of a true theory are true, but the truth of a theory can be put only in another theory.[152]

But while nothing interfered with Quine's relentless productivity, the situation at Harvard depressed him. On February 13, 1969, Quine wrote to his former student Follesdasl.

> The intellectual level of our graduate students in philosophy this year seems to me to be lower than I have previously seen it. The department's admission committee will now take matters more firmly in hand, I think, and admit for next year as few even as eight new graduate students in an effort to stem the tide of incompetence.[153]

This was a common enough lament among scholars as the sixties played out.

In 1970, Quine published *Philosophy of Logic*.[154] The book had been solicited by Elizabeth Beardsley for the series on Foundations of Philosophy that she and her husband were doing. The book is short –102 pages – but, as with *The Web of Belief*, Quine makes the most of the opportunity to preach his austere view of logic. The opening chapter is largely devoted to clearing away what Quine considered rubbish. This includes *propositions* and *statements* – the latter because Oxford philosophers had taken to using it for the "acts we perform in uttering declarative sentences." He therefore speaks only of sentences. He presents his view of observation sentences, and contrasts their responsiveness to observation with that of the more theoretical sentences of science. These of course are testable only in groups where there is considerable freedom for accommodating deviant observations. "Our theory of nature is under-determined by all 'possible' observations," Quine says, and hence there can be incompatible sets of hypotheses that account for the same observations. This thesis he would develop in a future paper. He also inveighs against the "myth of meaning" and the use of modalities. He explains the use of schematic letters and of semantic assent, quotation marks to form the names of expressions, and eternal sentences. He will talk of sentences in this book, and will introduce "truth" to fulfill "the purpose of reconciling the mention of linguistic forms with an interest in the objective world."[155]

The second chapter is entitled "Grammar." Having defined "phonemes," he puts the grammarian's task as deciding what strings of phonemes belong to the language. He does this by specifying a lexicon, a list of grammatical categories, and grammatical constructions in such a way as to yield a recursive specification of the infinite class of acceptable sentences in the language. The grammatical categories are substitution classes; members of the same category can be substituted for each other while preserving grammaticality. Thus Carnap's "This stone is thinking of Vienna" is a grammatical sentence despite its absurdity. Quine distinguishes between "immanent" and "transcendent" linguistic notions: the former being applicable within the language, the latter to languages in general. He then turns to logical grammar. Its categories are x-placed predicates where "x" goes from 1 to finite n, and an infinite category of variables, generated from a basic set by accentuation. The constructions are predication of one, two, three, etc. placed predicates. The remainder are constructions of sentences from sentences by negation, conjunction, and existential quantification. The other truth functional connectives – alternation, the conditional, and the biconditional, and universal quantification are definable in terms of the basic three. Quine emphasizes that schematic letters belong not to the object language but to the metalanguage. Names and functors can be used to produce singular terms, but Quine considers these "frills." Artificial languages constructed of these elements he terms *standard* languages, but within the class so defined Quine will omit the frills. He distinguishes the "particles," such as the negation sign, from the lexicon, but instead of making a separate category for them he takes them to be parts of the constructions in which they occur. He then takes up the propositional attitudes. He discusses various ways of dealing with them, but adopts none of them – propositional attitudes will not recur in the text. Similarly, he discusses modalities, but chiefly for the purpose of getting them out of the way.[156]

Chapter 3 is entitled "Truth." Having dealt with linguistic matters in the first two chapters, Quine says "the truth predicate then preserves his [the logician's or grammarian's] contact with the world." Quine gives a careful description of Tarski's definition of truth. Using the language already developed in terms of negation, conjunction, and existential quantification, and using sequences for the ordered arrays that satisfy sentences, he shows how the recursive definition of satisfaction is developed. Truth then is satisfaction by all sequences, falsity satisfaction by none. He then raises the question of the definition of "satisfaction"; is it possible to give an eliminative definition of "satisfaction"? The answer is yes, but only by using set theory, and Quine shows how this can be done. He also shows how Grelling's paradox can be avoided by the introduction of further devices from set theory and compares Grelling's paradox with Russell's.[157]

Quine is now ready to deal with logical truth, which he does in the next chapter. The truth or falsity of simple sentences is not settled by logic but presupposed; logic deals with the truth or falsity of the compound sentences constructed from the simple ones. Rather than dealing separately with logical implication, logical incompatibility, etc., Quine takes logical truth as his subject, since the other notions are definable in terms of it. The logical structure of a sentence is its construction in terms of truth functions, quantification, and variables. Hence logical structure and predicates are all there is to a sentence in the standard grammar that he has defined. This allows the following definition: a logical truth is "a sentence from which we get only truths when we substitute sentences for its simple sentences." Further "a logical schema is *valid* if every sentence obtainable from it by substituting sentences for simple schemata is true." Then a logical truth is "a truth thus obtainable from a valid logical schema." The notion of "validity" is thus introduced, and is developed as follows. The *set theoretic analogue* of a logical schema is

> a certain open sentence of set theory that we form from the schema in the following way. We change the predications "Fx", "Fy", "Gx", etc, to read "$x \in \alpha$", "$y \in \alpha$", "$x \in \beta$" etc., thus invoking variables "α", "β", "γ" etc. whose values are to be sets.

For two-placed predicates, we use ordered pairs, so "Hxy" becomes " x,y ϵ γ" correspondingly for three-placed predicates, and so on. This is an open sentence satisfied by some sequences of sets and not others.

Quine then introduces the notion of *model.*

> A model of a schema is a sequence of sets: one set corresponding to each schematic predicate letter in the schema, and, as the initial set of the sequence, a non-empty set "U" to play the role of universe or range of values of the variables "x", "y", etc. The set in this model corresponding to a one-place predicate letter of the schema is a set of members of "U"; the set corresponding to a two-placed predicate letter is a set of pairs of members of "U", and so on, ... The model satisfies the schema if, when we specify "U" as the range of values of the variables "x", "y", etc., and we assign further sets of the model to the respective set variables "α ","β", etc., the set-theoretic analogue of the schema comes out true.

This provides a new definition of validity: "A schema is valid if satisfied by all its models," and a logical truth is "any closed sentence obtainable by substitution in a valid schema." "U" must not be the empty set. We know that there are open sentences that do not determine sets, as Russell's paradox shows; we also know that there are sets that are determined by no sentences of the object language, as Grelling's paradox shows. Nevertheless, a schema that comes out true under all substitutions of sentences of elementary number theory is satisfied by every model, and if a schema is satisfied by every model, it comes out true under all substitutions of sentences. The requirement that the object language be rich enough for elementary number theory is essential for this theorem. The theorem itself follows from a theorem by Lowenheim, Hilbert, and Bernays, that if a schema is satisfied by a model at all, it becomes true under some substitution of sentences of elementary number theory for its simple schemata, and a theorem by Skolem, Herbrand, and Godel that "if a schema is satisfied by every model, it can be proved" – that is the theorem of completeness of the logic of quantification. But Quine, with his desire to stay as far clear of set theory as possible, then takes as his definition of logical truth "a sentence is *logically true if only truths come of it by substitution of sentences for its simple component sentences.*" The truth of this claim is guaranteed by the detour through set theory.

But this does not get us free of set theory, for sentences are sets of their tokens. Moreover, a sentence never uttered would have for its set of tokens the empty set, and all unuttered sentences would become identical. To avoid this, Quine says there are several ways to go. One is to take a string of signs as a sequence in the mathematical sense. The phonemes of the string can still be taken as sets of their tokens, but by using sequences we insure the unlimited supply required by the set theoretic theorems. A second is to take the signs and strings of signs as positive integers, since we have already assumed elementary number theory. A third way is, following the lead of the completeness theorem, to adopt some one of the various proof procedures and define a valid schema as one that can be thus proven. Any of these will do; all the ways of defining logical truth are extensionally equivalent.

Quine then introduces yet a fourth method of defining logical truth, and a more abstract one. Instead of specifying the grammar of the object language in terms of negation, conjunction, and existential quantification, he defines logical truth in terms of whatever grammar the object language contains. "Sentences have the same grammatical structure when they are interconvertible by lexical substitution," Quine says. So he defines a logical truth as "a sentence that cannot be turned false by substitution for lexicon." This allows Quine to conclude:

> What sentences of a language are to count as logically true is determined, on this theory, when we have settled two things about the language: its grammar and its truth predicate.[158]

One may already suspect that Quine has a particular reason for this emphasis on grammar.

Quine's preferred method of defining logical truth finds a difficulty in dealing with identity: a truth such as "x = x" is falsifiable when other predicates are substituted for "=". But Quine wants to keep identity as a part of logic, and his way of doing so is as follows. Suppose a language whose lexicon of predicates consists of a one-place predicate "A" and a two place predicate "B." Then define

$$(1) \quad \text{"}x = y\text{" for "}Ax \equiv Ay \cdot (z)(Bzx \equiv Bzy \cdot Bxz \equiv Byz)\text{"}$$

For any language with a finite lexicon, this method makes "x" and "y" indistinguishable by the predicates of the lexicon of the language. If, for example, people were identified only by their occupations, then any two people with the same occupations would be taken to be identical. All equations within the language are then taken as abbreviations for constructions on the order of (1).

Is set theory a part of logic? Frege, Russell, and Whitehead thought it was, but Quine does not. He first deals with the problem of attributes, which he believes is due to a confusion of use and mention,

> a confusion between mentioning a sign and using it. Instead of seeing "F" [of "Fx"] steadfastly as *standing in place of* an unspecified predicate, our confused logician sees it half the time as *naming* an unspecified predicate. Thus "F" gains noun status, enabling him to read "Fx" as "x has F" without offending his grammatical ear. Having got this far, he can round out his confusion by calling "F" an attribute.

And so of course the confused logician quantifies over "F", so we get "(∃F) . . . " As usual, Quine attacks attributes on the ground that there is no clear criterion for their identity; whereas sets are identical if they have the same members, multiple attributes may apply to the same things.

Quine shows how the rational numbers can be generated from the arithmetic of integers. But he then turns to the problems that challenge his definition of logical truth in terms of standard grammar. One such problem, raised by Davidson, is adverbs. Quine notes that this would require additions beyond the definition of logical truth in terms of truth functions and quantification, but he suggests several different way of accommodating it under his grammatical definition. A second problem, raised by Geach, is how to accommodate comparatives. Quine says "in a logically well-regimented language 'is bigger than' would figure as a simple two place predicate, and then the positive 'is big', where useful, would be paraphrased as 'is bigger than' followed by reference to some object chosen as a minimum standard suitable for the purpose at hand." Since predicates that take the comparative form are usually vague or elliptical, Quine thinks such a treatment would work with the grammatical theory of truth. A third problem is the propositional attitudes. Quine considers several ways of dealing with them but is not satisfied by any of them. And modality of course is a fourth problem that Quine sets aside. So Quine concludes

> The rewards of staying within the bounds of standard grammar are great. There is extensionality... There is, more generally speaking, the efficiency and elegance of the logic of truth functions and quantification. There is its completeness... There is the impressive concurrence of all those definitions of logical truth ...They all proved to demarcate one and the same class of sentences, as long as we adhere to standard grammar and allow ourselves a fairly robust vocabulary. There is a concurrence here that suggests that we have hold of something solid and significant.[159]

Quine then considers "deviant logics." The first example he picks is trying to translate from a radically different tongue. Could the native's logic be different from ours? Quine's answer is no because in translating it, we would foist our logic on him. If the native dissents from a compound but assents to its elements, we do not translate his compound as conjunction. When he assents to the compound and its elements, we can impute conjunction. Why then does logic turn out to be built into the translation when so many other things are not? Quine's answer is "it is in the incidence of obviousness that the difference lies." Quine adopts the rule "Save the obvious." So Quine holds that sentences that are obvious in any block of knowledge should be translated into English sentences that are also obvious. His examples are "1+ 1= 2" and "It's raining" uttered in the midst of a downpour. But on the score of obviousness, he claims that logic is peculiar: every logical truth "is either obvious as it stands or can be reached from obvious truths by a sequence of individually obvious steps." This is a very important claim for Quine, but before pursuing it, he considers some alternative proposals.

The next case considered is many-valued logics; these involve a rejection of the law of excluded middle and with it classical negation. Why would anyone want to do so? One reason, that Quine considers a bad one, is a confusion between knowledge and truth. It is often the case that we do not know whether a sentence is true or false. But Quine emphasizes that that does not show that the sentence itself is neither true nor false. A second reason is as a way of avoiding the paradoxes of set theory, by for example giving " $(\exists x)[(x \in x) \cdot -(x \in x)]$ " a middle truth value. Quine considers that this violates the "maxim of minimum mutilation." The classical logic of truth functions and quantification is free of paradoxes, and adding a third truth value would disrupt it. The paradoxes belong to set theory and should be dealt with there. Another reason is the Heisenberg indeterminacy principle. Quine objects to this on the same ground of minimizing mutilations. He comments "I do place the claims of physics somewhat above those of set theory, because I see the justification of mathematics only in what it contributes to our integral science of nature." But he notes that very few physicists have found three-valued logic useful in quantum mechanics. He considers the cost of going to a n-valued logic too high to justify such a move. The best known deviant logic is that of the intuitionists. The intuitionist demand for constructivism Quine notes, but he says that a good deal of what constructivism demands can be done in classical logic. He also points out that Weyl's constructive set theory uses classical logic; the constructivism appears only in the axioms for the existence of sets. Hence Quine concludes that no revision of classical logic is required. Similarly, he notes various forms of deviant quantification: quantified intuitionist logic, quantified modal logic, quantified multivalued logic, and even quantified propositional attitudes. Such deviations bring with them ontological consequences. To say just what an intuitionist quantified logic takes as existing, we would have to translate it into our (classical) logic since the intuitionist's meaning of "to be" is not ours.

Quine also deals with the problem of branching quantifiers. In a formula such as

$$(x)(\exists y)(z)(\exists w)Fxyzw$$

there is no way in classical logic to make the choice of "w" depend only on "z" and not also on "x". Hence the proposal

$$\begin{matrix}(x)(\exists y) & \\ & Fxyzw \\ (z)(\exists w) & \end{matrix}$$

The only alternative way to handle this, Quine says, is to introduce functions and quantify over them – e.g.,

$$(\exists f)(\exists g)(x)(z)(Fx(fx)z(gz))$$

Hence Quine concludes that this should be seen as a mathematical problem rather than a logical one. The final deviant that Quine considers is substitutional quantification, and here he reiterates the objections he has given before – namely, the inability of substitutional quantification to deal with nameless entities such as real numbers or sets not determined by a sentence.[160]

In the final chapter, Quine considers the grounds of logical truth. When one looks at the sciences as a whole, many have seen logic and mathematics as distinct from physics and the other natural sciences. Those, like physics, have been taken to be empirically testable, whereas logic and mathematics, being common to all the sciences, have been seen as different in kind from the others. But Quine holds that this division is a mistake. One ought not to think of the sciences as separate systems but as parts of one system of the world of which logic and mathematics are components. Between the more abstract parts of science and mathematics and logic, the division is gradual rather than sharp. Those who have seen logic and mathematics as truths of language while the natural sciences are truths of fact have distorted the picture. So have those, like Carnap, who have seen language as analogous to a formal deductive system. On the contrary, Quine says, one should look at how language is learned. Anyone who learns a language learns its grammar as well as its lexicon. Speakers of the same language may differ in their lexicons, but they all have the same grammar.

> So the logical truths, being tied to the grammar and not to the lexicon, will be among the truths on which all speakers are likeliest to agree... Naturally the habit of accepting these truths will be acquired hand in hand with grammatical habits. Naturally therefore the logical truths, or the simple ones, will go without saying; everyone will unhesitatingly assent to them if asked. Logical truths will qualify as obvious in the behavioral sense in which I am using this term, or potentially obvious.

One sees here why Quine has been so insistent throughout the book on the grammatical theory of logical truth. If Quine is to maintain his claim that logic lies at the center of the web of belief, and his claim that the truth predicate ties logic to the world, then he faces the problem of how this can be so since the connections among the sentences of science are logical connections of implication and deducibility that assume the truth of logic. To hold that science proves logic true would be circular, since the truth of logic is assumed in creating the science, while if the science is false then he would have the contradiction of assuming logic true and yet being compelled to find it false. The solution that Quine comes to is that logic is "obvious," since every member of the community knows the same grammar, however they may differ in lexicon, and so learns logic with their language. One wonders whether the students in Quine's elementary logic class found what they were learning "obvious."

Quine believed at this point that obviousness gave him a solution to the problem of the status of logic. In discussing logic and translation, Quine said that logic should be preserved in translation because it is obvious, and what is obvious for one language should be translated as obvious in another.

> The general policy of translating the obvious (that is, what is assented to as a matter of course) into the obvious is a policy that comes to a head in the logical truths, because of a combination of two circumstances. One circumstance is that the logical truths are either all obvious ... or else potentially obvious, in the sense of being derivable from the obvious by individually obvious steps. The other circumstance is that the translator can deal with them wholesale by abstracting shared skeletal forms. We see, then, how it is that "Save logical truth" is both a convention and a wise one. ... it gives logical truths no epistemological status distinct from that of any obvious truths of a so-called factual kind.[161]

That is, the general policy should be to translate what is obvious in one language into something equally obvious in the other. If logical truths are obvious then they ought to survive translation intact. And this, Quine thinks, enables him to preserve the claim that logical statements are as factual as those of physics. But this does not solve the problem of the circularity of using logic to construct the theory while also holding that it is defeasible like any factual statement.

The Philosophy of Logic is an exposition of Quine's particular view of logic. Its purpose was to inculcate his doctrines in philosophers generally and such laymen as were drawn to the subject. In ruling out modal logics and other deviant logics (e.g., many valued logics), as well as set theory, his aim was to make the term "logic" synonymous with first order logic with identity and to keep it purely extensional; no intensions need apply. Quine knew that any such book bearing his name would be widely read and he hoped to establish his view as canonical within the profession.

In 1970, Quine published a piece entitled "Grades of Theoreticity"[162] in which he introduced a new way of dealing with the homology problem that had plagued him since WO. Quine says he wants to examine the differences between grades of theoreticity, so he begins at the bottom with the notion of "data." As usual, he takes the starting point of knowledge to be sensory stimulations.

> We can look upon man as a black box in the physical world, exposed to externally determinable stimulatory forces as input and spouting externally determinable testimony about the external world as output. Just which of the inner workings of the black box may be tinged with awareness is as may be.

The stimulation of nerve endings is prior; when this input is processed to the level of awareness, we have observation sentences.[163] The latter are acquired by conditioning and are such that anyone who knows the language can verify or falsify them on the spot. This definition takes care of the objection that observation varies with interests and training since these factors are supposed to cancel out when the whole community is involved. There are, Quine admits, degrees of observationality that lead him to claim "the more observational a sentence is, the more nearly its use can be mastered ostensively." But Quine takes a new step when he says:

> We all have a sympathetic way of putting ourselves in the other fellow's place and sensing how the world would look from where he sits.[164]

This answers the objection that in ostensive learning the pupil and the teacher do not see precisely the same scene, and so far as I know this was the first occasion on which Quine used this argument. It also allows the native and the linguist to have the same perceptions. As usual, Quine invokes innate similarity standards, but he goes a bit beyond his earlier use in claiming "our learning of language thus depends heavily on the happy circumstance that similarity by our lights and similarity by the other fellow's lights go pretty much hand in hand."[165] Quine then moves to the compounding of observation sentences or terms, which he sees as "a step upward on the theoreticity scale, but a short one." "Individuation," Quine says, "is a big step, carrying us to another grade of theoreticity." Thus he continues to hold that individuation is done by language and comes after the learning of mass terms. Moving up, he considers the discovery of laws of nature at the near observational level. Since induction is learning, the discovery of the co-presence of the observed stimulus meanings of term A with that of term B counts as such a law. The means of representing generalities is quantification, which requires a major step up the scale, but Quine inserts substitutional quantification as an intermediate step before objectual quantification is attained. The move to full objectual quantification becomes evident when we begin to quantify over nameless things, such as electrons or real numbers. This brings us the objectual variable and general terms. Quine claims that if we had only substitutional quantification "then there is nothing recognizable as talk of objects; the question of what there is does not arise."[166] It is only with objectual quantification that we have objects, and "the values of the variables are what there are said to be." If we could stop here, Quine says, we would be "greatly relieved" but our drive for system and simplicity has driven us on to subatomic particles and transfinite sets. That our desire for system and simplicity should result in ever increasing complexity, he considers the greatest irony "of them all."

On August 18, 1970, Quine wrote to his former student Charles Parsons as follows:

> I like substitutional quantification because it is behaviorally more objective: as mentioned somewhere in "Existence and Quantification." (*Ontological Relativity*) it is almost free of indeterminacy of translation. I like it also for its implicit ontological economy; that is, when you do temporarily paraphrase it into terms of objectual quantification for the sake of an ontological assessment, you find that a denumerable ontology would have sufficed. If I could see my way to making do with substitutional quantification generally, in lieu of objectual, I'd jump at the chance. I'd not even temporarily paraphrase it into objectual terms for ontological assessment: I'd just turn my back on ontology, it being of a piece with the rejected brand of quantification. But of course I can't see how to do that.[167]

In *Roots of Reference*, Quine would again discuss substitutional quantification

I have remarked earlier on Quine's problem with "assent" and "dissent." Unless the linguist can determine the natives' terms for "assent" and "dissent," the process of translation cannot get started. Quine was aware of this problem and in his "Reply to Hintikka" in *Words and Objections* in 1969, he tried to deal with it. He wrote:

> The linguist's decision as to what to treat as native signs of assent and dissent is on a par with the analytical hypotheses of translation that he adopts at later stages of the enterprise: they differ from those later ones only in coming first, needed as they are in defining stimulus meanings.[168]

This is not a position Quine could hold to; it founds the more certain on the less certain. Quine had emphasized the lack of empirical control for analytical hypotheses; this is essential to his theory of the indeterminacy of translation. If the translations of "assent" and "dissent" have no better foundation than analytical hypotheses, then neither do stimulus meanings, and if that is so radical translation becomes so conjectural that it hardly rates as translation at all. Quine would soon seek another solution to the problem.

That Quine's behaviorism regarding language is inconsistent with Chomsky' s linguistic theory is obvious and it was certain that at some point they would collide. They did, in an article entitled "Methodological Reflections on Current Linguistic Theory."[169] Quine opens by distinguishing between rules that fit behavior and rules that guide behavior. Fitting, he says, is merely descriptive; all it claims is that the behavior happens to conform to the rule. "But the behavior is not *guided* by the rule unless the behaver knows the rule and can state it."[170] Actually, Quine's definition of guiding is wrong on two points: if taken literally, it forecloses the possibility of being guided by a rule one cannot state, which is part of the question at issue and therefore is question begging, and further guidance by a rule requires the behaver seeing to it that his behavior fits the rule. That Quine is wrong on these facts is shown by the fact that most speakers – including English speakers – obey grammatical rules they cannot state, but they can often recognize that they have made an error, as is shown by the fact that they correct these errors. But despite his definition, Quine is soon talking about unconscious rules. What Quine proposes is that there can be two systems of grammatical rules that provide recursive methods yielding the same infinite totality of sentences, but that are not consistent with each other.

What Quine challenges in Chomsky's doctrine is "it imputes to the native an unconscious preference for one system of rules over another, equally unconscious, which is extensionally equivalent to it."[171] That there can be such inconsistent systems of grammatical rules that are extensionally equivalent – that is, that determine the same infinite totality of sentences, Quine thinks is clear from the experience with axiom systems in mathematics. Quine takes Chomsky to hold that one of these systems is right and the other wrong. To which of course Quine replies, what behavioral evidence is there for their difference? Naturally, for Quine such evidence must be dispositions to verbal behavior. Quine also challenges the linguistic concept of when a sentence is well-formed, and uses Carnap's example of "This stone is thinking of Vienna." This is an odd example since it turns on the meaning of the sentence rather than its conformity to grammatical rules.

Quine also challenges the Chomskian thesis on linguistic universals. The argument here is the one from the indeterminacy of translation; if the subject-predicate distinction is found in all languages, that means all languages as we translate them, and we impose the subject-predicate distinction in the course of our translation.[172] But suppose we limit the language involved to English. Quine claims we do not understand the question of which grammar is right even in English.

> The essential problem remains; we do not really understand our English question. We are looking for a criterion of what to count as the real or proper grammar, as over against an extensionally equivalent counterfeit... and now the test suggested is that we ask the natives the very question which we do not understand ourselves: the very question for which we ourselves are seeking a test. We are moving in an oddly warped circle.[173]

Quine then tries to bolster his argument by comparing it to that of finding a definition for synonymy and analyticity. He cites work by Apostel and others to show that a test of synonymy similar to that proposed by Carnap led to indecisive results. But of course the issues concerning synonymy and analyticity have nothing to do with grammar. So after this detour, Quine comes back to grammar by challenging the notion of "deep structure." His objection is methodological. Can we distinguish between a grammatical structure supposed to be in the native's mind – i.e., innate, and an extensionally equivalent grammatical system? Quine holds that we cannot. Instead, Quine holds that every grammar is as "authentic" as every extensionally equivalent grammar. But Quine says that deep structure "and the transformations to and from it" might well be useful in the same way logical structure is. That is, each should be seen as a paraphrase of ordinary speech into a special form designed to serve particular purposes, of which he then presents verbal examples.[174]

The point of this article for Quine is clearly polemical. Given his own behavioristic psychology, he cannot accept Chomsky's view and wants to discredit it. He does not raise the issue of language learning of which he has made so much in other writings, and which Chomsky had used to show that a behavioristic psychology cannot account for the child's language acquisition. Nor does he mention repair data, which psycholinguists use extensively to show the presence of unconscious rules. The strategy is to externalize Chomsky's transformational grammar; to make it, like logic, a learned system useful for certain purposes. He still holds his earlier view that the learning of language is the learning of a skill, like riding a bike. Here at least he does appear to accept the notion of rule following but tries to externalize it.

Quine claims that in the translating of a foreign language, we will impose our linguistic system on the natives in making the translation, and that the so-called linguistic universals are the result of this sort of linguistic imperialism. But in fact, anthropologists and linguists do translations of other languages in which ontologies are very different from ours and which have grammatical differences from English. A trivial example will make the point. In Ojibwa, the third person singular pronouns distinguishes between animate and inanimate, but not between genders, whereas our language has three pronouns, one for inanimate things and one for each gender of animate things. How could we know this if our translations were as imperialistic as Quine claims?

Quine further claims that "we know logic when we know the logical constants." But learning the logical constants is not so easy. He thinks we can learn "negation" from "dissent." But learning "conjunction" poses a problem. We can know of two horses in a two horse race that they cannot both win, but we cannot say that one will not win nor that the other will not win. Hence we can know that the conjunction is false without knowing that the conjuncts are. We cannot, Quine says, learn categoricals from observing behavior since if "B" denotes "rabbit" and "A" denotes "undivided rabbit parts," one is not included in the other. Further, Quine admits that we can know what stimulations justify assent to a term, and dissent from it, and still not know what observable thing the term is true of. Quine further admits that identity, quantifiers, pronominal cross-reference, and pluralization can't be learned separately; they have to be learned together as parts of a system.

But there is a problem here. In *Philosophy of Logic* Quine argued that all speakers of the language know the same grammar. In this article ("Methodological Reflections on Current Linguistic Theory"), he argues that there are extensionally equivalent but incompatible grammatical systems. These were published in the same year. And as we will see, Quine later denied ever having said grammar was relative.

No one had a greater influence on Quine than Rudolf Carnap. When Carnap died in1970, Quine wrote a splendid eulogy that he gave at a memorial service in Boston, that was published as "Homage to Rudolf Carnap," and was later republished in the *Boston Studies in the Philosophy of Science.*[175] Quine says that "Carnap is a towering figure. I see him as the dominant figure in philosophy from the 1930s onward, as Russell had been in the decades before." Quine praised the *Aufbau.*

> Russell had talked of deriving the world from experience by logical constructions. Carnap, in his *Aufbau*, undertook the task in earnest. It was a grand project, and yet a self-effacing one, when so few philosophers understood technical logic. Much ingenuity went into the constructions, much philosophical imagination, much understanding of psychology and physics. If the book did not achieve its exalted purpose, it did afford for the first time an example of what a scientific philosopher might aspire to in the way of rigor and explicitness.[176]

Quine goes on to praise Carnap's *Logical Syntax of Language:* "it was the definitive work at the center, from which waves of tracts and popularizations issued in ever widening circles."[177] Carnap, Quine says, "more than anyone else was the embodiment of logical positivism, logical empiricism, the Vienna Circle."[178] Quine then describes Carnap's work in semantics and his *Logical Foundations of Probability.*

> Carnap was my greatest teacher. I got to him in Prague 38 years ago, just a few months after I had finished my formal studies and received my Ph.D. I was very much his disciple for six years . . . .But even where we disagreed he was still setting the theme: the line of my thought was largely determined by problems that I felt his position presented.[179]

Quine recalls his European trip in 1932 when having visited Vienna and the Vienna Circle, he then went to see Carnap in Prague.

> I was then an unknown young foreigner of 23 . . . it was extraordinary of anyone, and characteristic of Carnap, to have been so generous with his time and energy. It was a handsome gift. It was my first really considerable experience of being intellectually fired by a living teacher rather than a dead book.[180]

Quine recalls his Harvard lectures on Carnap, the Christmas trip to the APA meeting in Baltimore with Carnap, Goodman, David Prall, and Mason Gross, the wonderful year of 1939 when Carnap, Russell, and Tarski were all at Harvard, and his many meetings with Carnap and their correspondence. He closes with the comment "His death, while still at the height of his powers, marks a sad date in the history of philosophy."

Much attention has been given to Quine's disagreements with Carnap. In fact, Quine has been described as the man who destroyed logical positivism. What needs pointing out is how close he remained to Carnap, how much of Carnap's philosophy he retained. Quine was, in many respects, the child of the Vienna Circle – the last logical positivist.

We noted above that Quine had long been intrigued by Schoenfinkel's functional system of logic which he said provided a clarification of the nature of the variable. In a 1970 paper, "Algebraic Logic and Predicate Functors,"[181] Quine develops his own predicate functor logic, modeled on Schoenfinkel's, and shows that it is equivalent to quantification theory with identity. But he remarks that there is no proof procedure for this system.[182]

In 1970, Quine was invited to deliver the Carus lectures in 1971. He subsequently reworked the material from the lectures, and in 1973 published *The Roots of Reference.*[183] It hardly needs remark that Quine had been working on the problem of reference since the 1950s; *Word and Object* and "Ontological Relativity," which he gave at Columbia in 1968, were his most important publications on the subject before *The Roots of Reference*. The volume begins with the introduction that Goodman gave for the Carus Lectures – a fine example of Goodmanian prose.

> I have no idea what the roots of reference are, but I suppose that whatever has powers, as reference does, also has roots. But here I am probably guilty, in the language of *Methods of Logic*, of making, in the full sweep of a fell swoop, what amounts to a full swap or even a foul swipe.[184]

Quine begins by stating his problem: "Given only the evidence of our senses, how do we arrive at our theory of the world? Bodies are not given in our sensations but are only inferred from them." In seeking an answer, Quine reiterates his view that we should use the full resources of science, for he sees the problem as a scientific one – a problem of the science of science.[185] What science tells us, Quine holds, is that our contact with the world consists of sensory stimulations. Of course Quine insists that we must focus on dispositions to verbal behavior, but he also ponders briefly the possibility of analyzing behavior in terms of causes instead of dispositions. Quine takes causation to be the transference of energy from cause to effect. But he considers the notion of cause "out of place" in modern physics, since even the distinction between matter and energy is unclear. Further, what really interests us is contributory cause, and the attempt to analyze that in terms of energy transference he considers dubious. He therefore turns to dispositions, defining them as he has in a number of his earlier writings. The concept is imperfect, Quine admits, since it involves a reference to an unknown physiological state, but until happily that state is discovered, it can serve as a descriptive term. Quine includes here innate dispositions, whether learned in utero or coded in the genes.[186] As before Quine assumes an innate similarity relation, but here he adds refinements. By "receptual similarity" he means that two episodes of sensory stimulation are such that the second episode stimulates receptors that are in the neighborhood of the first, where "neighborhood" is used in its mathematical sense. Then Quine defines *perceptual* similarity as "*a* is shown to be perceptually more similar to *b* than to *c* when the subject has been conditioned to respond in some fashion to all episodes in the receptual neighborhood of *b* and to withhold that response from all those in the receptual neighborhood of *c*, and is then found to respond to those in the neighborhood of *a*." This is I believe Quine's first attempt at an analysis of perceptional similarity. It too rests on an innate similarity standard. Although perceptual similarity is subjective, it has behavioral consequences: a man's inductive expectations are reached by extrapolating along the lines of "perceptual similarity"; experiences that begin similarly are expected to turn out in similar ways. But if so, why should such induction based expectations turn out to be right? Quine's answer is natural selection: those who failed to form inductions that yield true expectations are no longer with us. But Quine faces a problem in explaining how it is that perceptual similarity standards change over time and experience. Perceptual similarity can be distorted by "internal states," Quine says; "what of his current purposes, passing memories, his interrupted train of thought? I speak mentalistically, but I refer to factors of his physical state."[187] The strategy of using mentalistic concepts and then claiming that he is referring to physical ones in doing so occurs repeatedly in this book. How is one to sort out what is due to perceptual similarity and what to "internal states"? Quine's answer is that since perceptual similarity has an innate base, it must be relatively stable, changing slowly, whereas the effects of the internal states will be variable. "It is a matter of detecting regular trends beneath the perturbations." Then there are the further problems of interpersonal agreement of judgments of similarity. Quine finesses this once again by an appeal to natural selection. "We may expect our innate similarity standards to be much alike, since they are hereditary in the race; and even as these standards gradually change with experience we may expect them to stay significantly alike, what with our shared environment, shared culture, shared language, and mutual influence."[188]

If perceptual similarity is to affect behavior, then Quine says there must be a connection between prior episodes and subsequent ones. Such a connection requires "traces" left by the earlier episode. Receptual similarity is not enough here because it accepts all stimulations equally, whereas "perceptual similarity hinges more on noticing." And noticing in turn depends on "salience." The conditions that determine salience are "focal position, motion, brightness, boundary contrast, gaudy color." These are innate determinants; similarity to traces of past episodes is an acquired determinant, which will both increase present similarities and revivify traces of the past ones. Quine assures us that he is using such mentalistic language only as a shorthand for the underlying physiology. Here Quine invokes pleasure and pain; traces encode the pleasure or pain of the prior episode, and "the drive to increase or decrease the similarity will therefore vary with the degree of pleasantness or unpleasantness of the earlier episode." The subject approaches that which gives pleasure and avoids that which gives pain. "To learn is to learn to have fun." Reward and punishment shape our behavior.

> The subject basks in present impingements and puts his best foot forward. Traces of past episodes tell him what to seek and what to avoid. Similarities point his strategy, which is that of exploiting the head starts, improving the advantages. The inductive method is implicit in that strategy, for in effect that strategy consists in reproducing some components of a past episode in the hope that other components will accompany them, or in averting some components of a past episode for fear that others may accompany them.[189]

Quine holds that this psychological description yields a reasonable "schema."

This is a more refined psychological model than Quine had previously used. It is also one that abounds in mentalistic terms. "The talk of a pleasure principle was a conspicuous instance of mentalistic idiom, however behavioral in intent. Similarly for salience and traces." But Quine tries to legitimize this usage as "heuristic" and posited only in the hope of physiological explanation "someday." He even expands this charity to "images." But Quine reminds us that the goal is to discover how we came to have science, and to do so without falling into the mentalism of "ideas." "How can we pursue such an inquiry while talking of external things to the exclusion of ideas and concepts? There is a way: we can talk of language." And so Quine focuses on language acquisition, where he holds that in its early stages ostension and conditioning suffice. Perhaps the mentalistic terms he has employed are reducible to physicals ones; at least he seems to think so. But he does not so reduce them and one doubts that he can. In fact, this whole account rests on a mentalistic psychology that Quine claims he rejects. But beyond language learning Quine finds the picture murky; the child talks of things not present, or the past and the future, and Quine admits that a behavioristic explanation of these developments eludes us.

Quine now turns to more familiar matters – observation sentences. Observations are basic to both language and science, and they are so through their "intersubjective immediacy".

> Ostensive learning is fundamental, and requires observability. The child and the parent must both see red when the child learns "red," and one of them must see also that the other sees red at the time.

Learning a language, Quine says, is learning the meaning of its sentences and so of what observations to count as evidence for them. But the verification theory of meaning went wrong in thinking that individual sentences have separate sets of supporting observations. Here of course is Quine's holism; only as parts of theories are sentences confirmed. To get started, Quine drops talk of observations in favor of talk of observation sentences such as "This is red." And what is that? "A sentence is observational insofar as its truth value, on any occasion, would be agreed to by just about any member of the speech community witnessing the occasion," and the speech community Quine says is defined as those among whom dialogue is fluent.[190] That it is sentences that are learned first rather than terms is one of Quine's most fundamental doctrines; he takes even one word expressions such as "red" to be sentences – that is, as meaning "This is red." Such sentences would seem to be confirmed by what they refer to, but since Quine has made the truth values of the sentence depend on agreement by witnesses, there is nothing to prevent subsequent retraction. In a lecture at Wayne State that year, Quine explicitly says that observation sentences are corrigible. The terms that occur in observation sentences occur also in theoretical sentences – that is how the theory is linked to observation. Conflicts within the theory can lead us to repudiate observation sentences. But this still assumes that witnesses sense the same, and the homology problem remains.

The child learns the sentences by conditioning – by reinforcement and extinction. In doing so, he learns the "similarity basis" of the sentence. This is a question of perceptual similarity: what tints count as red in his community. Salience helps the child learn, as does the ostending finger pointing at the patch. The pointing finger is soon replaced by "assent"; the child hazards "red" and others assent. "Assent " is thus learned in the same way; the child hazards "red" and the parent says "yes"; "what is learned here," Quine says, "is an equivalence: assent to a sentence entails the same rewards or penalties as a repetition of the sentence would entail." But this hardly works: if the child is rewarded for pointing at something red and saying "red," it hardly follows that he will be rewarded for pointing at the same thing and saying "yes." The ostension is to something red that the child sees; "yes" (or "assent") refers to the child's act in calling the perceived thing red. The two are obviously not equivalent. Further, on Quine's theory of language learning, the child is in the position of the linguist trying to master an unknown language. In Quine's theory, the child must learn "assent" and "dissent" before he can master other "sentences." He cannot volunteer a sentence he has not learned, and he can't learn the sentence without first learning "assent" and "dissent." Quine has yet to provide a convincing account of "assent." And the child must also master "dissent" which requires leaning that dissent is rewarded where assent is punished. Quine handles the learning of "good" similarly. "The two factors that make for learning, in general, are perceptual similarity and the pleasure principle; but in this case (red) pleasure does double duty, serving also as the similarity basis." But "good," like "sick" or "flimsy," is only partly an observation sentence; the child may not be able to tell that someone is sick by observation. But with "sick" and "flimsy," there is community agreement; not so Quine says for "good" in an aesthetic sense, for people disagree in their tastes. But on the morally good, the community does generally agree (although the community in this case need not be linguistic). For anyone who lived through the twentieth century, this is a bizarre statement.[191]

Quine then turns to the learning of masses and bodies. These are matters largely covered in WO. The difference between "mama" and "water" is that mama remains continuous despite distortions, whereas water comes in various amounts and shapes. But here Quine shifts his position. The type of similarity that persists through continuous distortion, Quine says, "is apparently a sort of similarity that we are innately predisposed to appreciate."

> It is no wonder that bodies, bodily identity and bodily persistence are the mainstay of ontology. Bodies, for the common man, are basically what there are; and even for the esoteric ontologist bodies are the point of departure. Man is a body-minded animal, among body-minded animals. Man and other animals are body-minded by natural selection; for bodymindedness has evident survival value in town and jungle.[192]

Quine has not said this before; why the change? He cites T. G. R. Bower's article in the *Scientific American*, "The Object in the World of the Infant"[193] – an early survey of studies of neonate cognition. Bower reports a number of experiments that support the claim that the prelinguistic infant has the object concept. Bower writes:

> This attainment is obviously one of tremendous significance. It transforms the perceptual world of the infant at one stroke into something very close to the perceptual world of the adult. According to these studies it seems that infants less than 16 weeks old live in a world articulated in terms of solids that are stably arranged in space according to their location, with a constancy of existence when they occlude one another.[194]

What then becomes of Quine's indeterminacy among rabbits, rabbit-slices, undetached rabbit parts, et al? If the individuation of objects precedes language, as it must if the child has it at sixteen weeks, and is a trait of the species, then Quine's jungle native does refer to an object when he points at a rabbit. Quine understands Bower's point; and in the remainder of the book rabbit slices, undetached rabbit parts, et al. are conspicuously absent. Instead, Quine turns to the learning of general terms. Given that the child has learned that the various appearances of Fido are those of one dog on what Quine calls a similarity basis, acquiring the general term "dog" requires a second order similarity – a similarity among similarities. "Thanks to his instinctive body-mindedness, he (the child) is an apt pupil when the general terms are terms for bodies."[195]

Quine then turns to the distinction between absolute and relative terms. "Small" and "mother" are absolute terms; "smaller than" and "mother of" are relative terms. These terms apply to objects, but Quine notes that relative terms apply to masses too – "darker than" is his example. And "same as" is an equivalence term. "Same as" may aid in individuating "dog," but it is to be applied to disjoint things; overlapping breeds confusion. But "same as" is a start on the identity predicate. Geach had argued that "identity" makes sense only when applied to general terms – i.e., "same dog." Quine agrees "as long as the sides of the identity sentence are demonstrative pronouns." But when we reach the point of identity statements concerning names, variables, and descriptions, this is no longer true.[196]

A crucial feature of human language is the "productivity of its combinations." Quine says, "We can imagine learning it [the combinations] as follows": "Given that the child has already learned the words 'yellow' and 'paper,' hearing them spoken enlivens traces of episodes in which yellow was salient" and similarly for "paper," and "these traces enhance the salience of any yellow or paper in the present scene." Hence, "all our mentor has to do to perfect our training in the compound 'yellow paper' is to discourage assent in those less striking cases where the yellow and the paper are separated." The same process accounts for other attributive compounds. Quine then takes "yellow paper" as an observation term, but he equates it to the sentence "the paper is yellow" We learn the "in" construction similarly. But these sentences are occasion sentences. Quine now introduces "standing sentences" and "eternal sentences." Eternal sentences are the stuff of science, or more exactly of the central portion of the web of belief.

> What makes occasion sentences less puzzling semantically than the eternal sentences is that we do not need to trace their systematic connections with theory in order to probe their meaning; they are at the periphery, where their meaning can be empirically assessed by checking them for assent and dissent, occasion by occasion. It is through the periphery that science and language imbibe all empirical content or meaning. To trace out the meaning of an eternal sentence deep inside the theory. . . . we have nothing to go on but its multifarious connections within the theory and ultimately, indirectly, with the periphery. Each of these strands being describable only by its interrelations with others, there ceases to be any clear sense in asking the meaning of a single such sentence at all.

But Quine has denied that the word "meaning" has any correct usage in a behavioristic psychology. How then can he talk of the "meaning" of occasion sentences? To pretend that somehow such usages are merely shorthand for behavioral descriptions is mere handwaving.[197]

Quine thinks the best way of coping with the "meaning" of eternal sentences is by retracing the process by which we learn "such language" in the first place. Using "snow is white" as his example, Quine holds that the child has been conditioned "to assent to the query 'white?' when snow is presented" and this conditioning is transferred to the word "snow." Learning an observation sentence, Quine holds, consists in learning the circumstances in which to assent or dissent. To learn an eternal sentence is to learn whether it is true forever. "First and last," Quine says, "in learning language, we are learning how to distribute truth values. I am with Davidson here; we are learning truth conditions." Apparently Quine identifies "meaning" with "empirical content or information," whatever that is. The learning process (all by conditioning) goes from "snow is white" relating mass terms to "Fido is a dog" relating the singular term "Fido" to the general term "dog." Quine says "the word ['Fido'] has induced an image of the integral animal." Quine of course claims that such mentalistic talk is reducible to conditioning, but it is worth noting that he finds it necessary to use the terms "image," "trace", "notice", and "salience." etc. to get his point across. These terms are not reducible to behavioral terms.

The next step in language learning is the categorical, e.g., "a dog is an animal," learned he claims by the transference of conditioning. But in accounting for the child's learning of such sentences, Quine says "he comes to sense similarity among the ways of coming to appreciate that a dog is an animal, that snow is white, and that Fido is a dog." Quine says, "I tread dangerous ground in speculating on an inward sense," but he claims "it is only a question of awareness of incipient drives toward specific overt behavior." "Drive" it should be noted is a term from Hullian behaviorism and refers to a postulated psychological state. Quine's behaviorism is here being stretched not only to the limit but past it. But Quine marches on, claiming transference of conditioning has enabled him to jump the gap between occasion sentences and standing sentences. He admits that his account attributes a confusion of use and mention to the child. Such errors must be corrected, just as we must eliminate the indicator words of occasion sentences if we are to reach the eternal sentences of science. "Language is conceived in sin and science is its redemption." The device of using transference of conditioning from the present object to the word for that object is essential for Quine's move from occasion sentences to standing and eternal sentences.[198]

Quine then turns to the learning of color words. "Red," as noted above, is learned as an observation sentence. But the general term "color" is not because there is color everywhere. So Quine proposes that from terms like "red-colored," "blue-colored," etc., we learn to abstract the general term "color." The particular color words are based on chromatic similarity, so "color" as a general term is based on a similarity of similarities. Shape words we acquire in a similar manner.

> As for the method of learning, we have already speculated on how the child would learn "α is a color"; it was a matter of his learning to distinguish color words from others. These speculations carry over in exact parallel to the learning of "α is a shape word."

One should note the word "speculation"; Quine says what he is presenting is a conjectural history of language learning. Whether he really regards it as speculative will appear later.[199]

Next, Quine takes up truth functions. Negation is obviously a sign of dissent and the child will learn it from occasion sentences and then will extend it to standing sentences. But conjunction is not so simple. For cases where both the conjuncts are true the conjunction is true. But there can be cases where the standard truth functional rule for dissent fails. While everyone agrees that a conjunction is false if its conjuncts are, if one or more of the conjuncts receives "abstain" as its value, the conjunction may receive either denial or abstention. Similarly, alternation follows the standard truth functional rule except where both its components receive an "abstain," when it will also receive an abstention. To deal with this, Quine uses a three valued logic with what he calls "verdict functions" that he regards as more primitive than the two valued truth functions. "Two valued logic is a theoretical development that is learned, like other theory, in indirect ways upon which we can only speculate." Surely, this is implausible; three valued logic is more primitive than two valued logic? And how did the child master "assent" and "dissent" if he is not thinking in terms of two values? But Quine carries on. These comments bring Quine to the subject of analyticity. Despite his comments about "meaning" above, Quine holds "no empirical meaning has been given to the notion of meaning, nor, consequently, to this linguistic theory of logic." This is claimed, one should note, despite the attribution of "meaning" to occasion sentences a few pages earlier. Nevertheless, since some people do learn sentences as true, he is willing to accept analyticity as the result of social consensus. "The *analytic* sentences are the ones whose truth is learned in that way [as previously described] by all of us; and these extreme cases do not differ notably from their neighbors, nor can we always say which ones they are."[200]

In Part III, Quine turns to reference. We have, he says, "reached the end of our speculations on the primitive steps." From here on, he tells us, he will speculate on how we learn to speak of objects. Quine employs his claimed indeterminacy of translation to make the point that "reference involves more than the simple ability to acknowledge a presence." But the indeterminacy of translation in the home language, from idiolect to idiolect, does not interfere with the learning of reference. "It all depends on what we are trying to do. It depends on whether we are going to make capital of relations of sameness and difference of meaning. In translation we do, whereas here I shall not." The child's problem is to learn the "referential apparatus" of English. The early learning of singular terms like "Fido" did not involve any distinctive objective reference; it is rather the notice of circumstances. The early individuative terms are general terms for bodies. "Bodies are the charter members of our ontology." Quine uses "square" to make his point. Whereas adding water to water just makes more water, adding one square to another does not make a square; square individuates. "Square" is a general term when predicated of something; when it is the subject of predication it is an abstract singular. "Here", Quine says, "is that first portentous step down the primrose path of abstract ontology." The line between concrete general terms and abstract singular ones is indistinct.

> Genetically what we have beforehand is just a play of grammatical analogies that mark differences in learning patterns. Centrally situated here is what we retrospectively classify as talk of bodies. Here is where the apparatus of objective reference gets its first development. Bodies are the prime reality, the object *par excellence*. Ontology, when it comes, is a generalization of somatology.

Physical objects are for Quine a generalization of body. Much of language does not have objective reference, but science does: "It is in imposing this referential pattern all across the board that scientific theory departs from ordinary language. We see the result: objective reference is central to our scientific picture of the world."[201]

Quine then turns to the relative clause, on the learning of which "I propose to speculate." The relative clause, Quine says, serves as a general term which, when predicated of an object, reaffirms what the sentence says about it. Geach has proposed an alternative that replaces the relative clause with a conjunction and simple pronoun. Quine admits that Geach's theory is plausible but he prefers the relative clause because he thinks it provides a more plausible process of acquisition. Further, Quine says, "I am not bent even upon a factual account of the learning of English, welcome though it would be. My concern [is] with the essential psychogenesis of reference." He modifies standard English by focusing on the "such that" construction. Quine emphasizes that the relative clause has further uses: it can put any sentence into the form of a predication, either objectual or substitutional. But it also plays a role in the universal categorical construction. Thus the sentence "Everything that we salvaged from the wreck is in the shed" becomes "Everything x that we salvaged from the wreck is in the shed." Quine particularly emphasizes the substitutional operator which he calls "the essence of the relative clause." Thus "Fα" becomes "α is a thing x such that Fx." The "x" is of course the variable that takes the place of the pronoun in this clause to keep the reference straight. So Quine holds that the relative clause is learned in predicative position, and then entered into the categorical construction. Quine's preference for this approach is clear when he points out the easy transition to

$$F\alpha \equiv (\exists x)(\alpha = x \cdot Fx) \equiv (x)(\text{if } \alpha = x \text{ then } Fx)$$

But this is not the order of acquisition. Quine shows that all the categorical forms can be generated by the use of the "such that" construction, and this introduces quantification.[202]

There are, Quine notes, two forms of quantification: substitutional and objectual.

> When its variable is conceived thus *substitutionally*, a universal quantification counts as true if and only if the open sentence following the quantifier comes out true under every substitution for the variable; and an existential quantification comes out as true if and only if the open sentence comes out true under some substitution.

Objectually construed, the variable refers to objects as its values. The two forms are not equivalent, having different truth conditions. Quine takes substitutional quantification as the form that is first learned: "I see this switch from substitutional 'x' to objectual 'x' as an irreducible leap in language learning." And it is objectual quantification that "becomes the distilled essence of ontological discourse." This is of course Quine's famous doctrine that to be is to be the value of a variable. Quine claims:

> By considering what steps could lead the small child or primitive man to quantification, rather than to the less tidy referential apparatus of actual English, we arrive at a psychogenetic reconstruction in skeletal outline. We approximate to the essentials of the real psychogenesis of reference while avoiding inessential complications.

So one sees here that what Quine has been calling his "speculations" are in his view the actual course by which both the individual and the race have come to quantification. "The relative clause and the categorical thus stand forth as the roots of reference."[203]

Quine's arguments so far are not clear. Apparently for him occasion sentences and observation sentences are learned by conditioning to neural stimulations. Sentences do not refer, but their terms do – to what? Quine says that they refer to "circumstances"; but given his discovery that infants see the world as one of "bodies," presumably a term like "ball," which for him is a sentence – "It's a ball," – should also refer to a ball as a body. But bodies for Quine are apparently not physical objects; the child does not refer to physical objects until he masters the relative clause and the categorical. One must guess, therefore, that the "bodies" perceived in infancy are developed into genuine physical objects by some process of construction and redefinition, though what this process is is not clear. Quine denies that the child has "objective reference" before he has the relative clause and the categorical. What then is the status of "bodies" or "circumstances" if they are not objective things? Apparently Quine is determined to hold to his claim that "to be is to be the value of a variable" no matter what.

But could the problem here lie in Quine's definition of "physical object" as "any material aggregate however scattered"? This involves the notion of a scattered four-dimensional physical object with temporal as well as spatial parts. This is not the standard concept of a physical object held by English speakers. Scientists may indeed think in four dimensions, but the idea that a physical object is "scattered," so that for example all dogs are parts of one physical object "dog," is one that few even among philosophers would endorse, nor would most scientists endorse it. Physicists do not consider all electrons as parts of a single physical object. One must therefore ask what language it is the acquisition of which Quine is describing? Since the relative clause and the categorical do not lead to a notion of objects as scattered, one must assume that what Quine means here by an "object" is the standard English concept, not his technical one.

Quine then turns to the acquisition of reference to abstract objects. He thinks that the first abstract objects to arise are attributes when they are taken as singular terms, as is the case with "square" in "square is a shape." This development is also facilitated by the relative clause, since it provides a general term for anything we can say about an object. But Quine now switches from attributes to classes, arguing that classes are as abstract as attributes but have a clear criterion of identity that attributes lack.

> In trying to imagine a psychologically feasible genesis of set theory, I shall start with general terms, including the relative clauses, and use substitutional quantification. The genesis will not be a matter of eliminative definition. It will proceed by irreducible leaps, but plausibly short ones.

Quine holds, as we saw above, that the categorical construction requires objectual quantification. But substitutional quantification does not require the categorical. The child learns that a universal (substitutional) quantification commands assent only if every substitutional instance commands assent. Should some of the instances receive abstention, though none receive dissent, he will face a situation like that of conjunction when one or both of the conjuncts receive abstention. The situation with existential (substitutional) quantification is parallel. Quine then summarizes his psychogenetic path to quantification:

> I shall sketch our pupil's past and present progress down this garden path. He learns his first variables, we saw . . ., by learning relative clauses; for, in my caricature, the relative clause has the form "thing x such that Fx." These first variables are substitutional, for he learns the relative clause in predicative position as a substitution idiom. Independently of this he learns also the categorical copula, without variables, as joining general terms "Every α is β." Next, we saw, he combines the two idioms, on the strength of an analogy between relative

> clauses and general terms . . . So, pursuing this analogy, our pupil slips relative clauses into the categorical. Thenceforward, the relative pronoun, or "such that" variable, figures as an objectual variable rather than a substitutional one.

Thus, Quine says, he first learns substitutional quantification, which he acquires "by learning its conditions of assent and dissent." The two forms of quantification look alike, and induce a resemblance between general and singular terms. This resemblance "makes the general terms feel like names of something or other" and lo, we have classes.[204]

Quine reaffirms his belief that his account of language learning is close to the actual process:

> In recounting this suppositious psychogenesis in terms of quantification, I am adhering to my caricature. My conjecture is that our actual learning of the ordinary idiomatic apparatus runs parallel to this caricature. But I gain visibility by bypassing the sinuosities of ordinary language.

Clearly Quine does believe that his account provides an answer to the question of how, from sensory stimulation, we have arrived at science.

He turns next to set theory. Just as substitutional quantification simulates objectual quantification, "so the 'such that' clause simulates a class name." We move from "thing x such that Fx," which is a general term, to the abstract singular term "{x:Fx}" – "the class of all things x such that Fx." The "such that" appears as a class abstraction. The substitutional transformation of "Fx" into "y is a thing x such that Fx" is taken as " $Fy = y \in \{x{:}Fx\}$" and the copula "is" becomes " $\in$." So given

$$\text{“}(x)(x \in \{y{:}Fy\} \equiv Fx\text{”}$$

we get the set theoretic law of *comprehension*

$$(1) \qquad (\exists Z)(x)(x \in Z \equiv Fx)$$

This assumes that there are no hidden free variables in "Fx"; if there are, what will be required will be a satisfaction condition, and such a condition is a value of the hidden free variable that satisfies some instance of (1) that results from putting some class abstract for "Z", and this Quine shows can easily be obtained, yielding for the hidden variable "w"

$$(2) \qquad (w)(\exists Z)(x)(x \in Z \equiv Gwx)$$

(1) looks like the axiom of abstraction that leads to Russell's paradox, but it does not do so here because "Z" and "x" are different types of variables; "we have two types in Russell's sense." But for all its similarity to standard objectual quantification, "all that is really afoot is substitutional quantification with concrete general terms as substituends." Can our pupil get "the benefit of all these classes without the onus of really assuming them?" Quine's answer is no because substitutional quantification fails to meet some standard laws of set theory such as the law of unit subclasses. As written, (1) has a substitutional variable, "Z" and an objectual one, "x." But Quine

shows that the derivation of the law of unit subclasses from (2) requires commutivity of existential quantifiers which cannot be done if one quantifier is substitutional and the other objectual. Accordingly Quine solves the problem by using objectual quantification everywhere. He goes on to examine further inadequacies of substitutional quantification and to make his case for objectual quantification.[205]

Quine then takes up identity; it was discussed earlier in the special case where it occurred restricted to a general term such as "apple" and flanked by demonstratives – e.g., "this is the same apple as this." The child comes to recognize "that whoever assents to sentences '$\alpha = \beta$' and 'α is a γ' will assent to 'β is a γ'." Thanks to the relative clause, "α is a γ" can in effect be any sentence containing "α." When we have class quantification, "$x = y$" becomes definable as "y" is a member of every class of which "x" is a member, including of course unit classes.[206]

What then of numbers? The usual set theoretic definitions are not available at the level of classes of individuals which is what has been defined so far. But Quine suggests an alternative:

> what perhaps does reflect the learning of number pretty well, when due allowance is made for the artificiality of the formalism, is numerical existential quantification as defined with the help of identity. We have
>
> $$(\exists_0 x)Fx \equiv \text{not } (\exists x)Fx$$
> $$(\exists_1 x)Fx \equiv (\exists x)(Fx \text{ and } (\exists_0 y)(Fy \text{ and not } (y = x)))$$
> $$(\exists_2 x)Fx \equiv (\exists x)(Fx \text{ and } (\exists_1 y)(Fy \text{ and not } (y = x)))$$
>
> and so on. There is no class quantification here at all; just objectual quantification over individuals.

But does this approach lead to the sums and products of arithmetic? Quine shows that in fact the standard recursions for addition and multiplication can be achieved by what he calls a "fantasy" picture of the learning process. Of course Quine recognizes that "children are taught prefabricated algorithms intensively in elaborate institutions." But Quine still believes that his "fantasy" account shows the "steps that would be required in learning arithmetic under ordinary conditions of language learning, if it could be done." Quine notes that a substitutional theory of numbers is possible only if one is prepared to abandon the principle that every number has a successor. Since the number of expressions is finite, "some number would be the last."[207]

Language, Quine says, "is learned in a succession of leaps, and the unconscious transition from substitutional to objectual quantification over numbers is one more such leap." Quine has now got his pupil into dealing with classes of individuals; he then needs to get him up to classes of classes of individuals, and so on. The mythical pupil has already learned how to use relative clauses in the construction "y is a thing x such that Fx" which is equated to "Fy." The same form will carry him to "Y is a class X such that FX" equals "FY." This progress requires only substitutional quantification, with a new single variable corresponding to each type. At some point Quine sees him going objectual "because of the law of unit subclasses or similar troubles." Although classes of higher types are scarce in ordinary discourse, Quine says his purpose in developing this "fictitious learning process" is to show how set theory might have emerged from "work at lower language levels." Quine does not, however, hold that Russell's theory of types is a natural development; if it has roots in the earlier stages of learning, it is nevertheless "primarily an artifice for blocking the paradoxes." The alternative systems of Zermelo, von Neumann, Bernays, and Quine himself, Quine thinks, are likely only after type theory has been created and the interpretation has gone objectual.[208]

Quine has now completed his review of "the genesis and development of reference," which he considers applicable both to the child and to the race, even though "it is imaginary in any event." Quine says, "I find the foregoing account plausible in its essentials, especially as a factual account of learning by the child." Quantification and variables are of course departures from ordinary language, as is the class vocabulary, but Quine says "my account of the learning of first-type set theory is meant as a schematized account of the child's actual learning of the ordinary language of abstract terms."[209] But he decides to address the question of reference to the past. Quine takes this to be a question about memory, not past episodes beyond living memory. For such a sentence as "I have seen a black rabbit," the child requires cues as to when to assent and when to dissent. The child is assumed to have already learned the term "black rabbit."

> In the idiom of images we might say that those words conjure up the right image, even failing the real thing. And now I fear we may have to assume yet a little more in the way of built-in faculties; namely, a discrimination on the child's part between two kinds of images, images of fantasy and images of memory.

An image, Quine claims, is "a neural event inducing a state of readiness for an appropriate impingement pattern." Quine claims that the child is aware "of that monitory neural event itself." The image of imagination is descriptive whereas the image of memory rests on acquaintance, using Russell's terminology.

> Now it is when the words "black rabbit" induce this latter, more vigorous sense of readiness that the child may rightly assent to the query "Seen a black-rabbit?" He is said to remember seeing one.

As Quine is aware this adds nothing to Hume's account of vivacity as the distinguishing trait of memory as opposed to imagination. And how does the child learn to respond to questions about the past? By the parent asking about events that he has seen the child witness and rewarding assent and punishing dissent. Such occasion sentences can be given dates and times and so become eternal sentences. "At last we have arrived at protocol sentences, as they were called in the Vienna Circle."[210]

Having as he believes taken care of the past, Quine turns to the future. "As the past tense hinged on memory, so the future must hinge on expectation." What interests Quine is the future tense, since that includes predictions. Suppose the child experiences the event "a followed by b." This will leave a memory trace, so that if the child experiences an a-like event, he will expect a b-like event. This process will lead to both true and false predictions which the parent will correspondingly reward or punish, until the child's use of the future tense "is eventually brought under control."[211]

Quine says that his process of psychogenesis has brought the child to the level of science represented by natural history, and no doubt measurement can be added since the child now has numbers. "But what of theoretical posits, hypothetical forces, hypothetical particles?" Boyle's law he says is at the level of natural history; it relates expandable tanks to thermometers. The expandable tank, he takes as similar to a "confined swarm, say, of bees." The molecules are imagined on analogy to the bees. But Quine does not push this further. "A perfected psychology of science would not aspire to keep causal track of the minds at the advancing front of natural science." It would be concerned rather with the "basic phenomena of natural science." And Quine adds that the minds at the forefront of natural science "are themselves aware of what they are doing."

But it remains to consider ontology. Quine thinks it a great improvement to conceptualize physical objects as four dimensional. Thus the Morning Star and the Evening Star are just Venus seen at different times. Stages of objects become real too. "Color" is accommodated as surfaces whose fine structure reflects light in "the appropriate band." Shapes are harder to fit into Quine's ontology, but he does so by taking them as aggregates of points, enduring in time. Spatio-temporal points may now be dealt with as quadruples of numbers; this can be applied to physical objects and to relations among such objects, such as their forming the corners of a square, a relationship that remains invariant independently of changing coordinate systems. But we still have to deal with the dual ontology of physical objects and abstract objects. And Quine laments that the higher reaches of set theory bring us a limitless supply of the latter. But we can also eliminate parts of the common ontology of attributes, propositions, possible objects, intensions, etc. When we regiment our language within the confines of truth functional logic and quantification theory we can at least reckon the costs and benefits of our ontology.

The last question Quine takes up is scientific method: "the question of how best to develop an inclusive scientific theory." The two principles he advances are simplicity and conservatism. We want the simplest theory possible, and we want to minimize the changes in our existing beliefs that we have to make to get it.

> Each of the leaps of language learning I have pictured is a private little scientific revolution, another step in the development of the system of the world. If the leap is one that conduces to simplicity in the child's evolving conceptual scheme, then normatively speaking it is good scientific method on his part, however unconscious. If it is a short leap, then again it is good, on the score of conservatism.

Scientific progress is governed by what Quine calls the maxim of *relative empiricism*: "Don't venture farther from sensory evidence than you need to." Quine has always had nominalist leanings, and gave up nominalism only when he was convinced that the alternative was to abandon large parts of mathematics. Substitutional quantification appeals to his nominalistic strain, but he finds it inadequate for the nameless things, physical or abstract, that there are. But he closes with a nod to Charles Parson's "semisubstitutional quantification" which combines substitutional and objectual quantification in a way that Quine says "does retain something of the desired nominalistic aura that is wholly lacking in the objectual version, and I see no reason to doubt that it meets the needs of set theory."[212]

The Roots of Reference is a strange book, stranger yet when one considers that this was his Carus Lectures. The problem he set himself was to explain how, starting from sensory stimulations, we have reached science. Quine takes this to be equivalent to asking how we have learned the language of science. The only science he "reaches" is set theory, but one must bear in mind that for him set theory is a part of physics. He views our learning to refer to objects, physical and abstract, as the indispensable way to set theory, and it takes most of the book to reach this point. The "psychogenesis" he describes is supposed to be that of the child and of the human race. Much of Quine's account is given in mentalistic terms such as "image," "notice", "salience", "pleasure", "pain," etc. Quine claims that all of these are reducible to behaviorism, but he does not so reduce them. If he were right in this claim, he could and should have written the book in behavioristic terms exclusively, but he did not, and I think could not. Moreover, Quine is aware, through the Bower article, that prelinguistic children have an object concept, and it is hard to believe that such children are not referring when they point at a physical object. If so, what is the point of this book? Further, learning a language – even learning the *language* of science, is not learning science. One can talk about fields, particles, forces, etc. without being a scientist or doing science, as science fiction writers have shown. But since Quine is wedded to the view that all thinking is in words and sentences, he cannot even frame his problem except as one of learning language.

But there is a further point. Quine sees his objective as a science of science – that is, a scientific explanation of how we have come to have the science we have. For this, he seems to believe that an account of how children learn the language of logic – truth functions, quantification, and identity, – is sufficient. Why? All children learn a language; not all, not even most, become scientists. At most, then, learning language is a necessary condition for learning science but not a sufficient one. But even telling the story of how language is learned, if this is to be a scientific account, one would expect the volume to be rich in references to the psychological and psycholinguistic literature in which the processes by which real children do learn languages are explored. Nothing of this sort appears here. What Quine actually provides is a conjectural account, one that he repeatedly says is speculative. And he then claims that it is, if not historically true, yet a true "skeletal" outline of the true process, though on what basis this claim is made is not apparent. Moreover, his claim that his account is behavioristic simply cannot stand. Repeatedly, he invokes mentalistic terms – notice, trace, salience, image, etc., that are not, and probably cannot be, reduced to behaviorism. Whenever the going gets tough, he invokes evolution and natural selection to resolve the problem. And if, as Quine claims, all this is required to learn logic, how does that square with his claim that logic is "obvious"? Finally, Quine refuses to modify his claim that "to be is to be the value of a variable." This requires him to deny that the child can refer to external physical objects until he has mastered the relative clause and categoricals. When a twelve month old toddler, innocent of language, points to an object and exhibits glee when the object is given to him, are we really expected to believe that the child was not referring to the object when he pointed at it? Furthermore, Quine claims that the relative clause and the categorical are the roots of reference. At what age does an average child master the relative clause and the categorical? Certainly not before ten; more likely fourteen or fifteen, if then. Are we really to believe that a ten year old child cannot refer to an external object? To so claim is to put dogma before fact.

Finally, it is obvious from this book that behaviorism does not provide Quine with a theory adequate for what he wants to say. He needs mentalistic terms to make his points. Quine was of course aware of this. It presents a problem that Quine would have to solve: either abandon behaviorism, or abandon the mentalistic terms he had found it necessary to use in the book. How he solved this problem we will see.

Chapter 4
Last Revisions

That Quine was distressed by the student uprisings of the 1960s has already been noted – distressed enough to consider leaving Harvard. He was not one to suffer in silence. In 1974 he published an article entitled "Paradoxes of Plenty."[1] He contrasts the heavy teaching loads and Spartan conditions of academic life in the 1930s with the largess brought by World War II and continued by the National Science Foundation and the National Endowment for the Humanities, the lightened teaching loads, the higher salaries and the increased secretarial help that followed. But he finds it paradoxical that these changes brought also less talented students, a proliferation of journals publishing poor fare, and a lowering of academic requirements. Instead of ushering in a renaissance of creative work, it brought a cheapening of academic values. "Mass subsidy," he says, "soon losses its luster and comes to be looked upon as each man's due."[2]

> The department that I know best has freed its graduate students of the requirement of general examinations, because these were said by student activists to induce anxiety. It has also ceased to require any history of philosophy for the Ph.D. in philosophy.[3]

But the prosperity of the 1960s had passed by the 1970s, and Quine thought that the passing of the affluence and the curtailing of funds for higher education might remove some of the ill effects that abundance brought.

With the publication of RR, Quine had brought his attempt at a genetic account of language learning to at least a tentative conclusion. As he told Bryan MaGee in 1978.

> In the few years since my book *The Roots of Reference* came out, all I have done is to produce numerous short pieces intended to clarify or defend or improve my philosophy at a variety of points.[4]

But this is not an entirely accurate statement. There were still problems that he needed to solve, lines of investigation that he had begun earlier that needed to be pursued, and there would be new problems that would arise. Among the holdover problems was Schoenfinkel's logic and its implications for the variable and Popper's claims for negative evidence.

M.G. Murphey, *The Development of Quine's Philosophy*, Boston Studies in the Philosophy of Science 291, DOI 10.1007/978-94-007-2424-2_4,

Quine had long been intrigued by Schoenfinkel's system of logic. In 1971 and 1972, he gave two papers exploring various aspects of this system. The first was presented in 1971, with the title "Truth and Disquotation,"[5] at a symposium in honor of Tarski at Berkeley. The paper was published in *The Proceedings of the Symposium on Pure Mathematics* in 1974, and reprinted in Quine's *The Ways of Paradox*. In this paper he explores what happens when a Tarski type truth predicate is defined in Schonfinkel's system and in his own predicate functor logic, which was adapted from Schoenfinkel's. As Quine puts it

> For Schonfinkel's language we get something that may be called *disquotation* – and in a stronger sense of that word than what Tarski's schema
>
> (1) ". . ..." is true ≡ ,,,,,,
>
> requires. We shall observe further why a general inductive definition of disquotation for arbitrary notations does not go through similarly. Finally, reverting to quantification of a sort, we shall consider what happens to the truth definition when quantification is reconstrued in terms of substitution rather than of objective reference.[6]

Quine shows that in Schonfinkel's system the result is precisely (1) but with the expression to the right of the biconditional being identical to the quoted expression on the left bereft of its quotation marks. In Schonfinkel's system this procedure cannot be generalized into an inductive definition of "disqx" (disquotation of x) for variable "x." The disquotation operator attaches to singular terms, but "the trouble is that the grammatical category of the resulting compound may be any or none, depending as it does on the reference of that singular term."[7]

In Schonfinkel's system, the well formed formulae are names. In Quine's predicate functor system, the well formed formulae are n-placed predicates, or, if $n = 0$, sentences. The problem here is that, if "satisfaction" becomes a predicate functor "satF," the degree of "satF" cannot be determined without knowing what "F" is true of.[8]

Finally, Quine considers the definition of truth in a system of substitutional quantification. He takes for his example protosyntax for the metalanguage with substitutional quantification. To avoid having the quantifications turn into finite conjunctions and alternations, he assumes an infinite number of constant expressions, built from a finite number of combinations and identity. This could be done taking "0" as a single term and then applying the successor function. Also as primitive predicates, he takes sum and product, which together with truth functions and substitutional quantification yields elementary number theory. For this language, truth is inductively definable. But whereas Tarski's definition can be fitted to "any specific theory, in the classical quantificational pattern, by just filling in the predicates and names and functors specific to that theory," in the case of protosyntax with substitutional quantification, much depends on the particular subject matter of the object language. If the theory "has atomic sentences whose truth is stubborn to hyperarithmetical degree, or is to be left open indefinitely for empirical determination, then there is in protosyntax no hope of an inductive truth definition for the theory."[9] Tarski's theory remains still the optimal one. But Quine adopts the term "disquotation" as the name for his truth theory.

The second paper, given the next year at the Boston Logic Colloquium, he titled "The Variable."[10] This was the aspect of Schoenfinkel's system that had interested Quine the most. He first dwells on the importance of distinguishing the variable from the schematic letter, which does not stand for an object but stands in for a predicate and is therefore not bindable. He also distinguishes the objectual variable from the bindable substitutional variable. In the case of the latter, the law of unit subclasses – viz.

(W)(W has members .⊃ (∃Z)(Z has a member of W as sole member))

would require that for any class "W" and any member "x," there be a membership condition that will pick out just "x." That this cannot be done where "W" is the class of real numbers is obvious. It amounts to requiring a unique membership criterion for every individual in the universe.[11] But Quine also wants to dissociate the variable from quantification. As he points out, quantification is present without the variable in the universal categorical, whereas the variable does its work in description, class abstraction, functional abstraction, and integration without quantification. At its most basic, the variable functions in the relative clause, and particularly in the "such that" idiom, "x such that Fx."

> Where the original sentence is thought of schematically as "Fa," the relative clause is the explicit segregation of the "F." The "such that" construction is the relative clause simplified in respect of word order, and fitted with a bound variable to avert ambiguities of cross-reference.[12]

Other uses, Quine says, are parasitic on this one. He then traces the history of the notion. Peano, he says, used "such that" expressions to designate classes, and quantified over them, whereas the relative clause is a predicate allowing at most substitution for a schematic letter. Russell took over Peano's use in PM and further confused "property, open sentence, and predicate." By 1930, when logicians were distinguishing first order theories from higher order theories, the use of schematic letters became common among continental logicians, though often taken as allowing quantification. Quine says that he himself was not fully clear on the matter until 1945 when "it dawned on me to call them predicates, thus recognizing at last that they were playing the role of the relative clause, the 'such that' clause." The notation of predicates and predicate functors can play some of the parts of classes without requiring classes to exist. Quine developed this in his theory of "virtual classes" in *Set Theory and Its Logic*. "We are noting how unready logicians have been to think directly in terms of a calculus of complex predicates."[13] This has been evident in the treatment of the Boolean calculus of classes, where there is no reason to introduce classes at all. Quine says it was only in the third edition of *Methods of Logic* that he used Boole's algebra in his presentation of monadic logic. The variable, Quine says, originated as the relative pronoun. It is clearly bindable in the relative clause using "such that." Bound variables disappear in the Boolean calculus of predicates, but they reappear when polyadic predicates are introduced.

> The basic job of the bound variable is cross-reference to various places in a sentence where objective reference occurs; and whereas monadic logic calls for this service only in the preparations, polyadic logic calls for it also within the ongoing algorithm, in order to keep track of permutations and identifications of arguments of polyadic predicates.[14]

Polyadic logic is decidable only so long "as there is no crossing up of argument places." So long as this does not occur, Quine notes that he has given a decision procedure that works not only for monadic predicates, but also for polyadic ones under certain conditions.[15]

The analysis of the variable was what had attracted Quine to Schoenfinkel's system. Ordinary quantification involves a combination of the separate ideas of quantity (all, some), bondage, cross-reference, and reference. These need not occur together – categoricals involve quantity without variables, class abstraction involves variables without quantity. Cross-reference involves an implicit use of identity signifying that different occurrences of the variable have the same reference. Quine was intrigued by the combination of these factors in quantification theory, which he credited to Frege.[16]

One of the implications of Quine's holism is that in an experiment to test a hypothesis, the finding of a negative result may not lead to the rejection of the test hypothesis. No one has emphasized the importance of negative results in hypothesis testing more than Karl Popper, and in 1974, Quine did a short piece entitled "On Popper's Negative Methodology"[17] that was published in the Library of Living Philosopher's volume on Popper. Popper is known for his view that evidence cannot confirm a theory but only refute it. Quine uses as his example the sentence "All ravens are black." For Popper, what others might cite as confirming evidence would mean only lack of refuting evidence, while one non-black raven would decisively refute the statement. Hence on grounds of conclusiveness, negative evidence is primary. "All ravens are black," Hempel had pointed out, is logically equivalent to "All non-black things are non-ravens," but this too is rejectable by one non-black raven. Quine holds that "raven" and "black" are projectable whereas "non-black" and "non-raven" are not. Hence he accepts "all ravens are black" as lawful, but not "all non-black things are non-ravens." This is an improvement on his earlier view in *The Web of Belief* that the logical equivalents of "All ravens are black" are also law-like. Existential statements however are difficult to refute even by many negative cases, whereas one positive case can confirm them. But since scientific laws are universal quantifications, Popper's doctrine holds for them.[18]

But, Quine says, consider "All men are mortal." To refute it, we require one immortal man. But to say "Jones is immortal" is the general statement "All future times are times in the life of Jones." Such a sentence cannot be confirmed or refuted alone; it is only as one of a group of sentences that the conjunction of the group's sentences may imply a testable consequence. Quine holds that the sentences of science often involve multiple quantifiers, some universal and some existential, and Popper's doctrine is difficult to apply in this case. But if the sentences of a theory are taken as a conjunction, then Popper's doctrine still makers sense; the refutation of any conjunct of a conjunction falsifies the conjunction, "while verification of such a consequence is as may be."[19]

In 1974, Quine published an article entitled "On Logically Equivalent Systems of the World".[20] This is major paper. He had already suggested in earlier papers that scientific investigation might not lead to a single final theory but to a number of equivalent ones owing to the underdetermined nature of science. The purpose of this article, Quine says, is to clarify the doctrine that science is underdetermined,

> under-determined not just by past observations but by all observable events. The doctrine is plausible insofar as it is intelligible, but it is less readily intelligible than it may seem. My main purpose in this paper is to explore its meaning and its limits.[21]

Quine emphasizes that the underdetermination thesis is not the same as holism. This gives him a chance to qualify his former claim that all science is at stake in every experiment. He says that there is considerable unity among the sciences due to the fact that all sciences share logic and mathematics, but science "is variously jointed, and loose in the joints in varying degrees." When a deviant observation shows up, we can choose what sentences are to be revised and what ones to let stand. How much of science is affected is our call, but not all of it at once.[22] But to return to the main theme of underdetermination, Quine modifies his definition of observation sentence. Previously he had defined an observation sentence as one that all "witnesses will agree to on the spot." But this he now says is only a "rough practical criterion." The definition he now gives is that those sentences are such that they could be taught by ostension, whether in fact they were or were not.[23] Further, "an observation sentence is an occasion sentence: it commands assent on some occasions and not others, depending on what is happening when and where the sentence is queried."[24] Sometimes Quine distinguishes observation sentences from occasion sentences, sometimes, as here, he identifies them. Observation sentences are not incorrigible, Quine says. Nevertheless, they supply the empirical content of science. To do so, they must be turned into standing sentences.

> Let us adopt, then, an arbitrary numerical system of spatio-temporal coordinates, and let us contemplate the infinite totality of what I shall call *pegged observation sentences*. Each observation sentence expressible in our language gets joined to each combination of spatio-temporal coordinates. The resulting sentences are standing sentences, some true and some false. The true ones do not depend for their truth on anyone's having made the observations; it matters only that the observable state or event in question occur, in fact, at the specified place-time. The time and place may be beyond the reach of all sentient life.[25]

But to bring the observation sentences into a form such that a scientific theory could imply them, we require conditional sentences whose antecedents specify boundary conditions and whose consequents are pegged observation sentences. These Quine calls "observation conditionals."

So then what is the theory that implies them? It is, Quine says, a "theory formulation" which is a conjunction of the axioms of the theory. Such theories can be axiomatized in multiple ways; there will be multiple formulations of a given theory, all of them logically equivalent.[26] But now Quine raises the question of what would be the case if in some theory formulation we were to interchange the terms "molecule" and "electron." The theory formulation so altered would not be logically equivalent to the one from which it is derived, but it would still imply the same observation conditionals. Hence the theories would be empirically equivalent. The molecule-electron case is a notational variant of the original theory, and Quine points out that there could be wholesale permutations of predicates that would still leave the theory formulations empirically equivalent though not logically equivalent. Hence he now proposes

> to individuate theories thus: two formulations express the same theory if they are empirically equivalent and there is a reconstral of predicates that transforms the one theory into a logical equivalent of the other.[27]

Quine assumes here that the language in use is English in regimented form – that is, truth functions, quantification and a finite lexicon of predicates. In these terms, "the theory expressed by a given formulation is the class of all the formulations that are empirically equivalent to that formulation and can be transformed into logical equivalents of it or vice versa by reconstrual of predicates." But even this definition of theory is not enough; Quine now takes theories to be mathematical sequences.

> Sentences ... and longer expressions, are to be taken rather as mathematical sequences of their component words or letters. An expression in this sense is a function, or class of ordered pairs; the first word or letter of the expression is paired with the number 1, the second with 2, and so on. In this way we can assure the existence of all expressions however long, all theory formulations as yet unconceived, all texts as yet unwritten; all "possible" expressions as one might say. Theories, finally, are classes of formulations, hence classes of expressions in this abstract sense; classes of functions.[28]

Having clarified his terms, Quine is now ready to state his thesis: "for any one theory formulation there is another that is empirically equivalent to it but logically incompatible with it, and cannot be rendered logically equivalent to it by any reconstrual of predicates." To prove this thesis Quine has to show that such alternatives are inevitable.

But stated in full generality, the thesis is untenable. The thesis fails where only a finite number of conditionals are implied. It also fails even with an infinite number of conditionals if they can be "encompassed by a single universally quantified conditional, or by finitely many." Accordingly, it requires that the observation conditionals true of this world be too ill assorted to be captured by any tight finite formulation. However, William Craig has shown that, given any formulation and any desired set of consequents, there is another class of sentences whose members are equivalent, one by one, to the sentences of the desired class and which admits of a mechanical decision procedure. Since a Craig's class is not a finite formulation, it does not directly contradict the underdetermination thesis, but since a theory formulation could apply to a recursive class of expressions, that would create trouble.[29]

But Quine claims that the underdetermination thesis remains important "in terms of what is practically feasible." "We, humanly, are capable of encompassing more observation conditionals in a loose theory formulation than in any tight system we might discover and formulate independently of any such loose formulation."[30] Then the underdetermination thesis applies to any such formulation. But this does not prove that logical equivalence might not be achieved by reconstrual of predicates.

> We might study two incompatible theory formulations, trying in vain to imagine an observation that could decide between them, and we might conclude that they are empirically equivalent; we might conclude this without seeing a reconciling reconstrual of predicates. This we might, but there still could be a reconciling reconstrual of predicates, subtle and complex and forever undiscovered. The thesis of underdetermination, even in my latest tempered version, asserts that our system of the world is bound to have empirically equivalent alternatives that are not reconcilable by reconstrual of predicates, however devious. This, for me, is an open question.[31]

So Quine proposes what he calls his "last ditch" version of the underdetermination thesis.

> Our system of the world is bound to have empirically equivalent alternatives which, if we were to discover them, we would see no way of reconciling by reconstrual of predicates. This vague and modest thesis I do believe. For all its modesty and vagueness, moreover, I think it vitally important to one's attitude toward science.[32]

Suppose, Quine says, that we had two such theories, logically incompatible, equally simple, each implying all true observation conditionals and no false ones, and irreconcilable by reconstrual of predicates. Which one would we say is true? Quine's answer is that we should take as true that theory that springs from our own theory of the world, for there is no higher authority to which one could appeal. But if we recognize the equivalence of the two theories, then Quine says we could adopt both as true and work within each, simply making it clear in each case which theory is in use. But Quine vacillated on this point. Gibson pointed out the vacillation in his paper in the Schilpp volume on Quine[33] and Quine acknowledged it. Replying to "Things and Their Place in Theories" and "Empirical Content" in *Theories and Things*, Gibson noted that in the former, Quine had held that only one of the theories should be taken as true while in the latter he took both as true. In his reply, Quine labeled the former his *sectarian* view and the latter his *ecumenical* view. Where the two theories conflict, Davidson had proposed to reconcile the conflicting sentences by changing the spelling of one of the conflicting terms and treating them as homonyms. But what if the theories cannot be reconciled?

> The remaining case, and the sticky one, is where the alien terms of the other [theory] are irreducible. The sentences containing them constitute a gratuitous annex to the original theory, since the whole combination is still empirically equivalent to the original. It is as if some scientifically undigested terms of metaphysics or religion, say "essence" or "grace" or "Nirvana," were admitted into the science along with their pertinent doctrine, and tolerated on the ground merely that they contravene no observations. It would be an abandonment of the scientist's quest for economy and of the empiricist's standard of meaningfulness. The sectarian position, then, is my newly recovered stance on these precarious slopes.[34]

There, for the moment, matters stood.

Quine followed up "Empirically Equivalent Systems of the World" with a paper that pushed his skepticism even further. In 1974, he gave "The Nature of Natural Knowledge"[35] as the Wolfson Lecture at Oxford. It was published in 1975 in an expanded form. Quine says the paper "is meant as a summary statement of my attitude toward our knowledge of nature." Doubt, as philosophers from Descartes to Peirce have claimed, is the mother of science, but Quine adds that science is the mother of doubt. In fact, the illusions of the skeptic are possible only against the background of knowledge of fact with which to contrast the illusion. Common sense knowledge is knowledge about bodies, which is already a step toward science. Epistemology, Quine says, is best looked upon "as an enterprise within natural science."[36] Quine then tells again his tale of how science has been acquired from observation sentences to general theories. But then Quine brings in a new consideration. The science we have acquired has "the arbitrary character of historical accident and cultural heritage; there was no hint of inevitability."[37] This is a consequence of Quine's claim of the underdetermination of science by all possible observations.

> What wants recognizing is that a physical theory of radically different form from ours, with nothing even recognizably similar to our quantification or objective reference, might still be empirically equivalent to ours, in the sense of predicting the same episodes of sensory bombardment on the strength of the same past episodes. Once this is recognized, the scientific achievement of our culture becomes in a way more impressive than ever. For, in the midst of all this formless freedom for variation, our science has developed in such a way as to maintain always a manageably narrow spectrum of visible alternatives among which to choose when need arises to revise a theory. It is this narrowing of sights, or tunnel vision, that has made for the continuity of science, through the vicissitudes of refutation and correction. And it is this also that has fostered the illusion of there being only one solution to the riddle of the universe.[38]

But how is this "tunnel vision" to be accounted for, if in fact science is as underdetermined by all possible observations as Quine claims? In the face of this "formless freedom for variation," why has our science not led us into a world of make-believe? Tunnel vision seems to imply a tunnel. What is it? Quine leaves us, as Emerson put it, "in the splendid labyrinth of [our] perceptions, to wander without end."[39]

There were at the time this was written multiple theories about the evolution of science, Thomas Kuhn's[40] being the most popular. Quine was apparently not a Kuhnian, he claimed that our science has developed in a particular way, and one would expect some explanation of why the development took the form it did. "Tunnel vision" is a nice metaphor but it explains nothing, and given this "formless freedom for variation," tunnel vision seems a very unlikely outcome. We are left in the position of a Darwinian equipped with fortuitous variation but without natural selection.

Quine followed up this article with a second in the same book; this one is entitled "Mind and Verbal Dispositions."[41] This is Quine's behaviorism in full bloom. On the first page he writes

> Most thought simply *is* speech, according to the pioneer behaviorist John B. Watson: silent, repressed, incipient speech. Not all thought is that. A geometer or engineer may think by means also of little incipient tugs of the muscles that are used in drawing curves or twirling cogwheels. Still, the muscles that play by far the major role, according to Watson's muscular theory of mediation, are the muscles used in making speeches.[42]

Quine does believe Watsonian behaviorism, though with refinements due to Skinner. He rejects what he calls "mentalism" – ideas, meanings, and the lot; for him, not only are there no such things but the belief in them is dangerous because it leads people to believe they have explained something by their use when they have not. To support his view, Quine turns as usual to the learning of language, which in childhood he takes to be a matter of conditioning and imitation. Quine is aware of Premack's work with chimpanzees and his success in teaching them "language," but he holds that what sets humans apart from other animals is their "unpredictable spontaneity of speech."[43] He rehearses his theory of how occasion and observation sentences are acquired, but he hazards a definition of "understanding a sentence: a man understands a sentence in so far as he knows its truth conditions." Just how a man could know what the truth conditions of a sentence are if he did not first understand the sentence is not revealed. Proof of a child's mastery of a sentence is based on query and assent. But when it comes to standing sentences, Quine says "I do not know how, in general, in terms of behavioral dispositions, to approximate to the notion of understanding at all, when the sentences understood are standing sentences." But this does not discourage Quine; rather, he takes it as meaning that standing sentences can only be understood in groups – that is, as theories that are tested by observation.[44] It is not obvious how taking standing sentences in groups enables us to understand them when we cannot understand them singly. From there he moves to the notion of equivalence: he writes "I am persuaded, indeed, that a satisfactory equivalence concept is impossible for standing sentences."[45] To support this, he cites the claimed indeterminacy of translation. Quine then says that language must be studied as a system of dispositions to verbal behavior. Objections that dispositions to assent or dissent involve a mental element leave him unmoved. "One partial criterion of what to count as a sign of assent is that the speaker would be disposed to produce that sign whenever a sentence is queried in circumstances in which he would be disposed to volunteer the sentence himself." He further says that "mind" ought to be construed "as a system of disposition to behavior,"[46] citing Ryle and Sellars as having held this or a similar theory.

There are, Quine holds, three levels of explanation: the mental, the behavioral, and the physiological.[47] The mental he dismisses. The only true explanations of behavior, he holds, are physiological. In between there is the dispositional. As terms like "soluble" are explicable by their microstructure, so he claims that dispositions to verbal behavior are explicable only by physiology. "The term 'disposition' has its significant application rather as a preface, each time, to an actual singling out of some physical trait." Talk of dispositions, therefore, is a way of referring to an as yet unknown physiological mechanism. So the statement, "Subject S is disposed to say 'red' whenever queried in the presence of something red," really means that he has some physiological mechanism as yet unknown that leads him to say "red" when queried in the presence of something red. If true, this reduces psychology to the status of natural history.

On November 12, 1975, Quine resigned from the Association for Symbolic Logic. "For years now the trend of my concerns has been such that I have utterly neglected the Journal."[48] Quine's shift from logic to semantics and ontology began in World War II. With the completion of his book on set theory in 1963, he largely abandoned the field of logic. He did continue to write on logic, but his interests were elsewhere.

In his contribution to the Schilpp volume on Quine in 1986, Hao Wang wrote a perceptive essay about Quine's work in logic.[49] Wang notes "It is Quine's habit to combine his book-writing with his course-teaching." In his reply to Wang, Quine says "the pedagogical motive has dominated my work in logic."[50] "New Foundations," Quine says, "...was my attempt at an efficient foundation of the subject for my course Mathematics 19; and it is characteristic that I have not felt impelled to try my mettle in those subsequent researches on NF, impressed though I am with what others have done."[51] It is not unusual for teachers to combine their teaching and their research, if they can, but as Wang shows, Quine's work in logic has been largely limited to exposition. Nor did Quine pursue the problems Rosser raised concerning *Mathematical Logic* beyond his article "Element and Number." It was Wang who correctly diagnosed where *Mathematical Logic* went wrong and proposed the revision Quine adopted, as Quine himself made clear. Wang remarks that Quine had developed few new ideas in logic since 1940. "Since 1940 Quine has tended to limit his attention to refining and deepening his understanding of those parts of logic which had interested him before."[52] Quine's contributions to logic have been largely restricted to simplification and clarification aimed at exposition, at which he has excelled. It is not at all clear that these retrospective descriptions correctly describe Quine's motives in the 1930s, when he may well have aspired to become more than a gifted expositor, but they do seem to be an accurate description of what he actually achieved in logic.

One of the problems facing philosophers of science is that science keeps changing, so they are always shooting at a moving target. Quine found himself in this position in 1976. The article in which he sought to deal with this change was entitled "Whither Physical Objects?" and was dedicated to the memory of Imre Lakatos.[53] Physicalism had been gospel for Quine ever since his days with Carnap, if not before; now he found the notion of physical object dissolving in his hands. Quine had previously defined physical objects "as the aggregate material content of any portion of space-time." But he had become increasingly nervous about the status of material objects. Einstein's "$E = mc^2$" showed that matter and energy are interconvertible, and on a number of occasions he had voiced the suspicion that quantum mechanics might undercut the notion of physical object. Here he describes a thought experiment involving electrons in which the ordinary criteria of object identity appear not to be applicable.

> All this is said to apply to other elementary particles as well as electrons. Matter evidently goes by the board. We are left rather with a field theory, a theory of the distribution of states over space-time.[54]

But this will require mathematics, since only so can the variation of such states as temperature be determined. Suppose we have a coordinate system for space-time, so that every point is specified by a quadruple of numbers. This is an idea he had gotten from Carnap in the 1930s, and had used himself in "Empirically Equivalent Systems of the World." Then let "Fα" mean "the mean temperature in degrees Kelvin of the region whose coordinates comprise the set α is x." The arbitrary character of the units of measurement can be met by choosing those that make for the simplest theory, and the arbitrariness of the coordinates is irrelevant since physical laws will quantify over all quadruples of numbers. It is clear that this involves a reduction to an ontology of sets, since all mathematics is reducible to set theory, but the sets will be ungrounded, there being no individuals to serve as ground elements. But Quine says that Fraenkel and von Neumann have shown that set theories need not be grounded. One can start with the empty set, then its unit set, and so on. Thus physics itself has led to the demise of physical objects and in its place a "hyper-Pythagoreanism."[55] So our ontology turns out to be that of set theory. But our lexicon of predicates, such as temperature, is not so reducible. "Our physical ideology remains unreduced to pure mathematics, for all the ontological debacle."

> We might most naturally react to this state of affairs by attaching less importance to mere ontological considerations than we used to do. We might come to look to pure mathematics as the locus of ontology as a matter of course, and consider rather that the lexicon of natural science, not the ontology, is where the metaphysical action is.[56]

Throughout his earlier writings, Quine had been insistent on the importance of ontology. His famous dictum – "to be is to be the value of a variable" – was but one expression of this; he deplored philosophers who were careless about their ontological commitments. But this article marks a change. Physics itself has undermined the ontology of physical objects. From here on, Quine would be less concerned about ontology.

In 1977, Quine published an article entitled "Facts of the Matter."[57] The early parts of the article contain little that is new. But Quine then moves onto new ground by invoking the arguments of "Whither Physical Objects?" Quantum mechanics dissolved the notion of bodies. "Thus at last bodies themselves go by the board – bodies that were the primordial posits, the paradigmatic objects most clearly and perspicuously beheld." Repeating his arguments from the "Whither Physical Objects?" Quine says "the brave new ontology is, in short, the purely abstract ontology of pure set theory, pure mathematics." "The lesson to be drawn from this debacle," Quine says, "is that ontology is not what mainly matters." And here ontological relativity is invoked to show that any ontology can be reinterpreted as another. What really matters, then, Quine holds to be sentences and their truth conditions. So what becomes of physicalism? It becomes thus clear that "there is no difference in matters of fact without a difference in the fulfillment of the physical state predicates by space-time regions,"[58] where space-time regions are defined by sets of coordinates. And how does this apply to the indeterminacy of translation? That now becomes "what excuse could there be for supposing that the one manual [of translation] conformed to any distribution of elementary physical states better than the other manual ... for supposing there to be a fact of the matter?" So if this argument is granted, ontology deals with pure sets only.

> The terms that play a leading role in a good conceptual apparatus are terms that promise to play a leading role in causal explanation; and causal explanation is polarized. Causal

> explanations of psychology are to be sought in physiology, of physiology in biology, of biology in chemistry, and of chemistry in physics – in the elementary physical states.[59]

Thus Quine remains what he calls a physicalist despite the evaporation of physical objects; his prior doctrines carry over under the new definition, but with the significant change that ontology now ceases to have the importance he had previously given it.

This ontological debacle dealt Quine's prior theory a severe blow. In WO and RR, he had devoted considerable attention to the process by which he believed children came to refer to physical objects. If now the notion of physical object was no longer tenable, what becomes of reference? What it is that the child learns to refer to? The only objects left standing are the abstract ones, which is the proper outcome if our ontology consists of sets. But was such Platonism something Quiner could live with? Is it really the objective of epistemology to establish Platonism? These were not conclusions that Quine could be happy with, but it took time for him to work out his answers. While he pondered his future course, he turned to other topics, but the problems raised by the death of physical objects were never far from his mind.

The first topic to which he turned was meaning. Despite his attacks on the notion of meaning, Quine knew he could not banish it by incantation. In "Use and Its Place in Meaning,"[60] he tried to find some way of using it.

> Words and phrases refer to things in either of two ways. A name or singular description *designates* its object, if any. A predicate *denotes* each of the objects of which it is true.[61]

He then turns to meanings and asks just what sort of thing a meaning is. As he remarks, "You would think we would know." But he thinks we do not, and sets out to tell us. First, however, he defines the word "expression" as "a string of phonemes – or, if we prefer to think in terms of writing, a string of letters and spaces."[62] There is a problem here because in standard linguistic usage a phoneme is taken to be a meaningful unit; Quine makes it clear that he is using the word without any such assumption of meaning. How then is "phoneme" to be understood here? "We can simply say that two sounds count as occurrences of the same phoneme if [the substitution of one for the other] has no effect on the speaker's readiness to assent to any string of sounds." So Quine decides to start with the notion of "means alike"; then the meaning is just "the set of the expressions that mean like it." This is the same strategy he had used in the 1940s in an attempt to define synonymy. Wittgenstein and Dewey, Quine says, thought that the meaning of a word was to be found in its use. This points toward a behavioral criterion but it does not provide it. We need, then, to consider not words alone but sentences. This leads him to the definition: "a word is synonymous to a word or phrase if the substitution of the one for the other in a sentence always yields an equivalent sentence." But this is still not satisfactory; Quine limits the equivalence to *cognitive* significance, meaning sameness of truth conditions. But this too is inadequate. The solution, Quine says, is Mill's notion of *concomitant variation* (correlation). And for this we must limit ourselves to occasion sentences – i.e., sentences the truth of which varies from occasion to occasion.

> If our interrogated informant is disposed to give matching verdicts on two such occasion sentences on every occasion on which we query the two sentences, no matter what the attendant circumstances, then certainly the two sentences must be said to be cognitively equivalent for him.[63]

Then the two sentences are said to be cognitively equivalent if cognitively equivalent for each speaker of the language, or, in certain contexts, for each speaker of a particular group. The "attendant circumstances" can be defined as sameness of nerve stimulations for the informant. Given this start, Quine defines "synonymy" for words and phrases as those which when substituted for one another always yield equivalent sentences for all members of the community. And given this much, the interchangeability criterion can be extended to standing sentences as well. There are of course restrictions: this notion of synonymy cannot be applied in quotations or to propositional attitudes.

Lexicography requires something more; it requires paraphrase into an equivalent sentence.

> Let us define a gloss of a sentence *s*, with respect to one of its words, *w*, as any cognitively equivalent sentence lacking *w* and containing only other words of *s* and words of higher frequency than *w*. A word may be called *reducible* if all occasion sentences that contain it admit of such glosses with respect to it.[64]

It is the lexicographer's task to find "glosses of occasion sentences with respect to all reducible words." This reduces less frequently used words to the more frequently used until one reaches a set of words that can be taken as common knowledge. Quine then proceeds to extend his criterion to standing sentences. These are sentences true for some substantial period; the shorter the period the closer they come to occasion sentences. "We might even extend the criterion to all standing sentences, provided that we take it only as a necessary condition of cognitive equivalence and not a sufficient one," whereas for occasion sentences it is both necessary and sufficient. For a sufficient condition "one standing sentence is cognitively equivalent to another if it can be transformed into the other by a sequence of replacements of words or phrases by cognitive synonyms." This is not quite a definition of cognitive synonymy for standing sentences, since a pair of sentences could meet the necessary condition but not the sufficient one, leaving its status indeterminate.[65] Moreover, Quine holds that his argument here has no bearing on the indeterminacy of translation.

This is a remarkable paper. In "Two Dogmas," Quine had rejected substitution in an extensional language as a criterion of synonymy because it was too easy to find countercases – "creature with a heart" and "creature with a kidney." Here he uses substitution in an extensional language but in a much better form. By limiting himself to occasion sentences, he can employ Mill's "concomitant variation" to determine which sentences are equivalent for an individual under the same stimulations, and then extend the test to the group. Further, he restricts this test to cognitive synonymy, employing the Positivist distinction between cognitive and emotive meaning, and taking cognitive synonymy as sameness of truth conditions – as verification. This gives him his definition for synonymy for words and phrases as those which, when substituted for each other, always yield equivalent sentences. Then the meaning is the class of all synonyms.

For lexicography we need also the concept of a "gloss" for a cognitively significant sentence *s* with respect to one of its words *w* as any cognitively equivalent sentence containing only other words of "higher frequency." This is Carnap's concept of reducibility applied in a different fashion. By "higher frequency," Quine means frequency of use by speakers of the language. His concept of a dictionary is one in which words are defined in terms of other words of higher frequency until we reach a set of words so frequently employed that their meanings may be said to be common knowledge.

Quine then extends his criterion of cognitive synonymy to standing sentences, but only as a necessary condition. The sufficient condition is that one sentence be transformable into the other by a sequence of transformations in which only cognitive synonyms are replaced by cognitive synonyms. This idea is one that he had explored in the unpublished paper that he had written while in the Navy.

This paper comes closer to providing a behavioral criterion of meaning than any other paper he wrote. It does not quite succeed since a sentence regarding "equilateral triangles" would be equivalent to one employing "equiangular triangle," and it furthermore assumes prior knowledge of "assent" and "dissent," as well as "means alike." If, as appears to be the case, Quine is taking the cognitive meaning of a sentence to be its truth conditions, then he is adopting a form of the old Positivist verification theory of meaning. If this is correct, he has then to explain how one can determine the truth conditions of a sentence without first knowing what the sentence means. Simply asking community agreement on assenting to a sentence, or to a pair of sentences, does not solve the problem, since such agreement could be reached concerning a sentence without the linguist having any idea what the sentence means. "Tresto arno vowe" and "Silba galon foss" might be judged synonymous by the community without our knowing what they mean.

In 1978, Quine published a paper entitled "On the Nature of Moral Values."[66] This is an unusual topic for Quine, who rarely wrote on anything but what he considered science, but his take on the subject is not surprising. He begins with the distinction between cognition and emotion. "It is the deep old duality of thought and feeling, of the head and the heart, the cortex and the thalamus, the words and the music." This division he sees reflected even in the origins of learning – namely, a similarity space and "an ordering of episodes along the valuation axis."[67] This equipment he takes to be innate. Similarity gives us a start on induction, e.g., the expectation that the future will be like the past, and our "innate likes and dislikes" give us a start on valuation, then further developed by induction. Just as similarity standards are developed into more advanced forms, reaching in due course the hypothetico-deductive method, so are our likes and dislikes. "We learn by induction that one sort of event tends to lead to another that we prize," and then we may come to prize the means themselves.[68]

"The transmutation of means into ends ... is what underlies moral training." Good behavior, at first, is technology. We then come to value the behavior as an end in itself, and it acquires a moral value. The transformation of means into ends Quine sees as basic to the acquisition of morality; this of course was a frequent refrain of Dewey's. Quine divides values into *altruistic* values – "values that one attaches to satisfactions of other persons," and *ceremonial* values – values attaching to practices of one's society or group. Note that Quine omits any mention of the prudential virtues. What is taken to be the moral code of a society is those moral values held by most members of the society, and is often implemented by rewards and punishments. Quine says he follows Schlick[69] in placing moral, aesthetic, and sensual values on a par. He also follows Schlick in relying on sympathy to account for one's pleasure or sorrow at seeing the pleasures or sorrows of others. Whether this sympathy is innate, as Schlick held, Quine does not say, but if so that would help to explain the infant's responsiveness to parental approval or disapproval.

Morality, while variable from one society to another, should exhibit common features in all since all societies have to cope with similar problems. In some societies, the members may believe in so comprehensive a code of divinely decreed values that their conformity is really prudential. But variation is intrasocietal too. Here one lacks the ability to test moral values empirically that one has in science; for morality a coherence standard is the best we can do. "It is a bitter irony that so vital a matter as the difference between good and evil should have no comparable claim to objectivity."[70] Hence the frequent recourse to divine decree to provide a justification for moral beliefs. Hence too the attempts at the reduction of moral doctrines to systems resting on premises that are expected to be universally agreed upon, as for example with Utilitarianism.[71] And in the last extremity, where disagreement is basic and irresolvable, we can fight. Since our moral values are generally vague, some disagreement is inevitable.

No doubt in earlier times, altruism was extended only to the members of one's own group. But today the borders have expanded and can embrace the whole species. One even thinks of moral obligations to the as yet unborn, as for example in the problem of the degradation of the environment.

> Thus we do what we can with our ultimate values, but we have to deplore the irreparable lack of empirical checkpoints that are the solace of the scientist. Loose ends are untidy at best, and disturbingly so when the ultimate good is at stake.[72]

By the late 1970s, Quine's celebrity was such that he had more requests for lectures and interviews than he could grant, or wished to. But in 1978, Quine agreed to an interview by Bryan Mcgee – the author of books on Wagner, Karl Popper, Schopenhour, and modern British philosophy – and the interview was published in Magee's *Men of Ideas*.[73] To the question as to what he regarded as philosophy's central task, Quine answered that he saw philosophy as continuous with and rounding out the scientific system of the world. Philosophy, he says, looks at "the abstract and theoretical end" of science. As to the perennial questions such as why does the universe exist, Quine says that there is no conceivable answer to them and so they are meaningless. Quine distinguishes two sorts of questions that are important to philosophy: ontological questions of what there is, and questions about what can meaningfully be said about what there is. Quine takes the materialist view: he believes that physical objects exist, but also that abstract objects exist. He holds that it is the position of physics that the world is a closed system; that all change is due to changes in the microphysical elements of space. He rejects dualisms, including the mind-body dualism: both are physical, and mental states are physical states of the body. And this leads of course to his behaviorism, since he sees behaviorism as the way to make "objective sense of mentalistic concepts." Given his physicalist view of the world, Magee wants to know how Quine can justify abstract objects. "The justification lies in the indirect contribution that they make to natural science," Quine says. It is mathematics to which he is referring here. Mathematics is essential to science. Quine is also asked what some of the established but mistaken ideas of philosophy are. His prime candidate is meaning. No behavioral sense can be made of the notion, Quine says. He also rejects "necessity," and the "fashionable philosophy of possible worlds." Quine denies any distinction between the laws of nature and "other true statements about the world." Both are known through the hypothetico-deductive method of science. Asked to specify his ontology, Quine asserts the existence of physical objects – "the content of any portion of space-time, however scattered is for me a physical object"; he rejects properties, propositions, and meanings. As for what can be said about objects, he rejects predicates that "have too little in the way of intersubjectively observable criteria, unless they compensate for that defect by contributing substantially to a well-knit system of the world that expedites prediction." Asked about his emphasis on language, Quine stresses the importance of semantic assent as a way of resolving disagreements; he also stresses the importance of discovering how language is learned along with the conceptual system learned with it. Asked about the major problems facing philosophy, Quine specifies three topics in philosophy where he would like to see breakthroughs. One is semantics, which he says is the theory of meaning. The second is the theory of the propositional attitudes. And the third is the minimum of mathematics necessary for natural science. "One effect to be hoped for, in such a minimization, is a more natural and conclusive solution than we now have for the antinomies of set theory."[74]

One should note that for the popular audience Quine stays with the ontology of physical objects instead of developing the ontology of pure sets, and that he does not discuss the status of logic at all. Magee did not ask about the latter and Quine did not volunteer. It is interesting that, having denied the existence of meanings, Quine thinks the semantic question one of the three most pressing in philosophy. But as we have lately seen, he is still seeking definitions of "meaning" and "synonymy."

In 1979, Quine published what amounts to an obituary for Kurt Godel,[75] who had died in January of 1978. Born in 1906, Godel had entered the University of Vienna in 1924. He received his doctorate in 1930; his doctoral dissertation was a proof of the completeness of the first-order predicate calculus. Every one of the valid formulae of the first-order calculus is provable by known methods. In the following year, he published his proof of the incompletability of elementary number theory. This proof, Quine says, "sealed his immortality." Whereas his dissertation had proven what many thought ought to be so, his incompletability proof stunned the mathematical and philosophical world. For any system containing elementary number theory, Godel showed that there is a sentence of the theory that "is true but unprovable."

Godel moved permanently to the United States in 1938 and became a permanent member of the Institute of Advanced Study at Princeton. By 1938 he had achieved what Quine calls "his third great discovery," which was published in 1940. Godel proved that both the special and the general continuum hypotheses are independent of set theory, and that the same holds for the axiom of choice. Any of these theses can therefore be added to set theory without danger of producing a contradiction. Moreover, the methods by which Godel achieved his proofs were original and have proven fruitful. Godel's numbering procedure from his 1931 proof has been used by many of his successors; he also contributed importantly to recursion theory. And the device of internal models which he used in 1940 has been important as well.

Quine lists some of the honors that came to Godel: election to the American Philosophical Society, the National Academy of Sciences, and the American Academy o Arts and Sciences, and some of his honorary degrees. Quine does not attempt to rank Godel against his contemporaries, but he was clearly the greatest logician of his time.

In 1980, Quine published an article entitled "What's It All About?"[76] This was a popular piece intended for a lay audience; it contains mostly material he had already published, but there is a paragraph on the inscrutability of reference that is worth quoting.

> I extend the doctrine [of the inscrutability of reference] to objects generally, for I see all objects as theoretical. This is a consequence of taking seriously the insight that I traced from Bentham through Frege and Russell: namely, the semantic primacy of sentences. Occasion sentences, not terms, are to be seen as conditioned to stimulations. These report the observations on which science rests. The scientific output is likewise sentential: true sentences, we hope; truths about nature. The objects, or values of variables, are just reference points along the way, and we may permute or supplant them as we please as long as the sentence-to-sentence structure is preserved.[77]

The abandonment of physical objects was forcing a revision of Quine's philosophy. One can see here the beginnings of the new direction he developed in response to the problem.

Also in 1980, Quine served as a commentator on a paper by Wilfrid Sellars. The paper is brief but it contains some important points, Quine says

> Mentalistic predicates can be tolerated in the manner of theoretical predicates in physics, e.g., electron spin, or even electron. For them there is no observational criterion, except as those predicates contribute to the coherence and simplicity of an inclusive theory for which there is observational support as a whole. In a word, we can admit them as hidden variables.[78]

This is the first occasion I know of on which Quine has admitted the possibility of such a status for mentalistic predicates. But Quine is no more prepared to adopt them than in the past. He comments that among hidden variables there are better and worse. The better ones are closer to behavior and amenable to explanatory hypotheses by neurology. His example – the only one he gives – is Watson's theory of thought as silent speech. This is no retreat from behaviorism. He also takes on the notion of events such as a man walking to the train station and whistling as he goes. Quine identifies events such as acts as states of the body; to the claim that this would make walking and whistling identical, he replies that the class of walkers and the class of whistlers are distinct, but can have a common member.[79]

Sellars suggests eliminating abstract objects such as classes by paraphrasing them into their members. As Quine puts it "the idea would be that when we quantify over sets we are to be considered to be quantifying ultimately over the tokens of the predicates that determine the sets." This Quine says would amount to substitutional quantification. But Quine points out that such a move will not do for quantification over impredictive sets, since those are sets whose "membership conditions quantify over the sets themselves." Since the proof of the continuity of the real numbers involves such sets, Quine holds that they are uneliminable, and the quantification involved must be objectual. Further, Sellars wants to use "meaning" in the claim that two expressions are alike in meaning. But Quine is not persuaded that sense can be made of such expressions. Quine adds:

> It is strange to find myself on the realist side of the nominalist-realist debate. I would be over there fighting the good fight shoulder to shoulder with Sellars were it not for the difficulties set forth in my earlier comments. I need no persuading of the attractions of nominalism if it can be got off the ground.[80]

One can see here that Quine's conversion to Platonism was one of necessity rather than choice.

The next year, Quine published an article entitled "What Price Bivalence?"[81] sparked, one assumes, by the attacks on bivalence by philosophers such as Michael Dummett. Quine sets the issue thus:

> We stalwarts of two-valued logic buy its sweet simplicity at no small price in respect of the harboring of undecidables. We declare that it is either true or false that there was an odd number of blades of grass in Harvard Yard at the dawn of Commencement Day, 1903. The matter is undecidable, but we maintain that there is a fact of the matter.[82]

So Quine defends bivalence despite the fact that it can lead to apparent paradoxes such as the sorites paradox. "If removal of a single grain from a heap always leaves a heap, then by mathematical induction, removal of all the grains leaves a heap." As Quine notes, the paradox is the result of the vagueness of ordinary terms such as "heap." One expedient is to resort to relative terms. Thus, using Russell's example, if the removal of one hair from a man's head does not make him bald, the removal of all his hairs should not make him bald. But we can take this in the form "balder than," with some stipulated point as marking "bald." Similar paradoxes are easily generated; what constitutes a mountain? How high? How much of the base? These are cases that can be settled by arbitrary stipulation. But there are some that cannot. Consider the surface of a table. There are marginal molecules that are not clearly either part of the surface or part of the atmosphere. Here, Quine says, "we must hold that there are physical objects, coincident except for one molecule, such that one is a table and the other is not." Hence, Quine says, if we hold to bivalence, we must do so even in the absence of objective fact. And Quine does; he claims that this is a paradox affecting common usage and one for which the "notions of austere physical theory remains in the clear." Evidently, Quine was prepared to pay the price for bivalence

In "Two Dogmas," Quine had declared for pragmatism, largely because he could not think of any other way to justify mathematics. But as he later claimed, he did so even though "it is not clear to me what it takes to be a pragmatist."[83] As noted above, the claim was absurd. In a paper he published in 1986 entitled "The Pragmatist's Place in Empiricism,"[84] he made clear his distance from pragmatism. He first defined "empiricism" in terms of five "turning points" that led to his own position. The five are: the shift from ideas to words (Tooke), the shift from words to sentences (Bentham), the shift from sentences to systems of sentences (Duhem), the adoption of methodological monism (presumably by Quine), and finally the abandonment of the goal of a first philosophy (also presumably by Quine). He then reviews in rapid order the work of Peirce, James, Dewey, and Lewis, and finds "little in the way of shared and distinctive tenets." His conclusion, which he states in his first paragraph, is "I suspect that the term 'pragmatism' is one we could do without."[85] His pragmatism had never been more than a temporary expedient, and now he had no further use for it.

Barry Stroud had been a student of Quine's and understood his doctrines very well. Stroud published a paper in which he maintained that, given Quine's theory, there is "the possibility that the world is completely different in general from the way our sensory inputs and our internal makeup lead us to think of it."[86] The problem is that Stroud is right and Quine knew it. But Quine has a way of answering it. Ontological relativity, proxy functions, and the rest, belong, he says, not to ontology but to the epistemology of ontology – to questions of method and evidence. Quine claims to be a robust realist: "people, sticks, stones, electrons and molecules are real indeed, on my view, and it is these and no dim proxies that science is all about." How does this realism fit with the epistemology Quine has developed? His answer is naturalism. Quine says his view is "immanent"; that he is working within science assuming the truth of science in order to determine how we come to have it. There is no transcendental point from which science itself can be judged. It could turn out that science is wrong and we could come to doubt it. "But our doubts would still be immanent, and of a piece with the scientific endeavor." But Quine goes on to consider the possibility of rationally reconstructing the world from sense data.

> My attitude toward the project of a rational reconstruction of the world from sense data is similarly naturalistic. I do not regard the project as incoherent, though its motivation in some cases is confused. I see it as a project of positing a realm of entities intimately related to the stimulations of the sensory surfaces; and then, with help perhaps of auxiliary realms of entities in set theory, contextually defining a language adequate to natural science. It is an attractive idea, for it would bring scientific discourse into a much more explicit and systematic relation to its observational checkpoints. My only reservation is that I am convinced, regretfully, that it cannot be done.[87]

Does this supply an answer to Stroud? We will see.

Quine's former student Follesdal wrote Quine on November 5, 1981.

> Semantics is like physics in being underdetermined by data. But why not then treat it like physics, positing meanings on a par with neutrinos and quarks?[88]

Quine's answer is "because, unlike neutrinos and quarks, they do not contribute to explaining experience." This of course means that Quine rejected the psychological work being done on concepts by people like Eleanor Rosch, assuming he knew about it. Watson and Skinner remained his authorities in psychology, regardless of the changes occurring in psychology.

In 1983 Quine published "Ontology and Ideology Revisited."[89] The article is a reply to various attacks on Quine's dictum "to be is to be the value of a variable." Most of the objections Quine dismisses. He does recognize that the "there is" locution is peculiar to English, and that it is conceivable that there could be a language no expression of which could be translated as "there is," and for this case Quine says that the attribution of an ontology would be impossible. But there is one objection that Quine takes seriously; namely, "that even a dumb animal reacts distinctively to particular things and kinds of things, thereby reifying them, recognizing their existence, incorporating them into his inarticulate ontology." Quine takes this as a valid point. In his 1951 paper "Ontology and Ideology" in *Philosophical Studies* 2:11–15, he drew a distinction between "ontology" and "ideology," the latter "being one's stock of simple and complex terms or predicates." This notion of ideology is now extended to include the responses of dumb animals.

> Ideology so construed may be called *perceptual* ideology, to mark both its breadth and its limits. It is broad in transcending the subject's lexicon, if any, and narrow in treating only of his direct responses to present stimulation. It accommodates dumb animals and remote aliens, thus supplying what was found wanting in my appeal to values of variables.[90]

From "perceptual ideology," Quine holds, we can derive "perceptual ontology." What this would include Quine says is our choice. If "it's a cat" belongs to the subject's ideology, we can include "cat" in his ontology. Similarly for "rain," if he reacts to rain so that it is counted in his perceptual ideology. But Quine considers the notion of "perceptual ontology" pointless since he claims perceptual ideology alone is sufficient. But Quine also says that ontology ought to be restricted to the values of bound variables.

> It is here, finally, that we come to appreciate the role of reification in the theory of scientific evidence: that it is a mere positing of neutral nodes in the structure of scientific theory, as shown by the proxy functions.[91]

Here the consequences of the theory of proxy functions emerge.

In 1984, Quine published "Relativism and Absolutism."[92] This is a very condensed presentation of Quine's views, but it records some important changes. Quine starts with the claim that scientific theory is supported on the one hand by systematic considerations, particularly simplicity, and on the other by "external information in the form of sensory stimulation." Observation sentences of course are the ones that are conditioned to sensory stimulations. These sentences are not *about* stimulations or observations; they are conditioned in the behaviorist sense that a dog is conditioned to salivate upon hearing a bell; the dog is not disposed to eat the bell but the food the approach of which he has been taught to expect when the bell rings. It is observation sentences that provide the basic evidence for science. And what makes a sentence observational is just that all speakers of the language will agree upon its truth under the same conditions of stimulation, or more precisely all stimulations within the generalization range of the first stimulation. Of course, this picture of observation sentences depends upon all community members having approximately the same standards of similarity, so it is relative to the human neural structure. So, Quine says, our training of animals such as dogs depends on their neural structure: their similarity standards must be "like" ours if they are to learn to obey our commands. Further, Quine says, "let us consider in what further way science may be said to be relative, supposing its empirical evidence fixed." One should note the appended clause, for the argument depends on it. In such a situation, Quine holds, science is underdetermined by observation sentences. That means that "different theories can be empirically equivalent." Where a "theory formulation" is taken in the sense of "Empirically Equivalent Systems of the World," that is, as the conjunction of the theory's axioms, many different formulation can be equivalent. Quine extends this to "drastically unlike" theory formulations, for he has shown that by using proxy functions the ontologies of theory formulations can be drastically altered. But suppose the theories are logically incompatible. And here Quine shifts his position. Suppose two theories are logically incompatible.

> There must be a sentence that is implied by one of the formulations and whose negation is implied by the other. It must be a sentence whose truth value is undetermined by all possible observations, since the two theory formulations are empirically equivalent. At least one term in the sentence, then, must be a theoretical term whose conditions of application are not wholly fixed by empirical criteria. We can then exploit its slack by treating the term rather as two inequivalent terms, one in the one theory formulation and the other in the other, and changing its spelling in one of the theory formulations. The two sentences cease to be contradictories. Similar adjustments elsewhere, as needed, render the two theory formulations logically compatible. I owe this expedient to Donald Davidson.[93]

All the "drastic" variants are therefore logically equivalent; if we take one as true, we can take them all as true, simply as complementary descriptions of our world. All this, Quine holds, is perfectly compatible with his "robust realism." It is science that tells us what is real and what is true. The truth of science and the reality of microphysical particles are not impugned by proxy functions or alternative formulations. "Those remarks had to do not with what there is and what is true about the world, but only with the evidence for what there is and what is true about the world."[94] They belong, not to ontology, but to the epistemology of ontology. But we can deal with the world only from within one scientific theory, and since all theory formulations are logically equivalent, it makes little difference which. "Such, then, is my absolutism. Or does it ring relativistic after all?"[95] Quine has here redefined his epistemology as a theory of evidence. This change had been developing since he gave up on physical objects; he needed a new way to deal with the reifications that had previously yielded objects and with the role of variables in his theory. His answer was to convert his epistemology into a theory of evidence, as is made clear in his next paper.

The next year, Quine published an article entitled "Sticks and Stones; or, The Ins and Outs of Existence,"[96] where he made apparent the consequences of proxy functions. The early portions of the article reiterate Quine's oft repeated account of language learning. He professes to be working within science to explain how, from initial stimulations, we reach the science that we have. Quine asserts "we are not aware of the triggering [the nerve stimulations]; it is rather what makes us aware of our surroundings." But does it? According to Quine, we first learn sentences, for the child's utterance "Mama" is to be understood as the sentence "Here's Mama" or "There's Mama." We learn these by conditioning – i.e., they are the responses we are conditioned to make upon receipt of certain nerve stimulations. They are not *about* what causes the stimulations. Quine says

> Observation sentences are not sentences about observations, nor sense data, nor stimulations, nor nerve endings ... It is inappropriate initially to regard them as about anything.[97]

Observation sentences (which are a subset of occasion sentences) need to be related to the standing sentences of science. Quine picks the "whenever" and "wherever" constructions to lead to the observational categorical. Yet "even here, where science is in the bud there is no talk of past or future, nor is there any call to impute reference to objects of any sort. The key idiom that clinches reference to objects is still in the offing." What relates these categoricals to science is the logical notions of implication and negation.[98] So Quine declares, as he has before, that the child can learn truth functions. Objective reference comes, Quine holds, with the relative pronouns and relative clauses. These Quine quickly converts to his preferred form of "such that." The bound variable emerges as the means of cross-reference. Quantification comes independently and then combines with the relative clause. At this point, Quine says, we have objective reference, for it is the bound variable that links us to objects as its values. As in other papers, Quine then leads his child prodigy to classes and numbers, and over these too we can quantify. Quine says

> I have been urging ... that observation sentences unformed in respect of reference to objects of any sort, are the best we can muster in the way of direct linguistic response to sensory stimulation. The positing of objects of any sort, from sticks and stones on up or down, is a sophisticated move that makes sense only after the mastery of the relative clause.[99]

But now, Quine says, suppose a proxy function that maps our ontology into another, so if "f" is such a proxy function, "dog" becomes "f of a dog." We can so transform our entire ontology into a different one. Would anyone notice?

> The old empirical evidence for the theory continues to support the theory when the ontology and terms are reinterpreted in this way. For all evidence stems from sensory stimulation and enters language through observation sentences. These are the final arbiter, and are neutral in point of objective reference.[100]

What matters to science and to discourse is structure rather than ontology. "The objects serve as mere nodes in the structure, and this is true of the sticks and stones no less than the electrons, quarks, numbers, and classes." Neither the logical nor the sensory links of the theory would be disturbed at all. "The whole cosmic upheaval ... has disturbed nothing."[101]

If one now feels that he has been led down the garden path, Quine's answer is "in a word, naturalism." What that means is that we must accept the ontology of our science as it is. For the ontological hijinks have not really been about ontology, Quine says, but about the epistemology of ontology. What our detour through proxy functions has shown is that the evidence that supports our ontology would equally support a totally different ontology. But this does not quite end the matter. Suppose one ontology actually switched for one of, let us say, numbers. "Yet many of the sentences will contradict sentences of our original theory." And how do we escape this? We can then consider our new theory as a foreign language and translate it into our old language by the converse of the proxy function. The conflict, we can then say, was merely verbal. The lesson from all this that Quine draws is: "look to the stimulatory input and the structure of scientific theory, and let the objects, the values of the variables, fall where they may."[102]

There are problems with Quine's new interpretation of epistemology. Quine holds that science determines ontology. But it is, according o Quine, quantum mechanics that shows that the notion of physical objects must be abandoned. This is therefore an ontological claim. But Quine continues to hold that the ontology of science contains sticks and stones, while it is epistemology that is deprived of physical objects. To justify this, Quine says that the answer is naturalism, which he takes as meaning that we must revise our theory of the world from within. But what has this to do with the locating of the demise of physical objects in epistemology rather than ontology? There appears to be no justification for Quine's saying that on the one hand science (quantum mechanics) shows that the concept of physical object must be abandoned, while continuing to hold that it is not in fact abandoned in ontology, and on the other hand making this scientific finding basic for epistemology.

In 1986, there was a meeting at Stanford of Quine, Davidson, Follesdal, and Dreben to discuss Quine's philosophy. Among the Quine papers there is a transcript of these discussions, and they are valuable in giving a picture of the problems Quine saw himself as facing at this time. The discussion was wide ranging. One problem raised was "what ontological relativity is relative to?"[103] It is relative to the language being translated and to the translation manual used. Different manuals construe ontology differently, relative to one's own language, taken as absolute. But is there relativity relative to the home language? If the home language is translated into itself by some transformation other than the identity transformation, it has to be treated as a foreign language. But then, Quine says, you get a regress, so the inscrutability of reference does not apply to the home language, but the doctrine of proxy functions does apply. Hence ontology can be switched, so long as truth is preserved. There are two separate problems here: one is to switch words for objects within the home language, the other is to change the ontology by proxy functions. Quine says "I consider myself a naïve realist when it comes to what there is." Quine holds that it is a matter of epistemology that the reification of objects helps to connect observation with theoretical changes by proxy functions, for that

> can be made without violation of the evidence. And it is not a theory to the effect that it is not a matter of nature and the reality of the world what objects exist and what don't. That is a question of science and it's not a question of methodology. And this discrepancy between the two is the same sort of discrepancy that we see again when we see that theories can be epistemologically equivalent, empirically equivalent, have the same evidence, and can be very unlike.[104]

All true theories should have the same reality. Truth theory, Quine says, covers more ground than translation theory because one could give a truth theory even of a language that couldn't be translated into the metatheory in which the truth was being specified.

"Assent" poses a further problem. Assent is a propositional attitude, a belief. How could one determine, even at a conjectural level, whether someone really does assent, and why? Quine says, "I'm not insisting on behaviorism, but I am insisting on physicalism." Since Quine refuses to accept mentalism, determining assent has to be a matter of conjecture. They are all agreed that the issue here is epistemological, but there is no further progress on "assent."[105]

Quine restates his position on world theories thus:

> I've come out in favor of the idea that there are mutually inconsistent and at least very different but empirically equivalent theories of the world but that it makes no sense to say they're all true. Only one of them is true, the one that we say rightly or wrongly is true is the one that we're working in.[106]

This is, Quine says, "a science fiction case." But in the case of incompatible theories, Quine opts for Davidson's idea of spelling to reconcile them, and he reiterates his sectarian position that Follesdal and Gibson had pressed him to take.

Discussion then focuses on Barry Stroud's question of "the possibility that the world is completely different in general from the way our sensory inputs and our internal makeup lead us to think of it."[107] What makes this a viable possibility is that for Quine occasion sentences are conditioned to stimulations rather than objects; hence if two objects cause the same stimulations they could not be discriminated. Quine's answer is "all we are claiming in our science is that the world fits this [theoretical] structure as far as it goes." The discussion then turns to warranted assertability. A warranted belief is one that fits all stimulations, but it could be false, as Stroud suggested. But truth is distinct from warranted assertability; it is not true of every meaningful sentence that either it or its negative is warranted assertable. Hence in dealing with warranted assertability, one cannot use a two valued logic. Nevertheless, every sentence is either true or false, regardless of whether or not we know which.

The discussion then comes back to the issue raised by Stroud. Quine is emphatic that "no observation sentence whatever describes the pattern of stimulations." Stimulations are causes, not what the sentence describes. This leads to the question of whether different observers have different stimulations. Clearly, they do. To get agreement among observers one must focus on the class of stimulations "that would prompt assent to members of the community." "All the stimulations that would prompt any member of the linguistic community to assent sincerely to the query 'Gavagai'? constitute the stimulus meaning or the affirmative stimulus meaning." They agree that a dummy rabbit could produce the same stimulations as a real one. Stimulations can also vary by lighting, angel, and color contrast. Hence classes of stimulations are the key. Perceptual similarity determines what stimulations are classed together. Quine remarks that he has been influenced by Carnap's *Aufbau* here. Davidson holds that the criterion is not stimulations but the shared scene. The reason his stimulations match Quine's is because they have the same causes – rabbit. "Why is this wrong?" he asks.

> The question is where in that [causal] chain we should look for the relevant similarity to give us correct translations. Your [Quine's] answer is right here. My answer is out there.[108]

Quine rejects Davidson's argument. The class of stimulations is the important thing. He refers to *Roots of Reference* and going from receptual similarity to perceptual similarity. We get the agreement by finding that they are the common response of these various people to the particular sentence: "And if that becomes the principle that holds, you have it right from the start." Davidson replies "I don't begin with what's here and build out. I begin out there and build in." But everything far away from our receptors turns out to be theoretical. This brings them back to the problem of empirically equivalent but incompatible world theories; are both true? Follesdal remarks that that would require an inclusive theory containing both the candidates, since truth only makes sense within a theory. Quine refers to a new paper he is publishing on indeterminacy of translation but without dispositions to verbal behavior. He says "The way to look at indeterminacy of translation really is just that we should recognize what there is to go on in translating, and how little there is to go on, and whether it results in indeterminacy, because we're taking a narrow sense of translation and not a wider one."[109] (The paper referred to is "Indeterminacy of Translation Again" where, as he remarked in the discussion, he made no use of dispositions to verbal behavior.)

Quine refers to holism and the freedom that it gives us to hold what we wish but that one must change sentences that imply a falsehood. That suggests a special status for logic. "It is interesting to think that these considerations may point after all to a difference in status between logic and the rest of science and furthermore this time a difference between logic and the rest of mathematics." That, Quine remarks, would please Carnap but not Hao Wang. Quine remarks that when he wrote "Two Dogmas" he had not worked out the relation between the periphery of the web and experience. The web, he says, consists of standing sentences, the periphery of observation categoricals which are standing sentences. Quine remarks of stimulations "That's my substitute for phenomena."[110] The antecedent and consequent are observation sentences. Observation sentences can be linked in turn by conditioning or whatever with stimulations outside the web. And that is where experience becomes stimulations. Occasion sentences are outside the web. That raises the question of whether observation sentences are corrigible; Quine says "no, because it is an occasion sentence that has no status except on that occasion and on that occasion it is assented to. If it does command assent otherwise, it does qualify as an observation sentence;" he means if it does command assent on the spot of all witnesses. "So that is not a case of corrigibility or incorrigibility as far as I can see. It doesn't apply. And then the record of this, the recorded observation, the protocol, is a standing sentence. And that's inside the web. And that is corrigible, and when we correct it when we change our minds about it, that's the occasion where we say we repudiate one of our purported observations." The text is not clear here, but apparently what Quine is holding is that occasion sentences are outside the web, whereas observation sentences, better observation categoricals, are inside the web. Observation sentences are not corrigible if they are assented to on the spot – Quine says corrigibility does not apply. But the record of the observation, i.e., the protocol, is inside the web. The text is not clear here, but Quine seems to be distinguishing occasion sentences from observation sentences ad claiming that the former are outside the web. This seems improbable and the transcription is likely in error.[111]

Truth, they agree, is within a theory; there can be multiple incompatible theories within one language. Dreben remarks, "Then Quine's notion of observation categorical is more important then ever because that's the only real link [to the world], and the rest takes on more and more the picture of idealism." They are agreed that there are no (true) tandem theories [two empirically equivalent but incompatible theories]: only one is true. (Presumably this is because of Davidson's spelling trick.) Suppose two grammars that produce the same totality of strings that are grammatical. Is there a fact of the matter between them? The choice would depend on what considerations were introduced – simplicity, genetic predisposition, etc. But Quine says "I've never talked of indeterminacy of grammar in the way I have of the other (translation)."[112] (But he has in "On Reasons for the Indeterminacy of Translation," and "Methodological Reflections on Current Linguistic Theory.") The problem, Quine says, of interlinguistic inputs is homology of nerve endings. Homology won't do it but something similar is required. The "burning problem" to come to terms with is intersubjectivity of stimulations; perceptual similarity is still in the individual.[113]

We saw earlier that Quine was forced to use mentalistic terms repeatedly in RR. This was not a satisfactory position for him, and in 1985 he tried to solve the problem in a paper entitled "States of Mind,"[114] a revised form of a 1980 paper he had given as a American Philosophical Association colloquium in Detroit. As one might expect, the piece is an attack on mentalism. The classic dualism of mind and body Quine considers redundant.

> The bodily state is trivially specifiable in the dualist's own terms, simply as the state of accompanying a mind that is in that mental state. Instead of ascribing the one state to the mind, then, we may equivalently ascribe the other to the body. The mind goes by the board, and will not be missed.[115]

We can, Quine says, continue to use the mentalistic terms, but interpret them as referring to states of the body; the mind is an unnecessary intervening factor best done without. Quine tries to bolster his case by arguing that the mentalistic terms are first learned from external influences. How after all do we learn to call our joys "joys"? "Clearly the answer is that such terms are applied in the light of publicly observable symptoms: bodily symptoms strictly of bodily states, and the mind is as may be." So Quine holds that without "outward signs" the mentalistic terms could never have been learned at all. Hence we can reject the mind as a make-believe entity, and take behavior as a sign of the underlying neural states. It does not follow however that mentalist terminology is or will be translatable into neural terms. Thus consider belief. It can be identified incompletely by verbal statements, and its strength perhaps by betting, though that test is rarely applicable outside the laboratory. The physical signs of belief are extremely heterogeneous, as is the case for all propositional attitudes. But Quine notes that some uses of propositional attitudes run deep. If Martha is teaching Tom the sentence "It's raining," she must hold the belief "Tom perceives that it is raining." All teaching by ostension requires such a belief on the part of the teacher, and the other propositional attitudes doubtless follow by analogy. So Quine does not succeed in eliminating mentalism after all. What is most striking here is on the one hand Quine's obvious animus against mentalism, while on the other hand his attempted reduction cannot be carried out, even on his own showing. Even if it could, it would be a reduction of the better understood to the less understood since we have no idea what the underlying physiological states are. In fact, Quine's claim that he can use mentalistic terminology while reinterpreting it to refer to bodily states masks the fact that without the mentalistic terminology he has no way of referring to the neurological states at all. This is a move that, in Russell's happy phrase, has the advantages of theft over honest toil.

Davidson had pressed Quine on the question of how adverbs can be incorporated into the language. In 1985, Quine answered in "Events and Reification."[116] We have already seen that Quine was puzzled by the problem of adverbs, and here he attempted a solution, and drew further lessons. First order logic can accommodate nouns, verbs, and adjectives; the problem is how to accommodate adverbs. Consider

(1) Sebastian walked slowly and aimlessly at t.

Davidson had shown that this can be reformulated as

(2) $(\exists x)$(x is a walk and x is slow (for a walk) and x is aimless and x is in Bologna and x is at t and x is by Sebastian).

This converts the adverbs to adjectives, but (2) posits the existence of a "walk" and hence of an event. If events are to enter our ontology, we need to know how they are individuated. Quine illustrates the problem as follows.

(3) $(x = y) \equiv (z)(x \in z \equiv y \in z)$

But classes are not individuated unless their members are. However,

> Since (3) explains identity of events by quantifying over classes of events. it individuates events only if the classes of events are already individuated, and hence only if events are already individuated.[117]

The lesson of (2) is that adverbs can be converted into adjectives by reification of "walk" – an event. Quine wants to absorb events into physical objects, the identification of which is already clear. Quine defines a physical object as the material content of some region of space-time. He takes events (i.e., Sebastian's walk) as the physical object Sebastian with changing coordinates over the time period of the walk. Then adverbs become adjectives applying to the physical object Sabastian. So, Quine says,

(4) A white cat is facing a dog and bristling

becomes

(5) (∃x)(x is a cat and x is white and x is bristling and x is dogward).

There are indeed questions about the individuation of physical objects. One may recall the problem of the table that Quine raised in an earlier paper. But Quine defines all of the various tables that differ only by the assignment of a molecule as the *extension family* of "table," and when it comes to quantifying over the table, any one will do.[118]

There is a further point that Quine wants to make here. Truth functional connectives join sentences, but rather loosely. Reification tightens these connections, as is evident in (5); the occurrence of the bound variable in each conjunct is what ties the constituent sentences into a single whole.[119] And the tightening of connections is crucial, not only within sentences but among them. Quine refers again to the lesson of proxy functions – that ontologies can be switched at pleasure. This, he says, "tells us that scientific evidence is a matter of sensory stimulation and the structure of the network of sentences." What the world is is defined by science; the epistemology of ontology shows that "a reshuffled ontology would fit all evidence just as well, but it would not fit it any better."[120]

In a letter to Gibson on April 18, 1987, Quine reflected on "Two Dogmas":

> In "Two Dogmas" I was not professing to prove anything. I was raising doubts and adducing possibilities ... in "Two Dogmas" I addressed the second dogma as one possible cause of people's belief in the first dogma. Today I see the second one as the important one.[121]

It was of course Quine's attack on the second dogmas that opened the way to holism.

There is among Quine's papers an exchange of letters between Davidson and Gibson that helps to bring out some significant points. Gibson writes "I took Quine's idea of the 'reciprocal containment' of epistemology and ontology as a crucial step toward getting a proper understanding of his philosophy."[122] Ontology, Gibson says, is fixed by science. And it is science that tells us that science is underdetermined by the totality of experience. But Gibson points out that since science is underdetermined, so is epistemology. Epistemology is not transcendent. Both ontology (science) and epistemology are immanent – that is, relative to our theory of science.

In 1990, Quine published a book entitled *Pursuit of Truth.*[123] Having discussed it at several conferences, Quine issued a revised edition in 1992. The objective of the work Quine describes as follows: "In these pages I have undertaken to update, sum up, and clarify my variously intersecting views on cognitive meaning, objective reference, and the grounds of knowledge."[124]

As often before, Quine posses his problem as, given the stimulations we receive from the external world, how have we managed to develop our science? That question he takes to be the question of evidential support for science. He starts his answer with observation sentences, taken here as including occasion sentences. He then describes observation sentences as he has before. It is observation sentences that provide the evidence for science; they are also the starting point for the learning of language. Taken holophrastically, however, observation sentences are conditioned to stimulations, and are, Quine says, theory free. Taken retrospectively, the terms of the sentence are theory laden and will reappear in theoretical contexts. A further advantage of this interpretation is that questions of reference are not raised. When observation sentences are used to test a theory, the hypothesis under test is never alone; in order to derive the testable sentences, auxiliary hypotheses are also required. The relations of the sentences within the theory are those of logical implication and deduction, and the testable sentence implied is an observation categorical, which Quine describes as "Whenever this, that," which is a standing sentence. What Quine calls the "free observation categorical" is a conjunction of observation sentences; a "focal observation categorical" rest on predication and binds its terms together with reference to something. Tests of such categoricals can refute them, but can only partially confirm them.[125] The rejection however is of the whole set of sentences that together implied it, meaning that one among them is false, but the choice of which one to reject rests with the investigator. This is of course Quine's holism. The decision of which hypotheses to withdraw is governed, Quine says, by the "maxim of minimum mutilation" – that is, changing as little of the theory as possible. He will not change the mathematics, because the consequences of doing so would be too disruptive. He will also not change any of the logic involved, but for a different reason from that which protects mathematics. "Any purely logical truth is thus exempted [from change], since it adds nothing to what S [the set of sentences involved in the test] would logically imply anyway." This comes to saying that logic has no empirical content.

But what about empirical content?

> Call an observation categorical *analytic* for a given speaker if ... the affirmative stimulus meaning for him of the one component is included in that of the other. Otherwise *synthetic*. Call a sentence or set of sentences *testable* if it implies some synthetic observation categoricals. Call two observation categoricals *synonymous* if their respective components have the same stimulus meanings. Then the *empirical content* of a testable sentence or a set of sentences for that speaker is the set of all the synthetic observation categoricals that it implies, plus all synonymous ones.[126]

Note that the terms "analytic", "synthetic", and "synonymous" are defined here purely with respect to stimulus meanings. This is not a departure from Quine's doctrine in WO.

Quine rejects the whole notion that there can be a foundation of certainty for science. But he does not abandon epistemology; he redefines it as "naturalized" – i.e., as the science that studies the relation of evidence to science. This does not however mean that his epistemology is no longer normative; rather, normative epistemology gets naturalized into a chapter of engineering, the technology of anticipating sensory stimulation. The fundamental norm is *nihil in mente quod non prius in sensu;* there is nothing in the mind that is not first in sensation. Further, naturalized epistemology is concerned with the general way of choosing hypotheses, as he made clear in *The Web of Belief.* There are "virtues" that guide the choice. But the claim that prediction is the checkpoint of scientific theory is not normative; rather, it is defined, in the "particular language game, in Wittgenstein's phrase: the game of science." But for all that, Quine says that science is not committed "to the physical, whatever that means." Quantum mechanics has rendered even that standby questionable, and one must consider the possibility that empiricism itself could be refuted. But science would still be science, "hinging still on checkpoints in sensory prediction."[127]

Quine said he was postponing the discussion of reference, but it was already implicit in the observation categoricals. "This is where I see bodies materializing, ontologically speaking, as ideal nodes at the foci of intersecting observation sentences." Quine then refers to the experiment of a toy car passing behind a screen and reemerging on the other side, and claims it reflects "the expectation of continuity of a present feature rather than the reification of an intermittently absent object." It should be noted that this is not the interpretation given to this experiment by psychologists, who see it as showing the child's recognition that the object persists while occluded. Quine's reason for his interpretation is that he wants to maintain his standard that to be is to be the value of a variable. Put into English rather than logic, this comes to saying that to be is to be the referent of a relative pronoun. Reification, Quine says, echoing his recent publications, serves to tighten the connections of truth functions. Reification "is a device for focusing observation sentences convergently." This brings Quine to the following claim.

> What particular objects there may be is indifferent to the truth of observation sentences; indifferent to the support they lend to the theoretical sciences, indifferent to the success of theory in its predictions.

To support this claim, Quine invokes proxy functions. Citing the Skolem-Lowenheim theorem, Quine holds that our entire ontology is reducible to the natural numbers. But since the numbers do not correspond one to one to the reifications that were our steppingstones in learning language, we could not have acquired our theory of the world with a Pythagorean ontology; we have to proceed in the old way. Furthermore, Quine notes that there are results in quantum mechanics that call the basic notion of physical particle into question. It may turn out that even the notion of existence no longer makes sense and some new alternative concept is required. But since observation sentences are bound only to stimulations, our science will not be disturbed by such an ontological upheaval. Our ontology is parochial; it is our contribution; but it is replaceable without falsifying science.[128]

We have noted before the consequences for epistemology that Quine drew from the demise of physical objects; here he addresses the consequences for science. His conclusion appears to be that all that matters for science is structure and stimulation. If so, this represents a very extreme form of empiricism in which the objects postulated in science are reduced to neutral nodes. Science would then consist of a structure of interlocking sentences with which, from prior stimulations, we predict future stimulations. But if science itself has overthrown the notion of physical object, how can he claim to be a naive realist and hold that sticks and stones are real? This is a problem that Quine needs to solve.

Quine then turns to meaning. He sets up the process of learning as follows. "We learn short sentences as wholes, we learn their component words from their use in those sentences, and we build further sentences from the words thus learned." He then points out that his thought experiment of radical translation should have yielded meanings if there were any; the result was the indeterminacy of translation, but no meanings. This thesis is a consequence of Quine's behaviorism, as he readily admits.

The starting point in language learning is observation sentences. They are also, he holds, the linguist's starting point in radical translation. To make this work, "the linguist must be able to recognize, if only conjecturally, the signs of assent and dissent in Jungle society." Quine claims that the linguist has much to go on here: a speaker will assent to an utterance in any circumstance in which he would volunteer it. But if the linguist must wait for volunteers, he will have a long wait. Quine notes that in earlier writings he had been troubled by the problem of intersubjective comparison of stimulations, as for example in "Propositional Objects." The issue was discussed at the Stanford conference in 1986. Quine says he rejected Davidson's idea that the stimulus should be the distal stimulus, i.e., "the nearest shared cause of the of the pertinent behavior of the two subjects." Instead, Quine introduces the notion of empathy. This is hardly a concept that one would expect from Quine, but he needs it for language teaching, since the teacher must be able to understand what the student is seeing, and for the problem of intersubjectivity. "We all," Quine says, "have an uncanny knack for empathizing with another's perceptual situation." So far as I know there is no behavioristic definition of "empathy." Quine reasserts his 1981 definition of observation sentences for single speakers, and then accounts a sentence observational for a group if it is observational for each member and if each would agree in assenting to it, or dissenting, on witnessing the occasion of utterance. "We judge what counts as witnessing the occasion, as in the translation case, by projecting ourselves into the witnesses position." This amended position has repercussions for translation: the alternative manuals of translation can no longer rest on shared stimulus meaning. "It is a matter of better or worse manuals [of translation] rather than flatly right or wrong ones." "What is utterly factual is just the fluency of conversation and the effectiveness of negotiation that one or another manual of translation serves to induce." This would seem to imply that one manual can be better than the other but Quine does not draw this conclusion. Granting now that stimulus meanings have become private, "what floats in the open air is our common language," which each internalizes in his own way. Quine then resumes his discussion of translation. But Quine now adds a further factor – belief. The linguist "will depend early and late on psychological conjectures as to what the native is likely to believe."

Practical psychology is what sustains our radical translator all along the way, and the method of his psychology is empathy: he imagines himself in the native's situation as best he can.[129]

Radical translation, according to Quine, exposes "the poverty of ultimate data for the identification of meanings." Given this situation, two translators, working independently on translations of Jungle to English, will almost certainly come up with different manuals of translation. Such is the indeterminacy of translation. But in fact what the translator will do is to impose his own views on the native, whenever there is doubt. But Quine makes it clear that this indeterminacy of translation does not extend to grammar.

What about reference? Quine says that in his earlier writings he had not clearly differentiated between "inscrutability of reference" and "ontological relativity." Now he says he can "say what ontological relativity is relative to . . . it is relative to a manual of translation." To say "gavagai" denotes "rabbit" is to choose one manual of translation over the other. Does this indeterminacy of reference hold for the home language? In *Ontological Relativity* Quine said that it did. He now rejects that conclusion since we can choose for the manual of translation the identity function.

Thus taking the home language at face value, the relativity is resolved. Reference is then explicated in disquotational paradigm analogous to Tarski's truth paradigm; thus "rabbit" denotes rabbits, whatever *they* are, and "Boston" denotes Boston.

And what of meaning? Philosophers have used meanings to serve as the essence of words, and Carnap used them to explain how mathematics could be meaningful without empirical content and why its sentences were necessarily true. Quine rejects both uses.

Holism lets mathematics share empirical content where it is applied, and it accounts for mathematical necessity by freedom of selection and the maxim of minimum mutilation.[130]

Quine then turns to intensions. If Martha is teaching Tom the language, then Tom's mastery of the sentence "It's raining," hinges on Martha's mastery, virtual if not literal, of the mentalistic sentence "Tom perceives that it is raining." The idiom is "x perceives that p," where "p" stands in for an observation sentence. And how does Martha do this? By empathy; by projecting herself into Tom's shoes. And this construction of "x perceives that p" can be extended beyond observation sentences to standing sentences. But Quine declines to extend it to "x perceives y," where "y" stands for an object. As previously noted, Quine does not allow objects to be directly perceived; presumably what the learner sees is a scene of which a surface of the object is a part though Quine does not say this. Instead, Quine assimilates "x perceives y" to "x perceives that p" where "p" stands in for another sentence. Perception, Quine says, is a momentary event; where the holding of perception lasts, it becomes "x believes that p." Ascription of belief, like ascription of perception, depends on evidence – for the latter, orientation and behavior plus empathy; for the former, the easiest evidence is obtained by asking "x" if he believes "p"; other clues may come from behavior, but the evidence becomes tenuous as the beliefs move away from observation sentences. Thus we come to propositional attitudes. "Empathy is why we ascribe a propositional attitude by a content clause." Having dismissed propositions, Quine takes "p" to stand in for a sentence. Ascription of an attitude can be applied to a cat; it does not require that "x" be a speaker. Such contexts are opaque; one cannot quantify into them. Quine uses the two sentences

(1) There are some whom Ralph believes to be spies.
(2) Ralph believes "(∃x)(x is a spy)."

(1) ascribes a belief *de re*; (2) ascribes a belief *de dicto*. The *de re* ascription assumes a relation in intension between thoughts and things intended; that, Quine rejects; but he considers *de dicto* ascriptions acceptable.

De Dicto ascriptions like (2) are acceptable because they can be rendered in predicate logic by converting the quotation to spelling, where the names of the constituent signs are combined by concatenation. But not all mentalistic talk is eliminable.

> What are irreducibly mental are the ways of grouping them: grouping a lot of respectably physical perceptions as perceptions that p, and grouping a lot of respectably physical belief instances as the belief that p. I acquiesce in what Davidson calls anomalous monism, also known as token physicalism: there is no mental substance, but there are irreducibly mental ways of grouping physical states and events.

Quine also takes modalities to be intensional and dismisses the lot. Physical necessity he takes to rest on a belief in essence that he rejects. And then, surprisingly, he turns on the notion of disposition. Although it admits of substitutivity of identity, it resists the predicate calculus. But the problem he considers solvable by taking the "ible" and "able" endings as vacuous, and counting such terms as "soluble" and "fragile" as physical predicates.[131]

Quine now turns to the question of truth. He takes truth as applying chiefly to eternal sentences, although it can also be applied to individual acts of utterance. Speaking then of eternal sentences, Quine dismisses the standard correspondence theory, and turns to Tarski's definition. Using the famous example

> "Snow is white" is true if and only if snow is white

Quine remarks "Truth is disquotation." To talk about a sentence of the language requires semantic ascent. Thus truth is predicated of the sentence named by "Snow is white," and what makes the sentence so named true is "the world being as the sentence says." This definition of truth can be extended to utterances. "I have a headache" is true if and only if the user has a headache while uttering it. Thus the truth predicate ties words to the world.

But truth has its paradoxes. The most famous is the liar paradox, which Quine gives in the form

> "yields a falsehood when appended to its own quotation" yields a falsehood when appended to its own quotation.

It is perhaps more easily recognizable in the form "This sentence is false" or various similar versions. Clearly, some restraint must be imposed. Quine first defines his term "disquote" as

> The truth predicate will be said to *disquote* a sentence S if the form
> "----------" is true if and only if ---------------
> comes out true when S is named in the first blank and written in the second.

The lesson of the paradox is that a language "cannot contain also a truth predicate that disquotes all its own eternal sentences." Quine then describes Tarski's famous theorem and how it was discovered. The language chosen for the proof is quantification with truth functions and the membership relation. It has a finite lexicon, and the means of quoting and appending. Truth applies to closed sentences; for open sentences the analogue of truth is *satisfaction*. "An assignment of objects to variables *satisfies* a sentence if the sentence is true for those values of its free variables." Thus an assignment is a function that relates one object to each free variable. Tarski first defined satisfaction for atomic sentences, each of which consists of a predicate joined to one or more variables, for each predicate in the lexicon. An assignment satisfies a conjunction if and only if it satisfies both conjuncts; it satisfies an alternation if and only if it satisfies one or both; it satisfies a negation of a sentence if and only if it does not satisfy the sentence negated; it satisfies an existential quantification "(∃x)(. . . x . . .)" if and only if some assignment, matching that one except perhaps for what it assigns to "x", satisfies " . . . x . . .". The definition is recursive, and therefore extends to all sentences of the language. Why does this not lead us right into paradox? The answer is that the definition of satisfaction is inductive; not direct. The definition explains satisfaction for each specific sentence, but it does not give a translation of "x satisfies y" for variable "y." Hence it does not yield "not(x satisfies x)," and so dodges Grelling's paradox. Thus we have the truth definition since "a closed sentence, having no free variables, is vacuously satisfied by all assignments or none according as it is true or false."[132]

To avoid the paradoxes of class theory, Quine says that some logicians (including Quine) have added classes that cannot be members of anything. Classes that are members of something other than themselves are called "sets" to mark the distinction, and classes galore can be added without danger. In such a system a direct definition of satisfaction and truth can be attained, but only for the sets without the added classes. This leads to a hierarchy of languages, each containing a truth predicate that applies at a level just below its own.

Quine then examines a number of challenges to the notion of truth involving the law of excluded middle. One is Aristotle's famous example: if it is yet to be decided whether there will be a sea battle tomorrow, then the statement that there will be such a battle tomorrow is neither true nor false. Theologians have found this doctrine very attractive since otherwise the future would be determined by God's foreknowledge and there would be no place for human moral responsibility. This doctrine Quine dismisses because he thinks it rests on two false premises: one is that there is an omniscient God; the other is the notion that determinism precludes freedom of action. "We are free and responsible . . . in that we act as we choose to; whether our choices are determined by prior causes is beside the point." It is charming to find Quine in agreement with Jonathan Edwards! Other challenges to the law of excluded middle raise the question of three valued logics, where the third value is "indeterminate." Quine of course admits the possibility but prefers to stay with the two valued logic; three valued logic he considers cumbersome. He also emphasizes the difference between truth and warranted assertability. A sentence may be true or false without our knowing which; if we have significant but not conclusive grounds for believing the sentence true, we are warranted in asserting it, but the sentence is still true or false, independently of our assertion. And what of mathematical truth, particularly the truth of sentences never applied? Higher set theory should be considered meaningful, since its sentences contain terms used at lower levels. Further, Quine cites Godel's axiom of constructability, and the efforts of others to show that the needs of science can be satisfied by predicative set theory. Quine sees these as simplifications that should be encouraged.

Finally, Quine takes up the problem of empirically equivalent but different global theories of the universe. If two such theories were logically compatible we would just incorporate the other into our own. If they were logically incompatible, Quine invokes Davidson's method of changing the spelling at the appropriate places to make them compatible. But what if the two theories are logically compatible "but, like Poincare's example, it hinges on some theoretical term not reducible to ours." Poincare's theory was that there is a central point in the universe such that as bodies move away from it, their length diminishes, so the boundaries of space are never reached. Hence, the central point is different from every other. Assuming there are no other grounds for choice, we can either take the "sectarian" position and reject the other theory, or the "ecumenical" position and welcome both. As we have seen, Quine vacillated on this issue until Follesdal and Gibson persuaded him to take the sectarian view. Quine compared the problem of indeterminacy between two such global theories with the indeterminacy of translation.

> What the indeterminacy of translation shows is that the notion of propositions as sentence meanings is untenable. What the empirical underdetermination of global science shows is that there are various defensible ways of conceiving the world.[133]

This small book contains some important shifts and clarifications of Quine's position. First, in holding observation sentences theory free, Quine asserts that there is something that is independent of our theories. Unlike Kuhn, for example, Quine holds that theories are not totally incommensurable; stimulations are real and theory free. Since observation sentences contain terms that are theory laden, one assumes that what Quine means is that observation sentences are theory free as holophrastically conditioned. But it is not clear why since it is our scientific theory that postulates stimulations and conditioning. Second, Quine finds the basis of mathematical truth in science; mathematics acquires its content from its applications; what to do about mathematics not as yet applied remains a problem. Logic however is taken as devoid of empirical content. But when Quine says that "any purely logical truth is thus exempted [from change], since it adds nothing to what S [the set of sentences involved in the test] would logically imply anyway," one is left wondering how an implication can operate when it is not included in logic. Third, Quine reasserts his doctrine that due to proxy functions ontologies can be changed at will. However, he does qualify this claim by saying that the *acquisition* of language requires the use of our standard ontology. Fourth, Quine introduces the concept of empathy to explain how teacher and pupil can share a scene; this is a partial answer to the problem of intersubjective stimulations that at the Stanford conference he had called the "burning problem." Fifth, he now adds psychology (empathy) as an important factor in translating Jungle to English, but again denies indeterminacy with respect to grammar. Sixth, he now attributes ontological relativity to the indeterminacy of translation, i.e., whether to translate "Gavagai" as "rabbit," whereas the inscrutability of reference rests on proxy functions. Seventh, Quine uses holism to account for the necessity of mathematics; those statements we refuse to abandon are the ones he calls necessary. Eighth, Quine continues to struggle with propositional attitudes; he is now willing to accept *de dicto* propositional attitudes, but rejects *de re* propositional attitudes. Ninth, he now resolves the issue of dispositions; since he believes that dispositions are caused by underlying physical or physiological states, he now takes the dispositional idiom as ascribing a physical property to the person or thing involved. Tenth, he clarifies his distinction between truth and warranted assertability. And finally, he softens his sectarian position regarding empirically equivalent systems of the world. Having found a way, through Davidson's device, of making all such systems logically compatible, he is willing to see them as just alternative views of the world that can be adopted at will.

In 1990, Quine published an article entitled "Three Indeterminacies,"[134] which represents an attempt at the clarification of his central doctrines. And some clarification is forthcoming. Quine gives the following definition: "An observation sentence is an occasion sentence that the members of the community can settle by direct observation to their joint satisfaction." He goes on to say that observation sentences "are reports not of sense data still, but of ordinary external circumstances." He then takes up the problem of matching observations of different observers. Quine says he was concerned over the problem as early as 1965; in 1981 he tried to by-pass the problem by accounting "a sentence observational for a whole community when it was observational for each member." But Quine notes that the inadequacy of this move was pointed out by Lars Bergstrom since a sentence could be observational for each member of the community yet not be assented to in the same situations. Quine rejects again Davidson's proposal that the stimulus be taken to be the distal one; Quine says "I remain unswerved in locating stimulations at the nearest point," since his goal is to explain how, from perceptual stimulations, we reach science. So here Quine proposes to do without intersubjective likeness of stimulations entirely. He now invokes empathy to explain how different witnesses can agree, and accounts "a sentence observational for a group if it is observational for each member *and* if each would agree in assenting to it, or dissenting, on witnessing the pertinent occasion."

This redefinition does not compromise indeterminacy of translation, though the elimination of like stimulations does shift the burden to conversation; "what is utterly factual is just the fluency of conversation and the effectiveness of negotiation that one or another manual of translation serves to induce." Quine remarks on the irony that the "gavagai" example is usually taken as demonstrating indeterminacy of translation when it was intended to show the relativity of reference. He continues to hold the latter doctrine, but emphasizes that ontology is relative to the manual of translation – e.g., one in which "gavagai" is translated as "rabbit." But he now says that ontological relativity does not apply in the home language since we can translate by the identity function.[135]

Quine then turns to reification. Proxy functions show, he says, that any ontology will do so long as it has a one to one correspondence to our usual one. "The gross bodies themselves, charter members of our ontology, could thus be superseded by proxies and not be missed." A *primitive* observational categorical of the form "Whenever this, that," is an irreducible generality prior to any objective reference. It is a "generality to the effect that the circumstances described in the one observational sentence are invariably accompanied by those described in the other." But if "circumstances" are described, surely they must be referred to. Quine distinguishes the *primitive* observation categorical from the *objectual* observation categorical such as "Tabby is white," which does involve objective reference.[136] It is not clear what this distinction means; "Whenever Tabby, white," which Quine calls a *primitive* observation categorical must refer to something if only the circumstances of Tabby and white; apparently what Quine means is that Tabby is not conceived as an external object in the primitive observation categorical but is so conceived in the objectual observational categorical. Just what the status of the "circumstances" is remains obscure.

Quine next turns to testing and refutation. He separates out sentences that are stimulus analytic; such Quine claims, is the status of "Robins are birds." Thus "analyticity" that Quine threw out the front door, seems to have returned through the back. It is only synthetic observation sentences that are testable. It is these that scientific theories imply. Then Quine gives his usual argument for holism, based on the need for multiple premises to imply an observation categorical. The revision of the premises will be based on the scientist's desire for simplicity and the maxim of minimum mutilation. Thus Quine says

> We exempt some members of S [the set of premises] from the threat on determining that the fateful implication still holds without their help. Any purely logical truth is thus exempted, since it adds nothing to what S would logically imply anyway.[137]

How the members of S can imply the test sentence without logic being involved is certainly not clear. But perhaps what he means is that logical truths can be eliminated because they are tautologies. Are they then analytic? Mathematics is also exempt from refutation for a different reason; its rejection would disrupt all science. All this is taken to illustrate Quine's holism. Testing is a question of "critical semantic mass." And holism also leads to the notion of empirical content.

> We can define it [empirical content] as the set of all implied synthetic observation categoricals, plus such further observation categoricals as may be *stimulus – synonyms* to any of these. Two observation categoricals are stimulus-synonymous if their component observation sentences have the same stimulus-meaning.[138]

So an untestable sentence or set of sentences has no empirical content. "These words," Quine notes, "echo an old verification theory of meaning, but they gain new force from our holistic standard of testability."

Quine then turns to empirically equivalent theories of the world. Quine describes briefly his own vacillation between the sectarian and ecumenical points of view. He discusses various alternatives, and finally concludes that "what appeals to me most" is

> Economy ... is imperative only as an ideal of theory construction and not of language. Meaningful application of the truth predicate, on the one hand, extends to the whole language and is not limited to any particular theory formulation. Empirically equivalent and logically compatible theories can be accepted as true descriptions of the world even if one of them uses terms irreducibly alien to the other. There is no call to fuse them into a single redundant theory. Our language can embrace the full vocabularies of both theories, and our truth predicate can then apply to each on its separate merits.[139]

He appears to have settled on the ecumenical position. But a total coherent theory of the world remains as a scientific ideal. The multiple theories Quine has just allowed will conflict in ontology, but this can be resolved by giving to each a particular predicate purporting to be true of only its objects.

Given its fame, "Two Dogmas" remained a center of contention. In 1991, Quine tried to clarify it in "Two Dogmas in Retrospect."[140] The article, he says, grew out of his difference with Carnap. The occasion for it was an invitation from the Program Committee of the American Philosophical Association to give a paper at the Toronto meeting. Quine remarks

> It is remarkable that my most contested and anthologized paper was an assignment. The response was quick and startling.[141]

As I have already noted, the paper was more a progress report than a finished essay. In retrospect, Quine faults the paper for having overstated his holism. While he says the claim that the whole of science is at stake in any experiment was correct "in a legalistic sort of way," he now sees what is at stake as the "critical semantic mass," meaning the cluster of statements sufficient to imply an observable experimental condition. Quine notes that when he wrote "Two Dogmas," he had not known of Duhem's work. He also says that implication is the relation between the cluster having semantic mass and the observation categorical, and that this gives logic a "special status." "We remain free here to adjust and to vary the limits of what to count as logic." The law of excluded middle could be dropped, as he had said in the original paper, but "it would just mean setting the limits of what to count as logic very narrowly and reckoning the law of excluded middle to the clusters of defeasible hypotheses that do the implying." His attack on the analytic/synthetic distinction was aimed at Carnap, who saw it as explaining the necessity of mathematics and its meaningfulness. Both, Quine says, are provided by his "moderate holism." The "empirical content" of mathematics comes from its applications; inapplicable mathematics survives because it shares the grammar and lexicon of applied mathematics. The necessity is due to our reluctance to change mathematics because of the disruption that it would cause in science. But Quine retains a modified form of analyticity. As he said in *Roots of Reference*, a sentence is analytic for a speaker "if he learned the truth of the sentence by learning the use of one or more of its words." Quine says that the logic of truth functions, quantification, and identity "would then perhaps qualify as analytic." Thus the analytic status of logic is here restored. The learning criterion, he thinks, suffices for logic, but since we rarely know how or when we learned a word, it cannot be extended. As we have seen, he found little that he could accept in the writings of the pragmatists. The "web" metaphor needed unpacking and "that was largely my concern in the ten years between 'Two Dogmas' and *Word and* Object." He has since clarified what in "Two Dogmas" he called "experience" as the "global *neural input* on that occasion." The sentences at the periphery of the web he subsequently defined as "observation sentences," connected to experience by "association with ranges of perceptually similar neural inputs," and observation sentences are connected to theory by shared lexicon.[142] Quine does not here remark that in retrospect it was the rejection of the second dogma that was really important, though he has made that point elsewhere.

In 1992, Quine published "Structure and Nature"[143] in the *Journal of Philosophy*. This brief article contains an interesting paragraph on observation sentences that is worth quoting.

> Grammatically, some [observation sentences] are indeed sentences, e.g., "It's raining," and some are nouns or adjectives, e.g., "Cat" or "Milk" or "White." To begin with, they are just expressions associated holophrastically with ranges of neural input. They are as if to say, with William James, "Hello, thingumbob again" – except that even "thingumbob" hints more of objectual reference than I could wish. The observation sentence names nothing, to begin with, neither neural input nor external object. But it is learned from adults who have learned the ways of reference, so it is indeed an expression, such as "It's raining" or "Cat" or "Milk" or "White", that is destined for eventual integration into a system of objectual reference on the child's own part as he matures.[144]

It is clear here that observation sentences as the child learns them have no objectual external reference. "The reification of bodies comes in stages in one's acquisition of language, each successive stage being more clearly and emphatically an affirmation of existence." This process is apparently not completed until the relative clause is learned. Quine then invokes proxy functions to show that the external reference of sentences can be changed at will. But he says this does not affect science – for two reasons: "First, implication hinges only on logical structure and is independent of what the objects, the values of the variables, may be. Second, the association of observation sentences with ranges of neural input is holophrastic. It is independent of reifications, independent of whatever objects the observation sentences or their parts may be taken to refer to as terms." Quine had evidently recognized that the claim that the child cannot refer to external objects until he has mastered the relative clause was untenable. He tries to remove the problem by making objective reference something gradually acquired, starting with the early "body-mindedness" of the infant but not reaching completion until the relative clause is learned.

In January of 1993, Davidson wrote to Gibson a very interesting letter.

> I never thought (or said) Quine *thought* observation sentences were about patterns of stimulation, nor did I ever say they were. What I argued is that his observation sentences weren't necessarily about rabbits either. Quine said they were about rabbits; but following out his instructions for determining meaning, it didn't turn out to be so. . . .For Quine (at least until *Pursuit*), both meaning and knowledge were based on "evidence" which we could at best infer was caused by rabbits, etc. If all rabbits disappeared, but our stimuli stayed the same,

our rabbit sentences and rabbit thoughts would (according to Quine) remain the same; but what would they be about? ... The question is, is the interpretation of observation sentences tied to the distal stimulus *directly* (which it certainly was not in the early days), or to something else? My *interpretation* is that the answer is now murky. My guess is that Quine remains an internalist (read Cartesian); if one takes the distal stimulus seriously, one is some sort of externalist. This is the difference between empiricists like Quine and me – whatever that makes me.[145]

Davidson here nailed a crucial problem in Quine's theory. Quine sometimes identified observation sentences with occasion sentences and sometimes not. He claims that observation sentences provide the evidence for science, but he also denies that they refer to anything, at least initially. Apparently they acquire reference gradually, becoming fully referential only with the acquisition of the relative clause. Just what this gradual process is is far from clear. After the 1986 Stanford meeting, Quine rejected the notion that observation sentences refer to the distal stimulus, but what they do refer to is not clear.

I stressed earlier the importance of Quine's concept of observation sentences. Quine dwelt on this in "In Praise of Observation Sentences," in the *Journal of Philosophy.*[146] He refers to the dispute within the Vienna Circle as to what constituted a "protocol" sentence, with Neurath on one side and Schlick on the other. Quine says that observation sentences are the starting point for knowledge. To the objection that an observation sentence such as "It's raining" assumes that the child already knows what rain is, he replies that observation sentences are bound by conditioning to neural input and require no prior knowledge of the world. Thus the child can acquire such sentences without prior knowledge of language. This point is important for Quine in explaining how first words are acquired. Observation sentences, Quine asserts, are occasion sentences; they are a verbal response to a particular neural intake which all competent witnesses will agree is correct. The observation sentence is Janus-faced, Quine says,

It faces outward to the corroborating witnesses and inward to the speaker ... It faces outward to its subject matter and inward to the range of neural intake that is keyed to trigger it. As a response to neural intake, the sentence is holophrastic: the neural intake is keyed to the sentence as a monolithic whole ... In its outward orientation to its subject matter, on the other hand, the sentence figures not holophrastically but piecemeal, word by word. Piecemeal is how the sentence relates to scientific theory, where its words recur in new combinations and contexts.[147]

The words of the observation sentences, Quine argues, are theory laden, and fortunately so because that is how observation is related to theory. "Bridge laws," as Reichenbach and Nagel defined them, are unnecessary; the common vocabulary of observation sentences and theories provide the required connection. And because they are tied to theory, they can be implied by theory, and so can serve as evidence for the theory. Quine uses this point to attack the notion, propagated by Kuhn and others, that different scientific theories are incommensurable. But, Quine replies, observation sentences are free of the underdetermination of translation since they are holophrastically conditioned to neural intakes. They are thus shared checkpoints for multiple theories and "should afford traces of commensurability insofar as the two theories are under empirical control at all."[148]

What comes through here better than in some of his other writings is Quine's insistence on an unmediated contact between the language learner and the external world. This is why he could not accept Davidson's view that observation sentences refer to the distal stimulus. What I think originally drew Quine to phenomenalism was the imperative of empiricism as he saw it that the learner have direct contact with the world. When Quine said at the Stanford Conference "stimulations are my phenomena," he was right. Like a phenomenalist, Quine holds everything beyond our neural receptors to be theoretical. Where he differs from the phenomenalist is in externalizing the phenomenal into the stimulatory. Of course the learner is not aware of his nerve excitations – no one is – so when Quine says observation sentences are conditioned to neural input, he is claiming that the learner can learn to associate the sentence with a scene. This is why he often describes the scene as rain or cold to avoid imputing reference to an object. This desire for immediate contact with the world often leads Quine to confusing occasion sentences with observation sentences. But in delaying the learners full recognition of objects until he masters the relative clause and the categorical, Quine delays it beyond plausibility.

Quine rejects the phenomenalist's epistemological program of the rational reconstruction of science, but he suggests that a physicalist reconstruction might be useful in riding science of what Eddington called "put up jobs" – parts of theories that can be dropped without being missed. Quine further holds that in the caregiver's teaching of language to the child, there must be empathetic ascription of perception to the child. Hence "I see the verbal ascription of perceptions as the primitive idiom of propositional attitude." Moreover, Quine notes the relation of observation sentences to ontology.

> The association with neural intakes, being holophrastic, is unaffected by any reassigning of objects to the terms involved. But also the logical relations of implication that connect scientific theory with observation categoricals are unaffected by one-to-one reinterpretation of terms; all that matters to logical structure is identity and difference. We conclude that the sensory evidence for science is indifferent to what things science says there are, so long as identity and diversity among them is preserved.[149]

Such ontological hi-jinks may seem reasonable with respect to abstract objects, but Quine says it is hard to take with respect to ordinary bodies.

> We were born body-minded. Natural selection implanted this bias in our innate standards of perceptual similarity, enhancing the salience of bodily contours and configurations. Bodies are our prototype for all reification, all existence. We could never, either as a race or as growing children, have worked our way up to objective reference without reifying bodies and substances, or something pretty much to that effect. But now we see, reflecting on the avenues of scientific evidence, that one could *retrospectively* swap even the bodies for arbitrary other objects without violence to the sentences affirmed in the theory or to the evidence for them, if there were any point in doing so.[150]

To this paper, Quine added an appendix on "Neural Intakes." Davidson had argued that the stimulus for perception is the distal stimulus. Quine opts for a proximal one. His reason is what he calls the "awkward heterogeneity" of the distal scene.

What matters, he holds, are the causal lines reaching the perceiver, and these are best described at the sensory surface. Current neurology shows that the processing of this intake is so complicated within the organism that its description there would be impossibly complex. Hence he takes the neural intake at the point where it strikes the perceiver's surface. Quine then compares his "global neural intake" to Carnap's "global experience" in the *Aufbau*. "It is uncanny," Quine says "how much better Carnap's thoughts are suited to the physicalistic, naturalistic stance than they were to his own phenomenalistic stance."[151] It is uncanny how closely Quine follows in Carnap's footsteps.

In 1994, Quine published "Promoting Extensionality" in *Synthese*.[152] The most interesting part of the paper is the section on propositional attitudes. When an ascriber attributes such an attitude to a subject, and speaks for the subject, the ascription is *de dicto*. But if the ascriber speaks for himself, it is *de re*. Thus in Quine's example

(1) (∃x)(Ralph believes that x is a spy)

where "x" refers to "Orcutt," we have a case of illegitimately quantifying into the belief context from outside. What is taken as *de dicto* or *de re* is the term or variable in referential position. Quine says that in 1956, he proposed analyzing the sentences in terms of quoted predicates; thus

(2) Ralph believes "is a spy" of Orcutt.

The referential position here is *de re* But making sense of (1), "which suggests that some trait or circumstance, some description, singles out the suspect in Ralph's mind" won't do, as Robert Sleigh's example of the shortest spy shows. Quine remarks "I now despair of a coherent theory of propositional attitudes *de re*."[153] The *de dicto* cases he can deal with by devices already covered, but the *de re* cases must be abandoned.

The other major threat to extensionality, Quine says, is modal logic. As before he rejects the notion of necessity and so all of the modal operators. But the problem of counterfactual conditionals cannot be so obviously dismissed. Quine takes it to be "elliptical and occasion-dependent." "A universally quantified truth-functional conditional is again implicit, I think, but with a complex antecedent some of whose clauses are left tacit, to be divined from the context and circumstances." Similarly, Quine takes disposition terms, not as involving possibility, but "that they just name ordinary properties in a special way, namely, by alluding to a fairly dependable and convenient symptom or test."[154] What they are symptoms of is the underlying microphysical state.

In 1995, Quine published "Naturalism; or, Living Within one's Means."[155] The article repeats much of what we have already covered, but Quine's remarks on mathematics, logic, and truth are worth noting. He says: "The basic laws of logic are internalized in learning the use of logical particles." Quine even suggests that logic so learned may be called analytic.[156] It is worth noting that C. I. Lewis had claimed that the laws of logic were true and analytic because they were determined to be so by the meanings of the logical constants.

The article contains a particularly full statement of Quine's view of mathematics:

> That leaves open the vast proliferation of mathematics that there is no thought or prospect of applying. I see these domains as integral to our overall theory of reality only on sufferance: they are expressed in the same syntax and lexicon as applicable mathematics, and to exclude them as meaningless by an ad hoc gerrymandering of our syntax would be thankless at best. So it is left to us to try to assess these sentences as true or false, if we care to. Many are settled by the same laws that settle applicable mathematics. For the rest, I would settle them as far as practicable by considerations of economy, on a par with decisions we make in natural science when trying to frame empirical hypotheses worthy of experimental testing.[157]

Science, as Quine conceives it, is a search for truth, but not just in the disquotational sense. Rather, truth is "an ideal of pure reason, in Kant's apt phrase, and transcendent indeed."

At the end of 1995, Follesdal wrote to Quine:

> It is . . . not enough to know what sentences the other assents to. It is crucial that we be able to find out what objects are referred to in these sentences and how the other's objects correlate with our own. Learning a language therefore depends on *de re* propositional attitudes. Reciprocally acquiring a language and using it to communicate consists in large extent in being better able to [do] the two key tasks that are required for mastering *de re* propositional attitudes: getting into the other's perspective on the world and its individuals, and correlating this to one's own. Mastering *de re* propositional attitudes and mastering a language are therefore inseparable. *De re* propositional attitudes are not useful solely for giving security agents a lead; they are vital for giving one another messages.[158]

Presumably the comment about security agents refers back to Ralph's problem with spies. Quine says that he had overlooked the point that "the propositional attitudes needed in handing language down have to be *de re*." But this has important implications for his theory of language learning. Quine had previously rejected propositional attitudes *de re*. He now found that he needed them, but just how he was to do this he does not say.

In 1996, Quine was eighty-six, but he was still hard at work and still writing. That year he published "Progress on Two Fronts."[159] The first front is the problem of translating between the jungle native and the linguist. Quine notes that in WO he accorded "sameness of stimulus meaning of the native's sentence for the native and of the English sentence for the translator." He took this as requiring homology of receptors. But Quine says he was uncomfortable with this theory. "How does mere sameness of distal cause, the jointly observed object, prevail over the proximal segments of the causal chain, inside the two observers, and still issue in agreeing response?" Quine says he got it wrong again in *Roots of Reference*, and did not finally find the answer until *From Stimulus to Science*. What he needed was harmony without interaction, and he found it in a pre-established harmony due to natural selection. The same cause serves to account for induction. "Natural selection has accordingly favored innate standards of perceptual similarity which have tended to harmonize with trends in the environment." "Natural selection," Quine says, "is Darwin's solvent of metaphysics."[160]

In the spring of 1996, Quine wrote out some of his thoughts on innatism in a document he entitled "The Innate Foundations of Endowments."[161] He poses the issue as Descartes versus Locke, and comes down on Descartes' side. Quine notes some of the responses of neonates, such as identifying different perspectives of objects as all aspects of one object, and then turns to induction. Since children can learn an action by reinforcement, they must have an innate proclivity to expect that under similar conditions, similar stimuli are followed by similar results. All learning, from humans to goldfish, has this character. Hence, Quine says, they have to have a hard-wired similarity metric, though one that can be further developed by experience. The similarity involved here, Quine holds to be perceptual similarity, as defined in *Roots of Reference.* He believes it testable in the form "*a* is more similar to *b* than to *c*." Why is the child so endowed? Quine's answer is natural selection; the reason we are so endowed is because all those not similarly endowed perished before they could reproduce. This of course does not explain how the endowment originated; that is presumably due to a prehistoric mutation. It only explains why, given some such endowment, we have it in the form we do. And that in turn assumes that the world presents regularities that are similar to our expectations. Descartes explained this by the goodness of God; Quine by the cruelty of natural selection.

> Natural selection also has an uncanny way of clarifying and integrating other familiar points of traditional philosophy too, quite apart from final cause. One such matter is induction. Another is Descartes' doctrine of innate ideas. A third is Leibniz's doctrine of pre-established harmony. Another is the Protagorean doctrine of Man as the measure of all things, and therewith the issue of realism vs. idealism.[162]

"Natural selection [is] the great solvent," Quine writes. He thinks it eliminates final cause by reducing it to efficient cause. This is of course true of the classical arguments from design, but it has to be further explained with respect to human conduct. It dissolves realism and idealism into each other; Quine writes "similarity is projected on the world from our associations." And what of the pre-established harmony? Once again, natural selection to the rescue; if our innate proclivities did not correspond to nature's regularities, our ancestors would have died out and we would not exist. One cannot avoid the feeling that Darwin has become a bit too convenient for Quine.

In 1995, Quine published his last book, *From Stimulus to Science*.[163] The book grew out of lectures that Quine gave in 1990 in Catolonia. Uncharacteristically, it begins with an historical sketch of the problem named in the title from Thales to Carnap. He gives an appropriate nod to Plato and Aristotle, skips the middle ages, acknowledges the Bacons, Roger and Francis, and then takes up Hobbes, whose view of knowledge he finds "strikingly modern." Descartes is treated next, then Locke, Berkeley and Hume. John Horne Tooke, a Quine favorite, comes next and then Bentham, Boole, and then, with nods to Frege, Peirce, Peano, Dedekind, and Cantor, we reach Whitehead and Russell, and from there to Carnap, who receives by far the most attention. There is no mention of Kant, or of any of the post-Kantians.

More surprisingly, there is no mention of Mill. But the history is focused on contributions to linguistics and logic, and reveals Quine's view of the significant figures who have pioneered on the path he has chosen to follow.

The most interesting of these sketches is the one of Carnap, which focuses on the *Aufbau*. Carnap was inspired by Russell's *Our Knowledge of the External World* which proposed "the explicit construction of the external world, or a reasonable facsimile, from sense impressions, hence from simple ideas." Russell did not attempt to carry out this proposal; Carnap did, and the result was the *Aufbau*. Quine describes in brief how Carnap undertook to do this, using logic and mathematics and one two-place predicate "remembering as similar." For basic elements he took "the individual's total experience at the moment, or perhaps during the specious present." Thus he started his project from a phenomenalistic basis – one that he later abandoned for physicalism. Quine sees his own naturalism as following in Carnap's footsteps, for "it is the rational reconstruction of the individual's and/or the race's actual acquisition of a responsible theory of the external world."[164] "Such is my option." He then says that "we can imitate the phenomenalistic groundwork of Carnap's *Aufbau*." Quine's "physical analogue" of Carnap's basic elements is "the temporally ordered class of receptors triggered during that specious present," which he terms a "global stimulus." Similarity of such global stimuli Quine equates to this "receptual similarity," while Carnap's "part similarity" he equates to his "perceptual similarity."

> Perceptual similarity is the basis of all expectation, all learning, all habit formation. It operates through our propensity to expect perceptually similar stimulations to have sequels perceptually similar to each other. This is primitive induction.[165]

And since it is the basis of learning, it cannot itself be learned, but must be at least partially innate. And of course Quine appeals to natural selection to guarantee that our standards of similarity match those of nature. Hence the pre-established harmony. He reiterates his view that dispositions are simply physical properties of the organism. The account of the observation sentence is Quine's standard one, and the interpersonal similarity agreement is again due to the pre-established harmony. A child learns sentences by ostension; he learns the particles "not," "and," etc., thus learning a bit of logic in the process. Then comes the "observational predication," but without reference or reification. The leap from observation sentences to observation categoricals is a major one, and Quine admits that he cannot explain it. But observation categoricals constitute a primitive theory of the world.

> At this point naturalism's modest simulation of an old epistemological quest is achieved, in a primitive way. We have a sketch of a causal chain from the impacts of rays and particles on our receptors to a rudimentary theory of the external world.[166]

Quine then introduces reification through what he calls "essential pronouns" – those that cannot be eliminated by substitution of the antecedent. This leads to variables and quantification; the "such that" construction follows with the relative clause and predicate abstraction or general terms. And Quine reiterates his slogan "to be is to be the value of a variable." But Quine then says, "long before reification in our clean-cut sense, bodies commanded the special attention of our remote ancestors as they do that of the modern man from early infancy." This is a bit misleading, for Quine does not attribute objective reference so early; our early perception of bodies is simply that of a repeatable similar "on a par with cold, thunder, and other unitary repeatable features of the passing show." Then, Quine holds, comes the discovery of the diachronic – of temporal sequence. Thus the child comes to recognize time and space and bodies continuous in both. Quine refers here to Piaget's work and to later unspecified work, but to what extent he is drawing on it is left unclear. But he is clearly drawing on Derek Bickerton's conjectures about the origin of language. Next, Quine admits abstract objects; the first such he thinks are properties, which he dislikes and converts to classes, then numbers, functions, etc. Physical objects no doubt are first thought of as "integral masses of matter," but Quine much prefers "the content of any portion of space-time, however irregular, indeed, however discontinuously scattered." But since there are cases where objects overlap, as do the states and counties of the United States, classes are necessary even in the "soft" sciences.[167]

Observation sentences, as Quine has defined them, can be learned in several ways; many are learned by ostension, but many are also acquired by verbal instruction. However learned, they are the checkpoints of science. In line with his holism, Quine emphasizes that a testable sentence is implied by a number of sentences jointly. This situation leads Quine to the following definition.

> A set of sentences that has *critical mass*, as we may say – that is, that implies some synthetic observation categorical – may be said to have those categoricals as its empirical content.[168]

It is of course observation sentences (observation conditionals and observation categoricals) that are the checkpoints for science, but Quine construes "science" broadly to include even history. That the normative aspect of epistemology is not lost in the turn to naturalism is made clear; it is concerned chiefly with the framing of hypotheses. Quine's maxim of minimum mutilation, which formulates his conservatism, and simplicity are norms that play this role; it is the art of framing worthwhile hypotheses, the technology of science.

Quine then turns to logic and mathematics. Implication, he says, "is the lifeblood of science ... it is what relates a theory to its checkpoints in observation categoricals." We learn logic, Quine says, in acquiring the use of the logical particles "not", "and", "or", "some", "every." Later we can formulate logic "in a complete formalization derivable from scratch in a couple of pages." But Quine holds that from this base we have predication, truth functions, and quantification.[169] Mathematics grades off from the purely formal to the applied where it merges into natural science, from which it acquires its empirical content. As for such fields as the higher reaches of set theory, since they share vocabulary with applied mathematics, they cannot be simply rejected. But the constraint of simplicity still applies. Godel has shown that any proof procedure will leave an infinite number of sentences of mathematics indemonstrable; what of these?

> I see nothing for it but to make our peace with the situation. We may simply concede that every statement in our language is true or false, but recognize that in these cases the choice between truth and falsity is indifferent both to our working conceptual apparatus and to

> nature as reflected in observation categoricals. It is like Kant's thing in itself, but seen as a matter of human usage rather than cosmic mystery.[170]

Quine then turns to "denotation and truth." He takes predicates as denoting every separate thing of which they are true. Two place predicates denote ordered pairs; n-placed predicates denote sequences of length n. But denotation can lead to paradox, as exemplified by Grelling's paradox. Unlike Russell's paradox, which it somewhat resembles, Grelling's paradox turns not on reification but on denoting; if heterological denotes itself, it is heterological only if it isn't. Tarski showed in 1935 how to resolve the problem. "First, we define denotation for each of the finitely many primitive predicates of the language, by disquotation." This gives an inductive definition for all predicates, primitive and definable in the formalized language. This language does not contain the predicate "denote." For that, we go to the metalanguage, where "denote" is defined as applicable to the object language. And so on up the meta... hierarchy. In non- formalized languages, one must simply beware the application, bearing the lesson of the hierarchy in mind. But if denotation is defined as applied to predicates of n-placed sequences, what of those applicable to zero length sequences? "In the zero case, there being nothing for the no place predicates to be true *of,* denotation reduces simply to truth outright." And truth, like denotation, must retreat up the hierarchy if we are to avoid paradoxes like that of the liar. And it is truth that science pursues.

> Science is seen as pursuing and discovering truth rather than as decreeing it. Such is the idiom of realism, and it is integral to the semantics of the predicate "true."[171]

Quine believes that he has dealt adequately with sameness of reference for bodies. But what of sameness of reference for abstract objects? Quine writes "I submit that intersubjective sameness of reference makes no sense, as applied to abstract objects, beyond what is reflected in successful dialogue." But, Quine says, this is also true of concrete objects once we leave ostension behind and talk of objects like elementary particles. The reason is proxy functions.

> Every object in the universe of discourse gets exactly one proxy, and no two objects get the same proxy. The reason such reinterpretations preserves truth values is that they preserve *sameness* of reference from mention to mention throughout discourse. Sameness of reference is what variables mark, and it is all that ontology contributes to science and truth.[172]

Here is ontological relativity with a vengeance, and Quine justifies it by saying "I conclude from [the indeterminacy of reference] that what matters for any objects, concrete or abstract, is not what they are but what they contribute to our overall theory of the world as neutral nodes in its logical structure." We can then, if we choose, "adopt" ostension as settling reference to observable concrete objects.

But now what of meanings? If we could define "alike in meaning," Quine says, we could, following Canap again, define meaning as the class of all sentences alike in meaning. For two occasion sentences, Quine defines sameness of meaning for a speaker at a time as "having the *disposition* to give the same verdict (assent, dissent, abstention) to both sentences on any and every occasion." As we saw above, for Quine a disposition is a physical property of a body. But now he says that "the subject *can* detect this state in himself and learn to express it as the two sentences having the same meaning". Quine adds "our healthy misgivings about introspective psychology must not lead us to underestimate this vital factor, whatever the obscurity of its neural mechanism."[173] For Quine, this is a startling exception to his vaunted behaviorism. And this is carried over to standing sentences when various pairs of words interchangeable in occasion sentences without disturbing sameness of meaning are taken as alike in standing sentences. This is however at best only a partial answer to the problem of sameness of meaning for standing sentences. Here the case of radical translation arises again, but Quine's answer this time is to use a bilingual; by doing so the problem of sameness of meaning reduces to that of a single individual. This he says works for occasion sentences but is problematic for standing sentences. But Quine claims that this in no way compromises the indeterminacy of translation. He moves from observation sentences to non-observational occasion sentences ("I have a headache") to standing sentences and analytical hypotheses. With all of these, the linguist tries to fit his translation to the native's beliefs, which he can guess, "since people are so much alike." But now Quine says that the linguist's goal is not translation but understanding or interpretation. Even such untranslatable sentences as "Neutrinos lack mass" are he claims interpretable. Just what this means Quine does not say but refers to Davidson as the source of the idea. But in the case of multiple sentences, non-observational occasion sentences are questionable, not to mention standing sentences.[174]

Quine then mounts his attack on mind. For him, there is no mind. Mental predicates should be treated as physical predicates applicable to the body. Quine is a monist who rejects Cartesian dualism. It is to physiology that we have to look for explanations of the so-called mental.

> The reduction of the mental to the physical. . .can be characterized in either of two ways: as *explaining* or as explaining *away.* There is no difference, but the first phrasing has a gentler ring. To have *repudiated* the life of the mind seems harsher than to have *explained* it in physical terms.[175]

Even Quine admits that we have at present no idea how to reduce the mentalistic predicate "thinking about Fermat's last theorem" to physiological terms. But having adopted Davidson's term "anomalous monism," he is content to assert that such a physiological explanation could be given had we "all pertinent information." Indeed, Quine goes back to Watson's thesis that we think partly with our muscles and that thinking is incipient speech. We can perceive another's unspoken thought by empathy, and Quine stresses the role of empathy in language learning. But "perceives that," like "thinks that," is a statement of propositional attitude which is intensional.

Quine's dislike of intensions is of long standing; he is an unwavering extensionalist. Intensions violate the substitutivity of identity and co-extensiveness. Quine is willing to accept propositional attitudes where they are construed as a relation between persons and sentences, where the "that" connecting "believes" (or thinks, etc.) with the sentence is taken as marking the beginning of a quotation. The sentences so quoted should then, Quine says, be spelled out in characters joined by concatenation. "Spelled out the words disappear and so do questions of substitutivity." The basic problem here is the confusion of the world of the person who has the attitude, whom Quine calls the "attitudinist," with the real world. Thus in

(1) (∃x)(Ralph believes that x is a spy)

the quantification lies outside the belief world but the variable it binds is inside. Compare

(2) Ralph believes that (∃x)(x is a spy)

where the "that" converts "(∃x)(x is a spy)" into a quotation and the quantification is within Ralph's belief world. In (1) the ascription is *de re*; on (2) *de dicto*. Quine accepts the *de dicto* attitudes, but not the *de re*. He says there are constructions that can render *de re* attitudes in external terms, but Quine does not do so here. There are other intensional idioms that Quine refuses to admit to science but that can be useful "outriders" – specifically the contrary-to-fact conditional. Modalities however Quine thinks we are better off without. But subjective probability he find acceptable when used with quotation in the manner of the propositional attitudes.[176] Finally Quine adds a very brief appendix on predicate functor logic.

From Stimulus to Science was Quine's last book, and it is clearly a summing up of his system. In the first chapter, he devotes more space to Carnap than to anyone else, but focusing on the *Aufbau*. In the appendix on neural intake that Quine added to "In Praise of Observation Sentences," he had remarked of Carnap's *Aufbau* "It is uncanny how much better Carnap's thoughts are suited to the physicalistic, naturalistic stance than they are to his own phenomenalistic stance." What is truly uncanny is how startlingly similar Quine's final system is to that which Carnap projected in the *Aufbau*. Both are purely structural systems. Both disavow ontology – Carnap because he thought the question one of metaphysics and therefore nonsense, Quine because proxy functions made external reference indeterminate. Both claim to preserve their empirical footing – Carnap through *foundedness* which he introduced as a concept of logic, Quine through the conditioning of observation sentences to stimulations. Was Quine aware of the similarity? Of course he was. Quine had always said that he had been Carnap's disciple in the 1930s and that the line of his own work had been in large part determined by Carnap's work. Moreover, in the parallels that he draws between Carnap's project in the *Aufbau* and his own attempt at a "rational reconstruction of the individual's and/or the race's actual acquisition of a responsible theory of the external world" one can see how closely he followed in Carnap's footsteps. Quine has been called the man who destroyed Logical Positivism; he might better be called the last Logical Positivist.

1997 brought the publication of "The Flowering of Thought in Language."[177] Quine opens the piece as follows

> Our first mental endowment is instinct. Then came thought, and later language. Thanks to language, thought then proceeded to flower. Such was our phylogeny. Ontogeny, then, true to form, recapitulates the sequence in the development of each child.[178]

This is Quine's most explicit endorsement of recapitulation theory, and it is surprising that even in 1997 he still held to such a theory that science has long since rejected. In fact, we know nothing whatever about the origin of language – not even when it developed or where. Furthermore, this is the only time that I know of when Quine said that thought preceded language. But he goes on to delineate a sequence of steps that may have brought us to where we are. There are few surprises in the description of the sequence, but Quine continues to draw parallels between phylogeny and ontogeny: "at some later point in the dim wastes of antediluvian time man achieved a more resounding breakthrough; and at some point our child recapitulates it, picking it up from us." This is the achievement of the observation categorical. And man, and child, march on until finally reaching science.

In 2000 – the year he died – Quine published an article entitled "Confessions of a Confirmed Extensionalist."[179] The article is interesting chiefly for the biographical material it contains. Quine wrote, "my first inarticulate hint of extensionalism may date from boyhood, when my liking for some Jewish schoolmates collided with someone's occasional derogatory remarks about Jews. I reasoned in effect that a class is to be evaluated, if at all, by evaluation of its members individually." He describes his love affair with PM at Oberlin. "I proceeded to Harvard for graduate work in philosophy because Whitehead was in philosophy there." When he went to Europe, he found that the economies over PM that he had achieved in his dissertation had "long been surpassed" by European logicians. Extensionality prevailed in Europe, but not at Harvard, where "Whitehead, Lewis, and Sheffer all swore by properties and propositions." With Carnap, Lukasiewitz, Lesnewski, and Tarski, extensionality was assumed "as a matter of course." Quine says he was an extensionalist at Oberlin and has been ever since.

Chapter 5
Conclusion

Quine has been hailed by some as the foremost American philosopher of the second half of the twentieth century. He is widely held to have refuted Logical Positivism, and to have established a number of skeptical theses, which may be summarized as follows.

1. There is no distinction between analytic and synthetic statements beyond our attitude toward them; no statements are true simply on the basis of the meanings of their terms.
2. The translation of a text from one language into another is not unique; there will always be alternative translations of the same text, and there is no fact of the matter as to which one is most accurate.
3. Scientific theories are underdetermined by empirical evidence. If we had a total scientific theory explaining all possible observations in the universe, there would be alternative theories that explained the same data equally well, and there would be no fact of the matter as to which one was the best.
4. Reference is inscrutable; one can change the ontology of any theory or language at will without affecting the truth of its sentences in any way; hence, there is no fact of the matter as to what a given theory or language is about.

There is nothing like a jaw-dropping skeptical thesis to excite the interest of philosophers, and Quine has contributed a full slate of such jaw-droppers. How did Quine arrive at these doctrines?

When Quine graduated from Oberlin, he already held some of the doctrines that would become the linchpins of his philosophy. The most important of these was behaviorism in the form given to it by Watson. Watson held that only behaviorism provided a scientific approach to psychology, and further that any doctrines that employed consciousness as an explanatory factor, or countenanced mental entities, were disguised forms of the psychology of the soul, and hence religious metaphysics. In accepting this view, Quine saw himself as a hardheaded empiricist and an opponent of any notion of a mental "substance,"[1] which he took to be another form of the psychology of the soul. It follows from this position that only what

M.G. Murphey, *The Development of Quine's Philosophy*, Boston Studies in the Philosophy of Science 291, DOI 10.1007/978-94-007-2424-2_5,

can be observed empirically can be real. Accordingly, Quine was strongly attracted to nominalism – that is, by the belief that only concrete particulars exist. He also considered talk of "ideas" and "meanings" as metaphysical nonsense, unless they could be behaviorally defined. Such a view made Kantian analyticity questionable, and led him to adopt an alternative definition that he derived from Lewis: analytic statements are those that we will never give up. But Quine was also a logician and a mathematician, and from Whitehead and Russell he had adopted the logistic thesis that mathematics is simply an extension of logic. Mathematics deals with abstract objects such as sets and functions. Empiricists have always found it difficult to justify mathematics, since its statements seem incapable of any form of empirical proof. Quine was confronted by this problem even before he graduated from Harvard.

Acting on Herbert Feigl's advice, Quine went to Vienna in 1932 and met the Vienna Circle. He also met Carnap, who was, he said, his greatest teacher. Carnap, whom Quine considered to be the embodiment of the Vienna Circle, had so profound an effect on Quine that for the next six years he considered himself Carnap's disciple. From Carnap, and other members of the Vienna Circle, Quine acquired a number of beliefs. Among these are (1) that philosophy should be the study of the language of science, (2) that all sciences are parts of a single scientific system of the world, (3) physicalism, (4) that philosophical controversy can best be dealt with through semantic assent, (5) holism, which Quine said he got from Carnap's *Aufbau*, (6) the distinction between cognitive and emotive meaning, (7) his great admiration for the *Aufbau.* He also derived from Neurath (8) the belief that he called "naturalism" – that we cannot transcend our own conceptual system but must work from within it to improve it.

His meeting with Carnap occurred at a time when the protocol debate was in progress between Neurath and Schlick, and Quine adopted the position that Carnap then held on that question. He was influenced not only by Carnap's *Logical Syntax of Language* but also, and more deeply, I think, by his *Aufbau*, which he took as a model of what a scientific philosopher ought to do. Quine himself has said that the line of his thought was in significant part determined by his agreements and disagreements with Carnap. One of these disagreements concerned Kantian analyticity. Carnap employed Kantian analyticity to guarantee the truth and necessity of mathematics. It took time for this disagreement to become clear; in his lectures on Carnap at Harvard, Quine employed both forms of analyticity – his own and Carnap's.

In Warsaw, Quine met Polish logic, and it must have come as a shock to discover how far ahead of his own work the Poles were. Although Quine never discussed this shock, his later letter to Tarski makes it clear that the Poles were far ahead of England and America. I think it is likely that Quine found in Tarski's doubts about the analytic/synthetic distinction a corroboration of his own. But he also learned Tarski's theory of truth, which he adopted as his own. And from this he drew the conclusion that sentences do not denote; their terms denote but the sentences themselves do not. It is the truth predicate that relates sentences to the world instead of denotation.

When Quine came back to Harvard as a Junior Fellow, he devoted himself largely to logic, and most of his writings of that period are on logical subjects. But he also lectured on Carnap's *Logical Syntax of Language*. The wine of logical positivism was running through his veins and it found expression in those lectures and particularly in the view of analyticity that he advanced. But the job he got at Harvard after his fellowship was as a logician, and he worked hard at it. Between 1934 and 1941, Quine published three books (SL, EL, and ML) and presented three logical systems (SL, NF, and ML). Productive though he was, his logical systems, like his dissertation, were all attempts at rewriting PM. It was PM that dominated his work in logic. Quine was a very good logician and expository writer, but he was not a great logician and he knew it. By the 1940s, his interests were shifting to semantics and ontology. He continued to work on logic, but the most important use he made of it was as a tool for the analysis of other subjects.

Nominalism is of course an ontological doctrine. To clarify just what the ontological commitments of a theory are, he coined the dictum that to be is to be the value of a variable. But could he be a logician-mathematician and a nominalist? Quine wrestled with the problem in the 1940s, and in 1947 he and Goodman published "Steps Toward a Constructive Nominalism." This work seems to have convinced Quine that he could not be a nominalist without giving up more mathematics than science would permit, and so he reluctantly abandoned the position.

His argument with Carnap over analyticity went public in 1940, but Quine did not abandon the notion; he had his own theory that he had derived from Lewis. But from the War on, Quine was thinking more and more about semantics, ontology, and epistemology. The position he held in 1943–1944 was that which Carnap had held when they had met in 1932. Ontology, Quine believed, was determined by science; what is real is what science says is real. Science, in which he included mathematics, is realistic. Epistemology, however, Quine called "idealistic" or "phenomenalistic." Quine clearly felt, as Carnap had, that an empirical epistemology must have direct contact with sensory experience. What he meant by calling it "idealistic" or "phenomenalistic" is that we project onto nature much of what we take as real. Secondary qualities, such as colors, do not describe the external world but are projected onto it by us. The same is true of similarity, which enables us to impose categories on continuously varying stimuli. Quine's view seems to have been phenomenalistic, as Carnap's had been in the *Aufbau*. But just how this phenomenalism was to be squared with his behaviorism and his realistic ontology Quine did not at that point know.

Carnap, and the Vienna Circle, had held that the Kantian analyticity of the statements of logic and mathematics gave them truth and necessity. Abandoning Kantian analyticity left Quine in need of a justification for these subjects. The logistic thesis guaranteed the truth and necessity of mathematics if the statements of logic were true and necessary, but without Kantian analyticity the best Quine could offer was that they are true and necessary because we think they are. Clearly, something more was required.

In the late 1940s, Quine had a number of problems that he was not able to solve. Abandoning nominalism left his ontological position unsettled; abandoning Kantian analyticity left logic and mathematics unsupported. In this situation, Quine began flirting with pragmatism as a possible way out. It was in these circumstances that he was invited by the American Philosophical Association to give a paper at the Toronto meeting. The paper he wrote was "Two Dogmas of Empiricism." His attack on Kantian analyticity was part of his ongoing argument with Carnap on the question and was aimed at Carnap's use of semantic rules to determine what sentences of his formal language were analytic. Quine did not mean that analyticity was indefinable, and he did not in fact abandon the concept. But in dealing with the second dogma, Quine espoused holism, which he stated in the paper that he had derived from Carnap's *Aufbau*; he did not know Duhem's work at that time. That auxiliary premises are involved in all scientific experiments and observations was not news; Lewis had discussed it,[2] and Cohen and Nagel had made a point of it in 1934.[3] What was new was the conclusion Quine drew from it – that statements cannot be tested singly and therefore that it is science as a whole that is at risk in any experiment. This claim badly overstated the matter as Quine later admitted, but if not all science, at least significant subsystems of science were at risk. He then declared for pragmatism, not only as a justification for logic and mathematical but for all science.

The response to this paper surprised Quine; more than that, it put him on the spot. He was far from certain about the claims he had made in "Two Dogmas," but the acclaim it brought him offered a strong incentive to push ahead on the lines he had laid out in the paper. Holism brought with it the underdeterminacy of science; if what is at risk in scientific experiments is the conjunction of the premises, then a negative result can be accommodated by rejecting any one conjunct, not necessarily the test hypothesis. Holism fits well with pragmatism, but it left Quine uncertain how to reconcile his realistic ontology with his phenomenalistic epistemology and his behaviorism. In the opening page of his 1952 *Methods of Logic*, Quine's phenomenalism is clear.

> The crucial point of contact between description and reality is to be sought in the utterance of a statement on the occasion of an experience which that statement utterance directly reports. The seeing of a green patch, and the simultaneous utterance "Green patch now," constitute the sort of composite event which, in its rare occurrence, gladdens the heart of the epistemologist.[4]

The relation of the statement to the observation here is phenomenalistic. But sometime in the next several years – probably in 1953 –he found a new way to solve his problem: he externalized his phenomenal report sentences by taking them, not as reporting observations, but at conditioned to the stimulations that cause the observations. And here his doctrine that sentences do not denote served him well. Sentences are conditioned to stimulations but they are not about stimulations. Since no one is aware of such stimulations of his nerves, this ploy allowed Quine to anchor observation sentences in the real world but to do so without having to have observation sentences refer to stimulations.

This move resolved some major problems for Quine. We know that our contact with the external world is through the stimulations of our nerves because science tells us so. Hence Quine can now take his epistemology as realistic rather than phenomenalistic since stimulations and conditioning are established by science. He could interpret the terms of his observation sentences as referring without having to claim that they are theory free; in fact, they are theory laden, and it is because they are theory laden that scientific theories can imply them. It also allowed Quine to redefine the function of epistemology. He could adopt what was all too clearly the case – that there is no justification for science beyond what science itself provides. Epistemology now becomes the scientific study of how we come to have the science that we do. There is no circularity here; one can apply science to the study of the acquisition of science since no claim of justification is involved. But what was such a study to be? Was it to be the historical study of how science has developed from antiquity to the present, or the study of how the child of our culture learns science? Quine chose the latter, which he sought to accomplish in WO.

There is an odd lacuna in Quine's theory of observation sentences. We are told that they are conditioned to stimulations holophratically, and that the sentences do not refer to the stimulations to which they are conditioned. It is the terms of the observation sentences that refer, and Quine claims that these terms provide the empirical content on which science depends. But nothing is specified concerning the relation between the stimulations to which the sentence is conditioned and the reference of the terms of the sentence. Suppose one were a Muslim child conditioned to utter "God is great" upon hearing the muezzin's call to prayers – a response to the correctness of which one's co-religionists would testify on the spot. Then is "God is great" an observation sentence? If not, why not? So far as I can see there is nothing in Quine's writings to rule out such a case. I think therefore that his theory of observation sentences is ω-inconsistent: if it covers the observation sentences on which science depends, it also covers a great deal more, some of which Quine would not approve.

Quine's theory of observation sentences solved the problem of how the "fabric" or "web" of scientific theory relates to what in "Two Dogmas" he had called "experience." The web of scientific theory includes both logic and mathematics. Distance from the periphery of the web to the center is taken to measure our degree of commitment to a sentence; hence he put logic and mathematics in the center of the web, since they are the sentences we will refuse to abandon, come what may. The structure of the scientific theory is supplied by logic which relates the sentences of the theory to each other and to the periphery – to observation categoricals. But Quine also came to the conclusion that logic and mathematics are separate disciplines – that is, he abandoned the logistic thesis. This move probably had several causes. One is the fact that there is no decision procedure for set theory whereas for logic there is: for the propositional calculus and for what Quine called "uniform quantificational schemata" there is a mechanical decision procedure. For general quantification theory there is no mechanical decision procedure, but any proof, once found, can be checked. Thus logic and mathematics (set theory) differ with respect to decidability. But I suspect a further factor was the development of alternative theories of number. Whitehead and Russell had defined numbers as equivalence classes; the number "n" is the class of all n-membered classes. But von Neumann defined numbers as sets of numbers; the number "n" is the set of all numbers preceding "n" and beginning with zero. These alternative definitions are equally adequate for mathematics. This had led Quine to the recognition that any progression will do for numbers, provided that it has a first member and that we stick to the same progression. Further, the way mathematics is derived from logic in the Frege-Russell tradition is by quantifying over predicate letters taken as bindable variables, but Quine considered these to be schematic letters that stand in for predicates and so are not bindable variables. Finally, Godel had shown that elementary number theory is incompletable, but that quantification theory is complete. All of these results suggest that set theory and logic are distinct. But Quine had taken Godel's incompletability proof as showing something further – that there is no line of separation between science and mathematics. By merging mathematics with science, Quine could endow applied mathematics with empirical content, and then allow non-applied mathematics to come into science on the coattails of applied mathematics. Mathematics therefore was part of science, but since Quine believed in the unity of science, he could hold that mathematics was analytic in *his* sense of analytic; it would not be changed because doing so would disrupt science as a whole.

That left logic. As Goodman pointed out, if as Quine held the structure of the scientific theory was provided by logic, then logic must have a different status than science. Quine tried many different ways to justify logic, none of which were satisfactory, and finally concluded that, after all, logic was analytic in the Kantian sense.

Quine defined epistemology as the scientific study of how we acquire science, and took this to mean how the child of our culture acquires science. But Quine, with his Watsonian ideas of thought and language, took this to mean how the child acquires the language of science. In WO he took the language of science to be first order logic. We have discussed in earlier chapters how Quine attempted to do this. He was not very successful. For one thing, Quine himself took epistemology to be a "chapter of psychology" by which he meant behavioral psychology. But Quine was not a psychologist. Worse yet, he undertook this task at just the time when psychology was leaving behaviorism behind and returned to the subject that had engrossed Locke and the Scottish Realists – human cognition. As Quine laid out what he thought was a logical picture of language acquisition, psychologists were hard at work studying the actual process by which children learn not only language but the character of the world around them. What psychologists found was not what Quine thought they ought to be finding. First, Quine overestimated the role of language. This seems like an odd claim, given the importance of language acquisition, but it is nevertheless true. Quine refused to grant the child objective reference – reference to things in the external world – until the he had acquired considerable linguistic sophistication. But psychologists have found that prelinguistic children refer, and furthermore that they see their world as one containing physical objects stably arranged in space and enduring even when occluded. But Quine, following Watson, did not grant such powers to children prior to learning language. Second, Quine thought that individuation was done *only* by language, but psychologists have shown that prelinguistic infants see a world of individuated objects.[5]

There is an irony here. Philosophy has been the mother of science. Many disciplines that we now recognize as scientific began as philosophical fields that broke away to become sciences. Psychology and cosmology are well known examples. Quine was quite right in his claim that epistemology is now a part of science – Cognitive Science. But Quine refused to recognize Cognitive Science as the new scientific epistemology. Instead he tried to carry out his epistemological investigation in terms of philosophy and the defunct psychology of behaviorism.

A further peculiarity of the book is its treatment of translation. Quine intended the gavagai argument to show the relativity of reference. But his readers saw it as showing the indeterminacy of translation, which it also did. The celebrity achieved by "gavagai" and "undivided rabbit parts" was astonishing. Quine recounts "a surprise I had in Spain: this young couple coming up a street in Granada, and the girl had two issues of a semi-annual journal on speech on philosophy (sic) of language entitled '*Gavagai*'. I hadn't even heard of it."[6] WO does of course claim indeterminacy of translation, though that is not what Quine intended to be its main point. Quine's real point in the argument was that there are no meanings, at least in any traditional sense, and that *reference* is indeterminate. Here again behaviorism determined Quine's conclusions – conclusions that non-behaviorists regard as bizarre.

There are a number of further problems with Quine's account. I have noted already Quine's inability to provide an adequate account of how the linguist learns the jungle terms for "assent" and "dissent." Equally suspect is Quine's claim that the child first learns sentences rather than words and that these do not refer to the world. There is, so far as I can determine, no evidence supporting these claims.[7] Infants acquire a "receptive vocabulary" – that is, words understood, before they produce words, and the words they do produce usually involve reference to things in the child's environment. Thus, "juice" is apt to be a request for juice, which obviously refers to something in the environment of the child. Quine here is imposing his philosophical doctrine on the infant. The denial of reference to the child's speech is carried to unjustifiable extremes; Quine seems determined to preserve his dictum that to be is to be the value of a variable even in discussing infancy. His tendency to identify language with thought (even though he admitted that they are distinct) and his denial of referential capacity come very close to denying that the prelinguistic child can think at all.[8]

Translation as Quine describes it depends upon the acquisition of "stimulus meanings." This requires homology of sensory receptors between the native and the linguist for which Quine presents no evidence. Further, once the level of stimulus meanings is transcended, we have to do with "analytical hypotheses" which are only minimally constrained by empirical data. Hence Quine repeatedly says that the translator will impose his own ontology on the native in the act of translation. If this were so, it would be impossible for us to discover an alien ontology if the native had one. But an examination of the anthropological literature will show that anthropologists have reported an astonishingly wide range of ontological beliefs in the different cultures of the world. Quine's claims for linguistic imperialism do not fit the data.

Quine does deviate from his behaviorism on one point; he endows the child with an innate similarity metric. Were this not done, there would be no way to explain how the child can learn different colors, or more generally impose categories upon continuously varying stimuli. This is no longer Skinner's empty organism. In subsequent writings, Quine has to endow infants with more and more innate equipment in an effort to make his claims plausible.

Finally, one of the most implausible claims in the book is that the indeterminacy of translation applies within the home language. This is taken to mean that members of the same speech community face indeterminacy of translation between idiolect and idiolect. Quine later abandoned this claim, but the fact that it was made at all illustrates very well his tendency to follow a doctrine once adopted down the barrel of a canon.

Having failed to get across his point about the ontological relativity in WO, Quine tried again in "Ontological Relativity." This paper marks the first use of proxy functions in Quine's work, though it took some time for him to develop the full implications of this argument. Quine treats language here as if it were a formal theory, and so argues that the reference of terms in the language can only be determined in a metalanguage. His examples are those of deferred reference, the proxy functions being one example. But although Quine was not entirely clear himself at this point, the inscrutability of reference due to proxy functions is a new argument, and is not the same as ontological relativity which rests on the indeterminacy of translation.

Quine returned to the problem of language learning in *The Roots of Reference* in 1973. His doctrine is the same as in WO, but it is now presented with greater psychological detail. The point of the book is to isolate the roots of reference, which Quine claims to be the categorical and the relative clause; the relative pronoun appears in the role of the variable of quantification. Quine holds that the child cannot refer to objects in the external world until this level of linguistic sophistication is reached, and he does this although he notes an article reporting studies on neonate cognition that directly contradict his claim.[9] His way out of this bind is to say that infants are innately programmed to perceive "bodies" but that these are not taken by the infant as external objects. The implausibility of this claim is clear when one considers at what age the ordinary child masters the relative clause. If he goes to a good school, perhaps by age ten. To hold that the child of nine cannot refer to an object in its environment is just absurd.

Why does Quine devote two of his major publications, one of which was his Carus lectures, to trying to impose a behavioristic scheme on the child's learning of language? I think that the *Aufbau* always remained for Quine a model of the sort of project that a scientific philosopher ought to attempt. He knew that the construction of the world from sense data that Carnap had attempted had failed, and as Carnap himself admitted, was impossible. But Quine undertook as his project "the rational reconstruction of the individual's and/or the race's actual acquisition of a responsible theory of the external world" – a project modeled on Carnap's. One should note the word *actual*. Although in both WO and RR Quine refers to his construction as "speculative," he also claims that it is an accurate depiction of the child's learning of language in "skeletal" form. He could not do it; his skeletal outline of the child's learning process was full of "leaps" and gaps (some based on analogy[10] and some unexplained) and unsupported claims. He was not a psychologist and he refused to discard his outworn behaviorism when the field of psychology moved on.

One of the most striking features of RR is that as Quine told his tale of the child's progress to the language of science, he is forced more and more to use mentalistic terms in order to say what he wants to say. Behaviorism was simply not adequate for Quine's needs. One might have thought that this would lead Quine to alter his behaviorism or abandon it, but in fact the opposite occurred. Quine claimed that language was a matter of dispositions to verbal behavior. He sought to overcome his problem by reinterpreting dispositions, dropping the implied possibility contained in the "ible" and "able" endings, and taking the terms themselves as referring to neural microstructures that cause the behavior. Disposition terms thereupon became physicalist terms; so did mentalistic terms, reinterpreted in the same way as physical. And this led Quine to deny the existence of mind. What had previously been taken as terms for mental states he now claimed referred to physical states of the body, and mind simply vanished. Quine's behaviorism won out.

As this should make clear, Quine's concept of "mind" was not at all that of psychologists. Quine saw the word "mind" as referring to a non-physical substance; it was "mind" in the sense employed by Idealists of an earlier era, or more recently Ryle's "ghost in the machine" made substantial.[11] For Quine, it was what Watson had called "soul". In getting rid of it, Quine reaffirmed his physicalism and materialism, and his theory of the categories.

For Quine had a theory of the "categories," as he called it, but not a traditional one. What Quine meant by this is that he recognized as real only what could be stated to be in terms of first order logic. This is the doctrine that underlay his repeated attacks on modalities, his rejection of propositions as objects, his refusal to admit intensions and mentalistic terms. Quine wanted the language of science to be as austere as possible; he would have liked to have barred all abstract entities, but found that he could not do without set theory if he was to keep science. For him, it was science that determined ontology, and that meant only those things that could fit in his model of the language of science could be real.

Of course this created problems with the abstract entities of mathematics. Quine had a profound distrust of set theory. In the opening page of "Steps Toward a Constructive Nominalism", Quine and Goodman wrote "We do not believe in abstract entities." Fundamentally, they say, this distrust of abstract entities "is based on a philosophic intuition that cannot be justified by appeal to anything more ultimate." But they go on to present what they consider supporting evidence for this view; the paradoxes of set theory.

> What seems to be the most natural principle for abstracting classes or properties leads to paradoxes. Escape from these paradoxes can apparently be effected only by recourse to alternative rules whose artificiality and arbitrariness arouse suspicion that we are lost in a world of make-believe.[12]

Quine would have been delighted if he could have done without set theory entirely. Repeatedly he said that he hoped that a way could be found to rid science of large parts of classical mathematics. There are no sets in quantification theory, and it was quantification theory that provided the categories. But he could not do it: mathematical entities remained integral to science.

There was one class of mentalistic terms that Quine could not fit into his scheme – propositional attitudes. He tried a variety of ways to solve this problem. Finally, he decided that he could accept propositional attitudes *de dicto* in the form "x believes that p", where "p" stands for a sentence that he could render by spelling and so without reference, but he could not find a way to accommodate propositional attitudes *de re.* The trouble was that he needed *de re* propositional attitudes for language learning. This problem he never solved.

Quine was also worried about the problem of how different individuals could have the same or similar stimulations. He tried a number of ways to solve this problem without success until at last he hit upon the idea of a pre-established harmony due to natural selection. As noted earlier, as Quine grew old and time grew short, he turned increasingly to Darwin for help.

Quine believed that science determined ontology. But as noted earlier, he became convinced by quantum mechanics that the notion of physical object was no longer tenable. This forced a drastic revision of his philosophical system. After Quine discovered his theory of observation sentences, he had defined epistemology as the science of science – the scientific study of how we have acquired science. Two of his major works – WO and RR – had been devoted to showing how our children came to master reference to physical objects. But abandoning the notion of physical objects, and adopting the theory of proxy functions, led him to redefine epistemology as a theory of evidence for science. He now claimed that science consisted of interlocking sentences held together by logical bonds where the variables are simply neutral nodes, and where some sentences are conditioned to stimulations of our senses. But it is very hard to see much content in this claim. That science consists of sentences bound to each other by logical connections and some of which refer to experience is so general a claim that it is almost vacuous. It is a claim none would deny because it says almost nothing

Quine said that it was findings in quantum mechanics that required the abandonment of physical objects. But he abandoned them only in epistemology. He claimed that science determined ontology – that what is real is what science says is real. If quantum mechanics required the abandonment of physical objects, why did it not do so in ontology? Quine claimed to be a naive realist in ontology, accepting the reality of ordinary physical objects. It is far from clear how he could hold that the finding of quantum mechanics apply only in epistemology and not in ontology.

Yet these remarks miss what seems to me to be the central thrust of Quine's work. To put it simply, Quine took Carnap's *Aufbau* as his model. He knew that the rational reconstruction of the world from a phenomenalistic base was impossible, as Carnap had also concluded, but he wanted instead a rational reconstruction of the language of science and of scientific theories that would achieve some of the same goals. His starting point was Carnap's view of epistemology and science when Quine had spent thirty-seven days with him in Prague. In one sense, Quine always remained a phenomenalist; his view of external objects as "posits" or "myths" is just what a phenomenalist would hold. By his ingenious trick of externalizing phenomenalism, his phenomena became stimulations. This allowed him to retain some of his phenomenalist doctrines but on a realistic basis, since it is science that tells us about stimulations. His rejection of Davidson's suggestion that one should start from the distal stimulus should be seen in that light.

As Quine developed his ideas from WO on, his views continued to change. The rejection of meaning, which is what the indeterminacy of translation was really about, his holism based on the underdetermination of science, the inscrutability of reference that was based on proxy functions, and the abandonment of physical objects, all led to a final view that, viewed epistemologically, science was a theoretical structure of interlocking sentences where the nodes (bound variables) were neutral as between different ontologies, but which rested on basic sentences conditioned to neural inputs. If one looks at the system Carnap created in the *Aufbau,* the similarities are striking. For Carnap too, the result of his epistemological construction was a structural system empirically grounded by "foundedness." Was Quine aware of the similarity? Of course he was. He and Carnap had followed different paths to their results, but as Quine said, the line of his thought had been largely determined by Carnap.

But what about the famous skeptical theses that Quine promulgated? Do they still stand, despite the problems in his system? Let us see.

1. The indeterminacy of translation. This thesis, as Quine himself has said, is a consequence of his behaviorism. But behaviorism is only one psychological theory, and not one currently enjoying high repute among psychologists. If instead one adopts a theory from Cognitive Psychology, Quine's thesis fails. Cognitive scientists have done extensive work on concepts, which many of them have taken as meanings. Eleanor Roach's work, published from 1973 on, advanced the idea of prototypes, which has been extensively studied ever since.[13] On this view, our linguist has a concept of rabbit and the native has a concept of Gavagai. How similar is the native's concept to the linguist's? This is a factual question, so there is a matter of fact as to which translation is most accurate.

One must hasten to point out that cross-cultural identity of concepts will surely be rare; the real question is about the degree of similarity. Every translator knows that exact translation between languages is virtually impossible. But not all translations are equal; some are more accurate than others.

2. Ontological Relativity. This thesis is a consequence of the indeterminacy of translation thesis, and fails if it fails.

3. The inscrutability of Reference. This thesis Quine based on deferred reference, as exhibited by proxy functions. It is well known that if a formal theory has a model, it has many models. It is also well known that the references of the terms of a formal theory cannot be entirely specified in that theory itself but only in a metatheory. Quine assumes that what holds for formal theories holds equally for natural languages. But does it? The answer is no. Quine assumes that the individuation of the world into individual objects is done by language. But there is compelling evidence that prelinguistic children perceive the world as largely composed of discrete and individuated objects.[14] Since this is true before the child has *any* language, it

is clearly an ability of human infants generally, – indeed, of animals generally – and therefore the linguist and the native can start from the recognition that each is observing a physical object. Further, when the child is taught by ostension that a word applies to a perceived object, the child takes it as referring to *that* object, not something else. But the heart of Quine's argument is the proxy function argument. Is it true that for any natural language, proxy functions can convert its ontology into an alternative one? In part, this depends on what we are talking about – the community of speakers of the language or the philosophers whose views are those of an elite set. Quine talks as though he is referring to speakers of the natural language in general. But if so, the claim fails. Take the case of ostensive reference. To fix ideas, let "A" be the speaker, let "x" be the object pointed at, let "y" be a different object that will serve as the deferred referent, and let "R" be a function from "x" to "y". Now in English, under what conditions is reference deferred from "x" to "y"? Assuming that the participants in the talk exchange intend serious communication, there are two: (1) where "x" cannot be the referent, then reference is deferred, and (2) where there exists a prior agreement or convention that determines the deferral of reference.

(1) It is a general rule for English that if "x" can be the referent, it must be. Consider the case where A points at a volume of Poe's poetry and says "He was a drug addict." Here the ostended object cannot be the referent, so reference is deferred. The referring function is that from books to authors, and the deferred referent is Edgar Allen Poe. But now consider the case where A points at the same book and says "That costs ten dollars." Here the referent must be the book because it can be. The principle involved is that if a thing can be the referent, it must be; if it cannot be the referent, then, assuming serious communication intended, the reference has to be deferred.[15]

(2) There are cases where by prior agreement, the referential principle just stated is abridged. The most obvious example is codes. During World War II, James Conant, then president of Harvard, received a phone call from Karl Compton, the message of which was "You'll be interested to know that the Italian navigator has just landed in the new world." Conant replied, "Were the natives friendly?" The answer was "Everyone landed safe and happy." What this exchange really meant was that Enrico Fermi had succeeded in building an atomic pile in Chicago – a sustained nuclear reaction – and that the results were as hoped.[16] But the agreement need not be explicit. In modern poetry, it is conventional that deferred reference is to be expected. Consider the following lines from Ezra Pound's "Hugh Selwyn Moberly:"

> There died a myriad, and the best among them
> For an old bitch gone in the teeth
> For a few dozen battered books.

The reader of the poem will know from the context that the first line refers to World War I by metonymy. The "old bitch" refers to England, and the "battered books" to English Victorian culture. But this sort of interpretative reading is largely confined to poetry and some fictional prose; it has little to do with ordinary life.

The alternative that Quine's doctrine is that of a small set of philosophers is true, at least for the unit set of Quine, but the argument would be flat out rejected by ordinary speakers of the language. For them, many of their words are linked to objects by direct ostension, and if a man points at a dog and says "That's a dog," and you reply that it is really the number eleven, you may find your sanity being questioned. For most people, the point of language is communication, and communication requires a shared, understood and fixed reference. Definitions do link languages to the world of perceived objects and events, and cannot be taken lightly as referring to something else. For formal theories, on the other hand, such switching of ontologies is possible, though it is not clear what that has to do with the problem Quine set out to answer.

It is of course true that for a natural language like English one could construct an isomorphic system, M, in which the referent of each referring expression in English corresponds to a different referent in M. But this does not affect the references of English expressions. The fact that such an ontological switch is *logically* possible does not render English reference inscrutable, since there are rules and conventions of English that prohibit such an alteration of reference. Constructing M might be an interesting technical feat, but the existence of such a system does not mean that the reference of English expressions could actually be deferred to those of M. It is possible to map the natural numbers (beginning with 0) into a denumerable series of nested concentric circles, but this does not mean that "2" refers to the third circle in the nest. The existence of such isomorphic systems has nothing to do with reference in the original system.

But it should be noted that there is an application of Quine's doctrine that he does not make. Nerve stimulations are shown to be real by our science. Suppose we were to apply proxy functions to them so that instead of stimulations we have psychedelic hallucinations. Then our observation sentences are conditioned to our visions. What sort of science would we have then?

(4) The underdetermination of science. Quine bases this claim on two arguments. The first, noted above, is that the testable consequences of a theory are implied by multiple premises within the theory; hence a negative experimental or observational result can be accommodated by rejecting any of the premises involved, and that what to reject is up to the investigator. This is true, but misleading as stated. When a scientist designs an experiment to test a hypothesis h, he employs as auxiliary hypotheses $a_1 \ldots . a_n$ only statements that he believes to be well confirmed – that is, much more probable than h. To do otherwise would be a mark of poor scientific judgment. Failure of the experiment to yield the predicted result will therefore lead to the rejection of h. But repeated failures of experiments in which a particular auxiliary hypothesis is involved may lead to new experiments to test the auxiliary hypothesis. To put it differently, going into the experiment, the scientist involved has a distribution of probabilities over the members of the set $\{h, a_1 \ldots a_n\}$ in which h has much the lowest probability. Failure of the prediction will therefore lead to the demise of h. But experimental outcomes can also affect the distribution of probabilities over the a's. If in a series of experiments in which a_i is a premise, repeated and unexpected failures occur, the scientist's probability assignment to a_i

may change and a_i may become the subject of a further experiment. It is also true that different scientists may have different probability distributions over the statements of the theory, and that may lead some to doubt what others do not. But the impression that Quine gives that the rejection of a premise is an arbitrary choice made by the defenders of *h* is quite false. There are indeed known cases where this has been attempted, the best known being the attempt by advocates of the phlogiston theory to explain away the gain in weight in combustion by claiming that phlogiston had negative weight. But what is striking is how rarely such events occur and how skeptically they are viewed by other scientists. The Fitzgerald contraction did not resolve the crisis in physics at the end of the nineteenth century.

Quine's other argument is the one from "Empirically Equivalent Systems of the World." As Quine himself said, this is a science fiction case, but thought experiments have their uses. What Quine supposes is the following situation. We have a coordinate system for the entire universe, so that to each and every point in the universe there correspond coordinates that fix the position of that point. Then Quine says "each observation [sentence] expressible in our language gets joined to each combination of spatio-temporal coordinates." This, Quine says, will constitute "all possible observation sentences." He defines a theory formulation as "a sentence – typically a conjunctive sentence comprising the so-called axioms of the theory." He then states his thesis: "our system of the world is bound to have empirically equivalent alternatives which, if we were to discover them, we could see no way of reconciling by reconstrual of predicates." This thesis admits of two versions which I have called the Weak form and the Strong form. The Weak form is: "suppose we have an infinite set of heterogeneous observation sentences and a finite theory formulation, T, from which all of the observation sentences of the set are derived. Then there exists an alternative theory formulation, T!, logically incompatible with T and such that it cannot be reconciled with T by any reconstrual of predicates, but from which all of the observation sentences are also derived." In this form the underdetermination thesis would surprise no scientist. He would simply point out that from the present empirical equivalence of T and T!(i.e., the fact that T and T! imply the same set of observation sentences), it does not follow that they will continue to be equivalent when future predictions are tested. Indeed, it is a truism that for any *fixed* body of data, alternative theoretical explanations can always be found; that is why scientists insist on prediction as the true test of any theory.

The Strong form of the underdetermination thesis is this: if we have a theory, T, that accounts for all of the observation sentences at every point in the universe of space-time – what Quine calls "all possible observation sentences" – then there will be an alternative theory, T!, that is empirically equivalent to T but logically incompatible with it, and there will be no reconstrual of predicates by which the two theories can be reconciled. (Quine later abandoned the requirement that the theories be logically incompatible in favor of Davidson's spelling trick.) What is different here is that *all possible* observation sentences are explained. That means that there can be no further predictions, no further observations; the data base is fixed and nothing further can be added to it. Under these circumstances, there is no fact of the matter as to which theory is right.

There are problems with Quine's assertion of the Strong form of the underdetermination thesis. As noted above with respect to the Weak form of the thesis, for any *fixed* body of data it is always possible to find alternative explanations. What the Strong form of the thesis does is to postulate a situation in which the total amount of data is fixed *and* in which no further predications can be confirmed that could distinguish between the alternative theories. It is obvious that the state Quine postulated is one which not only will not be realized but cannot be realized, but he does not suppose that it can – it is a science fiction state. But does this argument apply to science? It does not. The case Quine supposed is one that, even if it were realized, would prove nothing about *science*. *Science* is a process of inquiry by which we seek to discover truth. One can no more identify *science* with a particular scientific theory or a particular state of science at one time than one can identity music with Beethoven's fifth symphony. Quine's thesis concerns only a particular state, but one in which all science had ended, in which there is nothing more that *can* be discovered. In Quine's fantasy state, everything knowable is known, all observations are made, all experimental results are complete, and we have nothing better to do than to devise alternative ways of accounting for what we already know. In other words, Quine supposes a state in which the body of data is *fixed*, not just temporarily but forever; no further additions are possible. Inquiry here would not be science, for there is no truth left to be discovered. It is prediction and verification that define what Quine calls the "game of science"; without them there is no science. There could then be no decision among various theory formulations since no new predictions could be made or tested. In this situation, devising alternative theory formulations might be a pleasant pastime, rather like doing a cross-word puzzle, but it would not be science.

One should note here Quine's own statement regarding the relation of predictions to science.

> But when I cite predictions as the checkpoints of science, I do not see that as normative. I see it as defining a particular language game, in Wittgenstein's phrase: the game of science, in contrast to other good language games such as fiction and poetry. A sentence's claim to scientific status rests on what it contributes to a theory whose checkpoints are in prediction.[17]

Predictions in Quine's final system lose all point since their outcomes are already known; indeed, it is doubtful if they ought to be called "predictions" at all.

But there is a more fundamental problem. Quine specifies a state, K, in which all possible observation sentences "expressible in our language" are accounted for by the theories T and T! But in so specifying it, he makes K relative to a particular language of science. It is well known that the language of science changes over time as new discoveries are made, new instruments are created, and new terms are required for new results. Hence Quine's terminal state of science is fixed only relative to the specific language of science in use at K – L_k. But there is nothing in Quine's specifications to prevent someone at t_k (the time when K occurs) from constructing an instrument with which he is able to observe some phenomenon, z, never seen before. Since there is no term in L_k for the phenomenon z, he must coin a new one, and the addition of this term to L_k yields a new language L_{k+1}. There being no observation sentence in K mentioning z, at least one new observation sentence that does mention z must be added to K. But this contradicts Quine's claim concerning K that it contains all possible observation sentences, and shows that even if state K could be realized (which Quine does not think possible), it would not be the final state in which all possible observation sentences are accounted for. K is only final relative to L_k; it is not final relative to L_{k+1}. And of course the same process of discovery may be repeated in L_{k+1} to yield L_{k+2}, and so on.

The underlying problem here is Quine's attempt to limit the possible to the actual. He is not the first philosopher to try to do this; Peirce is a conspicuous predecessor. But it cannot be done. The actual cannot exhaust the possible. No matter how many observation sentences are accounted for at K, there is always another observation sentence possible that is not in K, whether it is accounted for by T and T! or not. Quine's rejection of modality leaves him no alternative but to try to translate the possible into the actual, but, as shown above, he fails in the attempt.

(5) The last of Quine's theses is the denial of Kantian analyticity, and the consequent claim that all sentences appearing in science are synthetic. If analyticity is defunct, there are two major issues that must be settled. The first concerns logic and has been described above. Quine's difficulty in finding a basis for logic led him to his final decision which is that the meanings of the logical constants are learned early and that they suffice to yield logic. What this does, whether Quine intended it or not, is to make all logical statements analytic in the Kantian sense; they are true by virtue of the meanings of the logical constants.

Furthermore, it is not clear that Quine's arguments are fatal to predicative analyticity. It is pretty well established in Cognitive Psychology that "bird" is a prototypical concept, and one of the features of the prototype in the case of "bird" is bipedality. Hence, the sentence "all birds are bipeds" is analytic.

Does this mean that Quine was wrong in his attack on analyticity? Yes and no. Quine's position rested on his behaviorism and the consequent belief that such mentalistic notions as "concepts" and "meanings" refer to nothing real. And in the 1940s, there was some ground for that: definitions of meanings and concepts were usually intuitive and vague, and the question of the nature of such notions was not viewed as empirically answerable. Quine's arguments in his lectures on Carnap assume that he can make any sentence analytic by an arbitrary redefinition. But what Quine failed to recognize is that "concepts" and "meanings" can be studied empirically. This has been one of the major thrusts of cognitive psychology. By the 1980s, the work of Eleanor Roach and others[18] had shown that such empirical study can be done, at least for some concepts. This is still a field of intensive research, but one result is that the prototypical character of some concepts, such as "bird," is well established. In this case, the establishment of bipedality as a characteristic of the prototype is now taken as empirically proven. Hence, one can say that "all birds are bipeds" is analytic. Quine was still very active in the 1970s and 1980s. If he did not accept this work, which was very widely discussed, it was because he did not want to accept it. So the situation is that for some concepts, we have definitions that are empirically based and well confirmed, but not (yet) for all. One can say that the idea of analyticity is sound where we know enough about the concept to say, on empirical grounds, what its components are, and that the number of such cases is certain to grow in the future.

It has been objected that there are alternative theories in Cognitive Psychology, such as Connectionist Theory, in which "All birds are bipeds" might not be analytic. I have not found in the connectionist literature a clear enough way of defining particular concepts to be able to say what the answer to that may be. But suppose the concepts of "bird" and "biped" turn out to be particular states of neural networks. Then the question of the status of "all bird are bipeds" will be determined by the relation between those network states. But that will be an empirical question, so there will be a fact of the matter. One day, I hope, we will know. It is also clear that some concepts are not prototypically structured; numbers are a good example.[19] But there is a good deal of evidence to support the claim that some concepts, including that of "bird," are prototypically structured.

The issue then is mathematics. Quine claims that all statements of mathematics are synthetic, that their empirical content comes from their applications, and that non-applied and non-applicable mathematics are accepted on "sufferance." This is thoroughly unsatisfactory situation. These claims are inconsistent with the way mathematicians think about their work and they leave unexplained why pure mathematics should be done at all. It seems to me that the question of the analyticity of mathematics ought to be reassessed. Lewis held all mathematics to be a priori analytic, but that any given mathematical theory may be replaced by some other mathematical theory on a pragmatic basis. The withdrawal of Euclidean geometry as a description of real space occurred because Riemannian geometry fit Relativity Theory better. Lewis' theory in fact gives a more realistic account of when and how mathematical theories are applied or rejected than does Quine's. The important thing is to recognize that this question is open, rather than being closed as Quine thought.

What then is one to say of Quine's "system"? It is a curious mixture of scientific realism and an externalized phenomenalism. Quine has little to say about science itself; he takes it for granted, except when he says that it requires the abandonment of physical objects. He was fully prepared to accept changes in physics, even though they required major revisions of his theories, but he refused to accept changes in psychology even when most psychologists abandoned behaviorism. The "action" for Quine is in epistemology. And the output, Quine said, was a theory of "evidence." But what is a theory of evidence? Since *anything* can be evidence for *some* hypothesis in some science (broadly construed, as Quine does), and since Quine's holism requires that "evidence" be related to theories, or sets of hypotheses, all one can really say at this level of abstraction is that the evidential statements must be empirical and must be implied by the theory. Quine's "theory of evidence" is nearly vacuous. The interesting questions here concern the verification of the testable consequences. Despite Quine's talk of occasion sentences, observation sentences, etc., he really tells us very little about how they are tested. We know that these sentences must be holophrastically conditioned to stimulations. But we know surprisingly little about these stimulations; as my example of the Muslim child indicates, there appear to be no constraints on what the causes of these stimulations must be. But we do know that observation sentences are not about stimulations. We are told that the terms occurring in observation sentences are theory laden, but not how they relate to sensory experience. Quine defines observation sentences as those upon the correctness of which all witnesses will agree on the spot. My example of the Muslim boy shows how dubious that definition is. His final definition is that a sentence is observational if it could be learned by ostension. But ostension depends in part upon the knowledge of the person to whom the ostended object is being pointed out. At one point Quine says that someone being a bachelor is an observation sentence for one who knows the individual's marital status. There are questions about the verification of observation sentences that are obviously important for a theory of evidence. The relations between occasion sentences and observation sentences needs clarification. Observation sentences are corrigible according to Quine, but the status of occasion sentences is unclear. At the Stanford conference, Quine said they were verified by what they report and so appear to be incorrigible. Surely a theory of evidence for science ought to deal with such matters. Instead what we have is a very general outline of a coherent system of sentences, some of which are conditioned to some sort of stimulations. If this is the net outcome of Quine's labors, his contribution to our knowledge is disappointing slight.

Why is Quine's contribution so slight? One reason was his stubborn refusal to abandon behaviorism when it became passé in psychology. Quine knew something of what was happening in psychology after 1960, but he refused to modify his position. As Gibson pointed out, his behaviorism is the Achilles heel of his philosophy[20] and Cognitive Psychology has furnished Paris' arrow. A second reason was his overestimate of the importance of language. Quine was not alone here; many philosophers who made what has been called "the linguistic turn" have become, in Wittgenstein's phrase, "bewitched by language." Claims such as that individuation is done only by language or that objective reference is solely due to language are so patently false that one is surprised to find them in the work of so gifted a man as Quine. A third reason, closely related to the other two, was Quine's refusal to abandon his pet doctrines, such as the dictum "to be is to be the value of a variable." This dictum is useful in analyzing the ontological commitment of developed theories, which was its original purpose. But to insist upon it in the child's learning of language was to deny the child objective reference until an absurdly late date. Finally, I think Quine always remained at heart a logical empiricist. He regarded Carnap as the embodiment of Logical Positivism, and despite his arguments with Carnap, he retained much of Carnap's view. His scientism, his behaviorism, his physicalism, his ambition to reconstruct our acquisition of science on a linguistic basis, all conspired to lead him into what appears now to be a dead end.

Why then did Quine hold the commanding position in American philosophy that he has been accorded? I think there are several reasons. One was the fact that he spoke from Harvard, traditionally the most eminent philosophy department in the nation. Second, he had a command of logic that most of his philosopher contemporaries did not. Few modern logicians have turned philosopher, as Quine did; one thinks of Godel, Tarski, Church, Rosser, Kleene, Fitch, etc. – for the most part they did logic and not much else. The outstanding exceptions are of course Russell and

Whitehead, but they were no longer active in Quine's later years. So Quine had an arsenal of weapons that his detractors could not match. Third, Quine's skeptical theses were indeed jaw-droppers; and whether other philosophers believed them or not, they could not stop arguing about them. Fourth, in a period of political turmoil, when Congressional committees were taking an unwholesome interest in what academics thought, Quine's work was so devoid of political implications that Congressional staffers, if they could have understood his writings, would have found little there to call subversive.

But I think Quine's celebrity came mostly from the fact that many believed he had destroyed Logical Positivism. One cannot overestimate the impact on American philosophy of the Logical Empiricists who fled to this country to escape Hitler in the 1930s and 1940s. Men like Carnap, Godel, Tarski, Hempel, Reichenbach, Frank, Feigl, and the rest were smart, thoroughly at home in logic and mathematics, well versed in science, and formidable in argument. Furthermore, they rapidly acquired distinguished professorships at eminent institutions in the United Sates where they continued to teach and publish – Carnap at the University of Chicago and U.C.L.A., Godel at the Advanced Institute, Tarski at Berkeley, Frank at Harvard, Hempel at Penn and then Princeton, Reichenbach at U.C.L.A., Feigl at the University of Minnesota, etc. Some American philosophers such as Charles Morris, Charles Stevenson, Ernest Nagel, and of course Quine, welcomed this infusion of empiricism and logical rigor. Many did not, though few were equipped to resist it. For these, Quine became the hero who slew the positivist dragon with its own weapons. The irony is that Quine was not so much the slayer of the dragon as its last offspring.

Notes

Introduction

1. Eiseley, Loren, *Darwin's Century* (Garden City: Doubleday, 1958).
2. D'Abro, A., *The Evolution of Scientific Thought* (New York: Dover, 1930).
3. Einstein, Albert, Boris Podolsky, and Nathan Rosen, "Can Quantum Mechanical Description of Physical Reality Be Considered Complete?" *Physical Review* 747:777–780 (1935).
4. Gilder, Louise, *The Age of Entanglement* (New York: Knopf, 2008).
5. Bochenski, I. M., *Contemporary European Philosophy* (Berkeley, CA: University of California, 1956), 95–96.
6. Ibid., 79–99, 123–128.
7. Allard, James, *The Logical Foundations of Bradley's Metaphysics* (Cambridge: Cambridge University Press, 2005).
8. Flower, Elizabeth, and Murray G. Murphey, *A History of Philosophy in America* (New York: Putnam's, 1977), Chapters 8 and 9.
9. Kuklick, Bruce, *Josiah Royce* (Indianapolis: Bobbs-Merrill, 1972). Flower and Murphey, Chapter 12.
10. James, William, *The Principles of Psychology* (New York: Henry Holt, 1910) – first edition 1890.
11. Kuklick, Bruce, *The Rise of American Philosophy* (New Haven: Yale, 1977); Bruce Kuklick, *Churchmen and Philosophers* (New Haven: Yale, 1985).
12. Murphey, Murray G., *C. I. Lewis, The Last Great Pragmatist* (Albany: SUNY, 2005).
13. Holt, Edwin et al., *The New Realism* (New York: Macmillan, 1912).
14. Bergman, Gustav, "Logical Positivism," in *A History of Philosophic Systems*, ed. Vergilius Ferm (New York: Philosophical Library, 1950), 472–473.
15. Moore, G. E., *Philosophical Studies* (London: Kegan Paul, Trebch, Trubner and Co., 1921), 1–30, 276–309.
16. Russell, Bertrand, *The Principles of Mathematics* (New York: Norton, 1950) – first published 1903.
17. Whitehead, Alfred North, and Bertrand Russell, *Principia Mathematica* (Cambridge: Cambridge University Press, 1950) – first published 1910.
18. Russell, Bertrand, *Our Knowledge of the External World* (London: George Allen and Unwin, 1952) – first published 1914.
19. Ibid., 79.
20. Ibid., 63.
21. Ibid., 68.
22. Ibid., 63.
23. Bubner, Rudiger, *Modern German Philosophy* (Cambridge: Cambridge University Press, 1981), 15–21. Bochenski, 129–139.

M.G. Murphey, *The Development of Quine's Philosophy*, Boston Studies in the Philosophy of Science 291, DOI 10.1007/978-94-007-2424-2,

24. Boole, George, *Mathematical Analysis of Logic* (New York: Philosophical Library, 1948) – first published 1847.
25. Lewis, C. I., *Survey of Symbolic Logic* (Berkeley, CA: University of California Press, 1918), Chapter 1.
26. Bell, Eric Temple, *Men of Mathematics* (New York: Simon and Schuster, 1930), Chapter 22.
27. Cantor, Georg, *Contributions to the Founding of the Theory of Transfinite Numbers* (New York: Dover, n.d.). Bell, *Men* Chapter 29.
28. Dummett, Michael, *Frege: Philosophy of Mathematics* (Cambridge: Harvard University Press, 1991).
29. Murphey, *Lewis* 80.
30. Kneale, William, and Martha Kneale, *The Development of Logic* (Oxford: Clarendon, 1962), 681.
31. Ibid., 673.
32. Fraenkel, Abraham, "Set Theory," in *Encyclopedia of Philosophy*, ed. Paul Edwards (New York: Macmillan, 1967), Vol. 8, 420–427.
33. Kneale, 682.
34. Fraenkel, 427.
35. Murphey, *Lewis*.
36. Wittgenstein, Ludwig, *Tractus Logico-Philosophicus* (London: Routledge, 1922).
37. Kneale, 720.
38. Wittgenstein, 4.21–4.221.
39. Ibid., 4.52–5.01.
40. Wang, Hao, "Quine's Logical Ideas in Historical Perspective," in *The Philosophy of W. V. Quine*, ed. Lewis Hahn and Paul Arthur Schilpp (LaSalle: Open Court, 1986), 623–648.
41. Tarski, Alfred, "The Concept of Truth in Formalized Languages," in *Logic, Semantics, Metamathematics*, ed. A. Tarski, trans. J. H. Woodger (Oxford: Clarendon, 1956), 152–278.
42. Godel, Kurt, *Consistency of the Axiom of Choice and the Generalized Continuum Hypothesis with the Axioms of Set Theory* (Princeton: Princeton University Press, 1940).
43. Lewis, C. I., and C. H. Langford, *Symbolic Logic* (New York: Century, 1932).
44. Heidegger, Martin, *Being and Time*, trans. J. Macquarrie and E. Robinson (New York: Harper, 1962).
45. Bochenski, Chapter 17.
46. Haller, Rudolf, "On the Historiography of Austrian Philosophy," in *Recovering the Forgotten Vienna Circle*, ed. Thomas Uebel (Dordrecht: Kluwer, 1991), 412–450.
47. Bolzano, Bernard, *Selected Writings on Ethics and Politics* (New York: Rodopi, 2007).
48. Brentano, Franz, *Descriptive Psychology* (London: Routledge, 1995).
49. Blackmore, J. T., R. Itagaki, and S. Tanaka, ed., *Ernst Mach's Vienna* (Dordrecht: Kluwer, 2001).
50. Waissmann, Friedrich, Foreword to *Moritz Schlick: Philosophical Papers* (Dordrecht: D. Reidel, 1979), Vol. II, xiii–xxxiii.
51. Murphey, *Lewis* 131.
52. Hempel, Carl, "Empiricist Criteria of Cognitive Significance: Problems and Changes," in *Aspects of Scientific Explanation*, ed. Carl Hempel (New York: Free Press, 1969), 101–119.
53. Carnap, Rudolf, *The Logical Syntax of Language* (London: Routledge, 1949).
54. Kraft, Victor, *The Vienna Circle* (New York: Philosophical Library, 1953); Joergen Joergensen, *The Development of Logical Empiricism* (Chicago: University of Chicago Press, 1951).
55. Joergensen, 48 ff.
56. Ibid., Jordan, Z., *The Development of Mathematical Logic and Logical Positivism Between the Two Wars* (New York: Oxford, 1945).
57. Kraft, 5.

Chapter 1

1. TML 14.
2. Ibid., 28.
3. Ibid., 38.
4. Ibid., 51.
5. Watson, John B., *Psychology from the Standpoint of a Behaviorist* (Philadelphia, PA: Lippincott, 1919).
6. F&Q/c, 390.
7. Stetson, Raymond H., *Motor Phonetics* (Amsterdam: North-Holland, 1951), 6.
8. James, *Psychology*, Chapter IX.
9. Kuklick, *Rise*.
10. Cohen, David, *J. B. Watson: The Founder of Behaviorism* (London: Routledge and Kegan Paul, 1979), 1–26.
11. Ibid., 27.
12. Ibid., 28.
13. Ibid., 35.
14. Ibid., 48–49.
15. Watson, John B., "Psychology as the Behaviorist Views It," *Psychological Review* 20: 158–177 (1913).
16. Ibid., 158.
17. Ibid., 161.
18. Ibid., 166–167.
19. Ibid., 174.
20. Watson, *Standpoint*.
21. Ibid., 320.
22. Ibid., 325.
23. Watson, John B., *Behaviorism* (New York: W. W. Norton, 1925), Chapter 10.
24. Ibid., 185.
25. Ibid., 185–195.
26. Ibid., 4–5.
27. Cohen, *Watson*, 218.
28. MSAm2587(1001); Quine to Skinner.
29. MSAm2587(2831–2832), 96.
30. TML 60.
31. Auto, 8–9.
32. MSAm2587 (2831–2832), 104.
33. Auto, 8.
34. Ibid., 8–9.
35. MSAM2587(1001); Quine to Skinner.
36. MSAm2587 (3190), 5; Auto, 9.
37. Auto, 9.
38. Lewis, C. I., *Mind and the World Order* (New York: Dover, 1929).
39. MSAm2587(3236).
40. MSAm2587(3236) "Concepts and Working Hypotheses", 1.
41. MSAm2587(3236) "Futurism and Conceptual Pragmatism."
42. MSAm2587(2391), 113–114.
43. PM, I, xiii.
44. Auto, 10–12.
45. Ibid., 11.
46. Ibid., 10.
47. F&Q/c, 501.
48. Lewis, *Survey,* 357.

49. LSeq, 34.
50. Ibid., 15–54.
51. Ibid., 64–74.
52. Auto, 11.
53. Ibid., 12.
54. Ibid., 13.
55. MSAm2587(3190), Preface.
56. Bailyn, Bernard, and Donald Fleming, *The Intellectual Migration: Europe and America 1930–1960* (Cambridge: Harvard University Press, 1969).
57. F&Q/d, 62.
58. Ibid., 24.
59. TML, 92–100.
60. Auto, 12.
61. Ibid., 12. Carnap, Rudolf, *The Logical Syntax of Language*, trans. Amethe Smeaton (London: Routledge and Kegan Paul, 1937) – published in German in 1934.
62. TML, 102–104.
63. MSAm2587(1060); Quine to Tarski 1/10/81. But see Auto, 15.
64. Carnap, Rudolf, *Der logische Aufbau der Welt* (*The Logical Structure of the World*), trans. R. A. George (Chicago: Open Court, 2003).
65. Russell, *Knowledge.*
66. Carnap, *Aufbau*, 60.
67. Ibid., 5.
68. Ibid., 98–101.
69. Ibid., 88.
70. Ibid., 88–89.
71. Ibid., 97–103.
72. Ibid., 23.
73. Ibid., 236.
74. Ibid., 236.
75. Ibid., 237.
76. Ibid., 237.
77. Uebel, Thomas, *Empiricism at the Crossroads* (LaSalle: Open Court, 2007), 2.
78. Ibid., 191–246.
79. Mancosu, Paolo, "Quine and Tarski on Nominalism," in *Oxford Studies in Metaphysics*, ed. Zimmerman (Oxford: Clarendon, 2008), Vol. 4, 22–24.
80. Ibid., 22–55.
81. Auto, 13.
82. MSAm2587(1001); Quine to Skinner 3/8/85.
83. MSAm2587(2831–2832), 35.
84. F&Q/c 62. "John Horne Tooke" *Wikipedia.*
85. SL, 5.
86. Ibid., 5.
87. MSAm2587(3286); Card file, For Nozick's seminar 4/4/90.
88. Creath, Richard, ed. *Dear Carnap, Dear Van* (Berkeley, CA: University of California, 1990), 48.
89. Ibid., 48–49.
90. Ibid., 59.
91. Ibid., 63.
92. Ibid., 63.
93. Ibid., 64.
94. Ibid., 65.
95. Ibid., 66.
96. Ibid., 69.

97. Ibid., 76–78.
98. Ibid., 79–80.
99. Ibid., 80.
100. Ibid., 86.
101. Ibid., 86.
102. Ibid., 89.
103. Ibid., 92.
104. Ibid., 92.
105. Ibid., 93.
106. Ibid., 95.
107. Ibid., 97.
108. Ibid., 97.
109. Ibid., 99–101.
110. Ibid., 30.
111. Ibid.,103.
112. Ibid., 28.
113. Murphey, *Lewis*, 159.
114. Lewis, C. I., "The Pragmatic Conception of the A Priori," *Journal of Philosophy* 20:169–177 (1923).
115. Carnap, *Syntax*, 153.
116. Quine, W. V., "Truth by Convention" in WP, 77–106.
117. Ibid., 77.
118. Ibid., 104.
119. Ibid., 105–106.
120. Peirce, Charles S., *The Collected Papers of Charles Sanders Peirce* (Cambridge: Harvard University Press, 1932), 2.204.
121. Quine, "Truth by Convention," 102.
122. Ibid., 106.
123. Auto, 16.
124. Russell, Bertrand, *The Principles of Mathematics* (New York: Norton, 1950), vi.
125. Joergensen, Joergen, *A Treatise of Formal Logic* (Copenhagen: Levin and Murksgaard, 1931); W. V. Quine, "A Note on Nicod's Postulate," *Mind* 41:345–350 (1932).
126. Quine, W. V., "A Theorem in the Calculus of Classes," *Journal of the London Mathematical Society* 8:89–95 (1933).
127. Quine, W. V. "Ontological Remarks on the Propositional Calculus," *Mind* 43:472–476 (1934).
128. WP, 267.
129. MSAm2587(2831–2832), 157.
130. TML, 123–124.
131. Ibid., 119–120.
132. Ibid., 130.
133. MSAm2587(2831–2832), 177.
134. Quine, W. V., "A Reinterpretation of Schoenfinkel's Logical Operators," *Bulletin of the American Mathematical Society* 42:87–89 (1936).
135. Auto, 15.
136. Ibid., 16–17.
137. PM, Section B, 12.
138. Kneale and Kneale, *Development of Logic* 669. Frank Plumpton Ramsey, *The Foundations of Mathematics*, ed. Braithwaite (New York: Humanities Press, 1950), 1–61.
139. Quine, W. V., "The Axiom of Reducibility," *Mind* 45:498–500 (1936).
140. Quine, W. V., "Set-Theoretical Foundations for Logic," *JSL* 1:45–57 (1936); SLP, 83-00.
141. Wang, Hao, *A Survey of Symbolic Logic* (Peking: Science Press, 1963), 387.
142. Quine, "Set-Theoretical," SLP, 83–99.1.

143. Ibid., 852.
144. Auto, 173.
145. SLP, 884.
146. SLP, 905.
147. Quine, W. V., "Toward a Calculus of Concepts," *JSL* 1:2–25 (1936).
148. Mancosu, 25–28.
149. Ibid., 28.
150. Ibid., 29.
151. Carnap, Rudolf, *The Philosophy of Rudolf Carnap*, ed. Paul Arthur Schilpp (LaSalle: Open Court, 1963), 34.
152. TML, 128–129.
153. Quine, W. V., "Logic Based on Inclusion and Abstraction," *JSL* 2:145–152 (1937); SLP, 100–109.
154. Ibid., 102.
155. Quine, W. V., "New Foundations for Mathematical Logic," *American Mathematical Monthly* 44:70–80 (1937). reprinted in FLPV, 80–101.
156. Ibid., 89.
157. Ibid., 89.
158. Rosser, Barkley, "On the Consistency of Quine's 'New Foundations for Mathematical Logic,'" *JSL* 4:15–24 (1940), 23.
159. Rosser, Barkley, *Logic for Mathematicians* (New York: McGraw-Hill, 1953), 204–206.
160. Specker, E., "The Axiom of Choice in Quine's 'New Foundations for Mathematical Logic,'" *Proceedings of the National Academy of Sciences* 1953:366–368 (1953).
161. Jenses, Ronald, "On the Consistency of a Slight (?) Modification of Quine's *New Foundations*," in *Words and Objections*, ed. D. Davidson and J. Hintikka (Dordrecht: D. Reidel, 1969), 278–291.
162. Quine, W. V., "Reply to Jensen," in ed. Donald Davidson and Jaakko Hintikka, 349.
163. Quine, W. V., "On Cantor's Theorem," *JSL* 2:120–124 (1938).
164. Mancosu, 29.
165. WP, 199–202.
166. Creath, ed. *Dear Carnap, Dean Van* 260.
167. TML 131–137.
168. Quine, W. V., "On the Theory of Types," *JSL* 3:125–139 (1938).
169. Quine, W. V., "Logic based on Inclusion and Abstraction," SLP, 100–109.
170. Quine, "Theory of Types," *JSL* 3:125–138 (1938).
171. Ibid., 134–136.
172. Auto, 18–19.
173. MSAm2587(1001); Quine to Skinner 11/18/37,2.
174. Carnap, *Philosophy*, 35.
175. Auto, 19.
176. Carnap, *Philosophy* 64–65.
177. MSAm2587(1060); Quine to Tarski.
178. MSAm2587(655); 2/20/41.
179. *Journal of Philosophy* 84:29–32 (1987). Tarski to White.
180. Tarski, Alfred, "On The Concept of Logical Consequence," in *Logic, Semantics, Metamathematics*, ed. A. Tarski, trans. J. H. Woodger (Oxford: Clarendon, 1956), 409–420.
181. Ibid., 418–419.
182. Carnap, *Philosophy,* 30.
183. Quine, W. V., *Mathematical Logic* (Cambridge: Harvard University Press, 1940); W. V. Quine, *Elementary Logic* (Boston: Ginn and Co., 1941).
184. *Elementary Logic,* 40.
185. Ibid., 91.
186. Ibid., 116–117.
187. Ibid., 56.

188. Ibid., 146.
189. Ibid., 116. Cf. also 119.
190. Ibid., 147.
191. Ibid., 166.
192. Auto, 27.
193. Tarski, Alfred, *Introduction to Logic* (New York: Oxford, 1946) – reprint of first English edition of 1941.
194. MSAm2587(499).
195. MSAm2587(2831–2832), 180.
196. TML, 130.
197. *Mathematical Logic,* Chapter 1.
198. Murphey, *Lewis*, 320.
199. Quine, *Mathematical Logic*, 24, 71, 112, 133, 147, 226.
200. ML, Chapter 1.
201. ML, Chapter 2.
202. ML, Chapter 3.
203. PM, 150.
204. *Mathematical Logic,* 150–152.
205. ML, Chapter 4.
206. ML, Chapter 5.
207. ML, Chapter 6.
208. ML, Chapter 7.
209. MSAm2587(2831–2832), 215, 231.
210. TML, 146.
211. Rosser, Barkley, "The Burali-Forti Paradox," *JSL* 7:1–17 (1942).
212. Auto, 21–22.
213. PM, III, 75.
214. Quine, W. V., "Element and Number," *JSL* 6:136–149 (1941); SLP, 121–149.
215. SLP, 124–132.
216. Quine, W. V., *Mathematical Logic* (Cambridge: Harvard University Press, 1947), 215.
217. ML, 246.
218. MSAm2587(2831–2832), 294.
219. Wang, Hao, "Quine's Logical Ideas in Historical Perspective," in *Philosophy of W. V. Quine*, 631.
220. For an overview of Quine's work in logic, see Joseph Ullian, "Quine and Logic," in *The* Cambridge *Companion to Quine*, ed. Roger Gibson (Cambridge: Cambridge University Press, 2004), 270–286.

Chapter 2

1. Quine, W. V., *O Sentido da Nova Logica* (Sao Paolo: Martins, 1953).
2. Quine, W. V., "Whitehead and the Rise of Modern Logic," in *The Philosophy of Alfred North Whitehead*, ed. Paul Arthur Schilpp (New York: Tudor, 1951), 127–163.
3. Ibid., 139.
4. Ibid., 142.
5. Ibid., 146–149.
6. Ibid., 144–145.
7. Ibid., 153.
8. Ibid., 156–158.
9. TML, 183–192.
10. MSAm2587(2831–2832), 247–264.

11. Mancosu, Paolo, "Harvard 1940–1941; Tarski, Carnap and Quine on a Finitist Language of Mathematics for Science," *History and Philosophy of Logic* 26:327–357 (2005).
12. *Journal of Philosophy* 40:113–127 (1943).
13. FLPV, 170.
14. Quine, "Notes on Existence and Necessity," *Journal of Philosophy* 40:113–127 (1943).
15. Ibid., 113–114.
16. Ibid., 118.
17. Ibid., 115–127.
18. Ibid., 121.
19. Ibid., 120.
20. MSAm2587(3169), 20–35, 131.
21. Ibid.
22. MSAm2587(420); Quine to Goodman 12/19/44.
23. Wang, *Philosophy of Quine*, 632.
24. MSAm2587(420); Quine to Goodman 8/14/45.
25. MSAm2587(3181).
26. Ibid.
27. JSL 10:1–12 (1945); SLP, 181–195.
28. SLP, 185.
29. Ibid., 188.
30. Quine, W. V., "Concatenation as a Basis for Arithmetic," SLP, 70–82; 71.
31. Auto, 26.
32. MSAm2587(420); Quine to Goodman 11/30/46.
33. Ibid., Quine to Goodman 2/19/47.
34. Ibid., Goodman to Quine 2/24/47.
35. Ibid., Quine to Goodman 2/28/47.
36. Ibid., Goodman to Quine 3/22/47.
37. Ibid., Goodman to Quine 3/29/47.
38. Ibid., Goodman to Quine 4/5/47.
39. MSAm2587(2831–2832), 269.
40. JSL 12:43–48 (1947).
41. Ibid., 44.
42. Ibid., 45.
43. Ibid., 46.
44. Ibid., 47.
45. MSAm2587(420); Quine to Goodman 7/6/40.
46. Ibid., Quine to Goodman 7/26/46.
47. Ibid., Quine to Goodman 2/19/47.
48. Ibid., Goodman to Quine 2/24/47.
49. Goodman, Nelson, and W. V. Quine, "Steps Toward a Constructive Nominalism," *JSL* 12:97–122 (1947).
50. Quine, W. V., "Nominalism," in *Oxford Studies in Metaphysics*, ed. Dean Zimmerman (Oxford: Oxford University Press, 2008), 3–21.
51. Ibid., 6.
52. Ibid., 14.
53. Ibid., 16–17.
54. Ibid., 21.
55. Ibid., 21.
56. JSL 12:74–85 (1947).
57. Ibid., 81.
58. Ibid., 82–84.
59. Goodman and Quine, "Steps."
60. Ibid., 105.

61. Ibid., 106–107.
62. Ibid., 108.
63. Ibid., 108–109.
64. Ibid., 110.
65. Ibid., 111.
66. Ibid., 118.
67. Ibid., 117–118.
68. Ibid., 122.
69. Mancosu, "Tarsli and Quine on Nominalism," 43.
70. Tarski, Alfred, *Introduction to Logic*, trans. O. Helmer (New York: Oxford, 1941).
71. MSAm2587(1060); Tarski to Quine 12/8/39.
72. Quine, W. V., *Mathematical Logic* Revised Edition (Cambridge: Harvard University Press, 1951).
73. Ibid., Preface.
74. Ibid., 162–166.
75. Ibid., 165.
76. Wang, Hao, "A Formal System of Logic," *JSL* 15:25–32 (1952), 26–27.
77. ML, 162.
78. Wang, "Formal System," 27.
79. Creath, *Dear Carnap, Dear Van,* 240–241.
80. Ibid., 245.
81. Ibid., 247–248.
82. White, Morton, *A Philosopher's Story* (University Park: Penn State University Press, 1999), Appendix.
83. Ibid., 337–338.
84. Murphey, *Lewis*, 261–262.
85. Ibid., Rudolf Carnap, *Meaning and Necessity* (Chicago: University of Chicago Press, 1956), 55.
86. White, *Story*, 339–340.
87. Ibid., 341.
88. Ibid., 343.
89. Ibid., 342.
90. Ibid., 345.
91. Ibid., 346.
92. Ibid., 348.
93. White, Morton, "The Analytic and the Synthetic: An Untenable Dualism," in *John Dewey: Philosopher of Science and Freedom*, ed. Sidney Hook (New York: Dial Press, 1950), 316–330.
94. Ibid., 330.
95. Quine, W. V., "On What There Is," *Review of Metaphysics* 2:21–38 (1948); FLPV 1–19.
96. Ibid., 11.
97. Ibid., 15.
98. Ibid., 16.
99. Goodman and Quine, "Steps."
100. Quine, "On What There Is," 16.
101. Ibid., 17.
102. Ibid., 19.
103. Quine, W. V., "Identity, Ostension, and Hypostasis" in FLPV, 65–79.
104. Ibid., 71.
105. Ibid., 74.
106. Mancosu, "Quine and Tarski on Nominalism," 43.
107. Quine, W. V., "Where is Logic Going?" *F&Q/c* 148–151.
108. Ibid., 151.
109. Quine, W. V., "The Entangled Philosophies of Mathematics," (2008) *F&Q/c* 157–162.

110. Ibid., 158–159.
111. Ibid., 162.
112. MSAm2587(3181) "The Ontology of Logic."
113. Ibid.
114. Ibid.
115. W. V. Quine, "Two Dogmas of Empiricism" FLPV 20–46.
116. Ibid., 20.
117. Ibid., 22.
118. Ibid., 27.
119. Ibid., 32.
120. Ibid., 33.
121. Ibid., 34.
122. Ibid., 41.
123. Ibid., 42.
124. Ibid., 43.
125. Ibid., 43.
126. Ibid., 44.
127. Ibid., 46.
128. MSAm2587(499); Hempel to Quine 5/21/51.
129. MSAm2587(3236); Lewis to Quine 3/13/51.
130. Creath, *Dear Carnap Dear Van*, 427–432.
131. For an excellent study of the Quine–Carnap relationship, see Daniel Isaacson, "Quine and Logical Positivism," in *The Cambridge Companion to Quine*, ed. Roger Gibson (Cambridge: Cambridge University Press, 2004), 214–269.
132. Murphey, *Lewis,* 238.
133. Cohen, Morris R., and Ernest Nagel, *An Introduction to Logic and Scientific Method* (New York: Harcourt-Brace and Co., 1934), 219–221.
134. Duhem, Pierre, *La Theorie Physique: Son Object, Sa Structure* 1905: English translation *The Aim and Structure of Physical Theory* (New York: Atheneum, 1962).
135. Quine, W. V., "Two Dogmas in Retrospect," F&Q/c 390–400.
136. MSAm2587(407); Quine to Gibson 4/19/87.
137. Quine, "Two Dogmas," 42.
138. TML, 226.
139. Quine, W. V., "The Problem of Meaning in Linguistics," FLPV 47–64.
140. Ibid., 49.
141. Ibid., 52.
142. Ibid., 56.
143. Ibid., 63.
144. Ibid., 63.
145. FLPV, 102–129.
146. Ibid., 103.
147. Ibid., 120–121.
148. Ibid., 123.
149. Ibid., 125.
150. Ibid., 127–129.
151. Ibid., 129.
152. FLPV, 130–138.
153. Ibid., 138.
154. FLPV, 139–159.
155. Ibid., 142–143.
156. Ibid., 148.
157. Ibid., 153.
158. Ibid., 159.

159. FLPV, 160–167.
160. Ibid., 167.
161. FLPV.
162. MSAm2587(2831–2832), 269.
163. MSAm2587(287); Quine to Davidson 8/13/65.
164. Quine, W. V., *Methods of Logic* (London: Routledge and Kegan Paul, 1952).
165. Ibid., xi.
166. MSAm2587(420); Goodman to Quine 1/2/51.
167. *Methods of Logic*, 23–37.
168. Ibid., 100–113.
169. Baars, Bernard, *The Cognitive Revolution* (New York: Guilford Press, 1980) 66–74.
170. Ibid., 62–63.
171. Boring, Edwin G., *A History of Experimental Psychology* (New York: Appleton-Century-Crofts, 1950), 650.
172. Baars, 210–211.
173. Ibid., 273–284.
174. Chomsky, Noam, "Review of Skinner's *Verbal Behavior*," *Language* 35:26–58 (1959).
175. Baars, 175–177.
176. TML, 229.
177. Quine, W. V., "On ω-inconsistency and the So-Called Axiom of Infinity," SLP, 114–129.
178. Quine, W. V., "Three Grades of Modal Involvement," WP, 158–176 (1953).
179. Quine, W. V., "On Mental Entities," WP, 221–227.
180. Ibid., 222.
181. Ibid., 225.
182. Ibid., 225.
183. Ibid., 226.
184. Ibid., 227.
185. Quine, W. V., "The Scope and Language of Science," WP, 228–245.
186. Ibid., 234.
187. TML, 247.
188. SLP, 212–219, 1954.
189. SLP, 205–211, 1954.
190. SLP, 224–226.
191. SLP, 220–223.
192. SLP, 37–51.
193. WP, 107–132.
194. Schilpp, P., ed., *The Philosophy of Rudolf Carnap* (LaSalle: Open Court, 1963), 915–921.
195. Ibid., 916.
196. Ibid., 916.
197. Ibid., 920.
198. Ibid., 921.
199. Quine, W. V., "On Frege's Way Out," SLP, 146–158.
200. Ibid., 150.
201. Ibid., 156–158.
202. Quine, W. V., "A Way to Simplify Truth Functions," *American Mathematical Monthly* 62:627–631 (1955).
203. Quine, W. V., "Posits and Reality," WP, 246–254.
204. Ibid., 252.
205. Ibid., 253.
206. TML, 262.
207. Quine, W. V., "Unification of Universes in Set Theory," *JSL* 21:167–279 (1956).
208. Specker, Ernest, "The Axiom of Choice in Quine's New Foundations for Mathematical Logic," *Proceedings of the National Academy of Sciences* 39:972–975 (1953).

Chapter 3

1. Quine, W. V., *Word and Object* (Cambridge: Harvard University Press, 1960).
2. Ibid., Section 2.
3. Ibid., Section 5.
4. Ibid., Section 6.
5. Ibid., Section 7.
6. Ibid., Section 8.
7. Ibid., Sections 9–10.
8. Ibid., Sections 12, 14.
9. Ibid., Section 13.
10. Ibid., Sections 12, 14.
11. Ibid., Section 15.
12. Ibid., Section 15.
13. Ibid., Section 16.
14. Spelke, Elizabeth, "Perception of Unity, Persistence and Identity; Thoughts on Infants' Conceptions of Objects" in *Neonate Cognition*, ed. J. Mehler and R. Fox (1985), 89–113; Renee Baillargeon, "Representing the Existence and Location of Hidden Objects: Object Permanence In 6- and 8-Month Old Infants," *Cognition* 23: 21–41 (1986); Claes von Hofsten, and Elizabeth Spelke, "Object Perception and Object-Directed Reaching in Infancy," *Journal of Experimental Psychology: General* 114: 198–212 (1985).
15. Chomsky, Noam, "Review of Skinner's *Verbal Behavior*," *Language* 35:26–58 (1959).
16. WO, Section 17.
17. Ibid., Sections 17–18.
18. Ibid., 92. Although there is no acknowledgement, this phrase is surely adapted from James' famous line: "A polyp would be a conceptual thinker if the feeling of 'Hello! Thingumbob again!' ever flitted through its mind" James, William, *Psychology* I: 463.
19. WO, Section 19.
20. Ibid., Section 20.
21. Ibid., Sections 22–23.
22. Ibid., Section 24.
23. Ibid., Section 25.
24. Ibid., Sections 17, 26.
25. Ibid., Sections 17, 26–29.
26. Ibid., Sections 30–31.
27. Ibid., 133. Maria Luisa Della Chiara and Giuliano Toraldo Di Francia, "Quine on Physical Objects," in *On Quine*, ed. Paolo Leonardi and Marco Santambrogia (Cambridge: Cambridge University Press, 1995), 104–112.
28. Ibid., Section 34.
29. Ibid., Sections 35, 37–38.
30. Ibid., Section 40.
31. Ibid., Section 43.
32. Ibid., Section 45.
33. Ibid., Section 46.
34. Ibid., Section 47.
35. Ibid., Section 47.
36. Ibid., Sections 49–52.
37. Ibid., Sections 53–55.
38. Ibid., Section 56.
39. MSAm2587(420), 12/15/58.
40. MSAm2587(420), 12/24/58.
41. MSAm2587(420), 12/30/58.
42. MSAm2587(420), 12/30/58.

43. MSAm2587(420), 5/10/60.
44. Quine, W. V., "Simple Theories in a Complex World," WP, 255–258.
45. Ibid., 255.
46. Ibid., 258.
47. Ibid., 258.
48. Quine, W. V., "On The Application of Modern Logic," WP, 33–39.
49. Ibid., 37.
50. Ibid., 37.
51. Ibid., 39.
52. MSAm2587(499), 5/8/62.
53. MSAm2587(3281), Philosophy 148 card file, 1956–1961.
54. Quine, W. V., *Methods of Logic* (New York: Holt, Rinehart and Winston, 1963).
55. Quine, W. V., "The Ways of Paradox," in WP, 1–18.
56. Ibid., 17.
57. Quine, W. V., *Set Theory and its Logic* (Cambridge: Harvard University Press, 1963).
58. Ibid., viii.
59. Ibid., 87.
60. Ibid., 81.
61. Ibid., 329.
62. Quine, W. V., "Ontological Reduction and the World of Numbers," WP, 212–220.
63. Ibid., 214.
64. Ibid., 215–217.
65. Ibid., 218.
66. MSAm2587(22831–2832), 404.
67. MSAm2587(2831–2832), 440–441.
68. MSAm2587(2831–2832), 401–402.
69. Quine, W. V., "Foundations of Mathematics," WP, 23–32.
70. Ibid., 28.
71. Ibid., 31.
72. Ibid., 31–32.
73. SLP.
74. WP.
75. Jean van Heijenoort, ed., *From Frege to Godel* (Cambridge: Harvard University Press, 1967).
76. Ibid.
77. Quine, W. V., "Russell's Ontological Development," TT, 73–85.
78. Ibid., 73.
79. Ibid., 73.
80. F&Q/d, 111.
81. Quine, "Russell's Ontological Development," 74.
82. Ibid., 75.
83. Ibid., 77.
84. Ibid., 77.
85. Ibid., 80.
86. Ibid., 83.
87. Ibid., 83–84.
88. Ibid., 84.
89. Ibid., 84.
90. Ibid., 85.
91. Auto., 38.
92. Quine, W. V., and Joseph Ullian, *The Web of Belief* (New York: Random House, 190).
93. Ibid., Ch. 1.
94. Ibid., Ch. 2.
95. Ibid., Ch. 3.

96. Ibid., Ch. 4.
97. Ibid., Ch. 5.
98. Ibid., Ch. 6.
99. Ibid., Ch. 7.
100. Ibid., Ch. 8.
101. Stevenson, Charles, *Ethics and Language* (New Haven: Yale, 1974).
102. Quine and Ullian, Ch. 9.
103. Quine, W. V. "Linguistics and Philosophy," WP, 56–58.
104. Ibid., 57.
105. Ibid., 58.
106. MSAm2587(287).
107. MSAm2587(2831–2832), 452. Auto, 452.
108. Quine, W. V., "Philosophical Progress in Language Theory," *Metaphilosophy* 1:2–19 (1970).
109. Ibid., 8–9
110. Ibid., 9.
111. Ibid., 12.
112. Ibid., 13–14.
113. Ibid., 18.
114. Quine, W. V., *Set Theory and Its Logic* (Cambridge: Harvard University Press, 1969).
115. Ibid., 164.
116. Ibid., vii.
117. Parsons, Charles, "A Note on Quine's Treatment of Transfinite Recursion," *JSL* 29:179–182 (1964).
118. Quine, W. V., *Ontological Relativity and Other Essays* (New York: Columbia University Press, 1969).
119. Ibid., "Speaking of Objects," OR, 1–25.
120. Ibid., 1.
121. Ibid., 8.
122. Ibid., 9–11.
123. Ibid., 12.
124. Ibid., 12–13.
125. Ibid., 14.
126. Quine, W. V., "Ontological Relativity," in OR, 26–68.
127. Ibid., 27.
128. Ibid., 28.
129. Ibid., 30.
130. Ibid., 40.
131. Ibid., 45.
132. Ibid., 64.
133. Quine, W. V., "Replay to Chomsky," in *Words and Objections*, ed. Donald Davidson and Jaakko Hintikka (Dordrecht: D. Reidel, 1969), 309–310.
134. MSAm2587(2831–2832), 455.
135. Quine, W. V., "Epistemology Naturalized," in OR, 69–90.
136. Ibid., 72.
137. Ibid., 82.
138. Ibid., 83.
139. Ibid., 87.
140. Ibid., 8–89.
141. Quine, W. V., "Existence and Quantification," in OR, 91–113.
142. Ibid., 107.
143. Ibid., 112–113.
144. Quine, W. V., "Natural Kinds," in OR, 114–138.
145. Ibid., 125.

146. Ibid., 126.
147. Ibid., 134.
148. Ibid., 138.
149. Quine, W. V., "Propositional Objects." in OR, 139–160.
150. Ibid., 1157.
151. MSAm2587(499).
152. MSAm2587((499), 11/26/69.
153. MSAm2587(363), 2/13/69.
154. W. V. Quine, *Philosophy of Logic* (Cambridge: Harvard University Press, 1970), second edition 1986.
155. Ibid., Ch. 1.
156. Ibid., Ch. 2.
157. Ibid., Ch. 3.
158. Ibid., Ch. 4.
159. Ibid., Ch. 5.
160. Ibid., Ch. 6.
161. Ibid., Ch. 7.
162. Quine, W. V., "Grades of Theoreticity," in *Experience and Theory*, ed. Lawrence Foster and Joe W. Swanson (Amherst: University of Massachusetts Press, 1970).
163. Ibid., 2–3.
164. Ibid., 3.
165. Ibid., 5–6.
166. Ibid., 7.
167. MSAm2587(824), 8/18/70.
168. *Words and Objections*, 312.
169. Quine, W. V., "Methodological Reflections on Current Linguistic Theory," *Synthese* 21:386–396 (1970). F & Q/c 215–227.
170. Ibid., 215.
171. Ibid., 216.
172. Ibid., 220.
173. Ibid., 221.
174. Ibid., 225.
175. Quine, W. V., "Homage to Rudolf Carnap," WP, 40–43.
176. Ibid., 40.
177. Ibid., 41.
178. Ibid., 42.
179. Ibid., 42.
180. Ibid., 43.
181. Quine, W.V., "Algebraic Logic and Predicate Functors," WP, 283–307.
182. Ibid., 305–307.
183. Quine, W. V., *The Roots of Reference* (LaSalle: Open Court, 1973).
184. Goodman, Nelson, "Introduction" to RR, xii.
185. RR Section 1.
186. RR Sections 2–4.
187. RR Sections 5–7.
188. RR Sections 7–8.
189. RR Section 9.
190. RR Section 10.
191. RR Sections 11–13.
192. RR Section 14.
193. Bower, T. G. R., "The Object in the World of the Infant," *Scientific American* 225:30–38 (1971).
194. Ibid., 37.

195. RR Section 15.
196. RR Section 16.
197. RR Sections 16–18.
198. RR Sections 17–18.
199. RR Section 19.
200. RR Section 20.
201. RR Sections 22–23.
202. RR Sections 24–25.
203. RR Section 26.
204. RR Section 27.
205. RR Section 28.
206. RR Sections 28–29.
207. RR Section 30.
208. RR Sections 30–31.
209. RR Section 32.
210. RR Section 33.
211. RR Sections 34–35.
212. RR Sections 35–36.

Chapter 4

1. Quine, W. V., "Paradoxes of Plenty," *Daedalus* 103:38–49 (1974). TT 194–198.
2. TT, 197.
3. Ibid., 197–198.
4. "Ideas of Quine," Interview by McGee, Brian, *Men of Ideas* (London: Oxford, 1978), 168–169. F&Q/d 5–17.
5. Quine, W. V., "Truth and Disquotation," WP, 308–321.
6. Ibid., 309–310.
7. Ibid., 315–316.
8. Ibid., 317.
9. Ibid., 320–321.
10. Quine, W. V., "The Variable," WP, 272–283.
11. Ibid., 272–273.
12. Ibid., 275–276.
13. Ibid., 278–280.
14. Ibid., 281–282.
15. Ibid., 282.
16. Ibid., 277.
17. Quine, W. V., "On Popper's Negative Methodology," in *The Philosophy of Karl Popper*, ed. Paul Arthur Schilpp (LaSalle: Open Court, 1974), 218–220.
18. Ibid., 219.
19. Ibid., 220.
20. Quine, W. V., "On Logically Equivalent Systems of the World," *Erkenntnis* 9:313–328 (1975). F&Q/c 228–243.
21. Ibid., 313.
22. Ibid., 314–315.
23. Ibid., 315–316.
24. Ibid., 316.
25. Ibid., 316–317.
26. Ibid., 318–319.
27. Ibid., 320.

28. Ibid., 321.
29. Ibid., 324–326.
30. Ibid., 326.
31. Ibid., 436–327.
32. Ibid., 327.
33. Gibson, Roger, "Translation, Physics, and Facts of the Matter," *Philosophy of Quine* 139–154, 152–153 (1986).
34. Quine, W. V., "Reply to Gibson" in *Philosophy of Quine*, 157.
35. Quine, W. V., "The Nature of Natural Knowledge," in *Mind and Language*, ed. S. Guttenplan (Oxford: Clarendon, 1975), 67–82.
36. Ibid., 67–68.
37. Ibid., 80.
38. Ibid., 81.
39. Emerson, Ralph Waldo, "Nature," in *Nature, Addresses and Lectures*, ed. James Elliot Cabot (Boston: Houghton Mifflin, 1885).
40. Kuhn, Thomas, *The Structure of Scientific Revolutions* (Chicago: University of Chicago Press, 1962).
41. Samuel Guttenplan (Oxford: Oxford University Press, 1975).
42. Ibid., 83.
43. Ibid., 85.
44. Ibid., 89.
45. Ibid., 90.
46. Ibid., 91.
47. Ibid., 92.
48. Quine, W. V. to Parsons MSAm2587(824), 11/12/75.
49. Wang, Hao, "Quine's Logical Ideas in Historical Perspective," *Philosophy of Quine* 636.
50. Quine, W. V. "Reply to Wang," *Philosophy of Quine* 644.
51. Ibid., 645.
52. Wang, "Quine's Logical Ideas," 632.
53. Quine, W. V. "Whither Physical Objects," in *Essays in Memory of Imre Lakayos*, ed. Robert Cohen, Paul Feyerabend and Mark Wartofsky (Dordrecht: D. Reidel, 1976), 497–504.
54. Ibid., 499.
55. Ibid., 501–503.
56. Ibid., 503–504.
57. Quine, W. V., "Facts of the Matter," in *American Philosophy from Edwards to Quine*, ed. R. W. Shahan and K. R. Merrill (Norman: University of Oklahoma, 1977). F&Q/c 271–286.
58. Ibid., 189–192.
59. Ibid., 196.
60. Quine, W. V., "Use and Its Place in Meaning," TT, 43–54.
61. Ibid., 43.
62. Ibid., 43–44.
63. Ibid., 49.
64. Ibid., 52.
65. Ibid., 50–54.
66. Quine, W. V., "On The Nature of Moral Values," TT, 55–66.
67. Ibid., 56.
68. Ibid., 56–57.
69. Schlick, Moritz, *Problems of Ethics* (New York: Prentice-Hall, 1939).
70. Quine, "Moral Values," 63.
71. Ibid., 64.
72. Ibid., 66.
73. D. Follesdal and D. B. Quine, ed., *Quine in Dialogue* (Cambridge: Harvard University Press, 2008), 5–17.

74. Ibid., 17.
75. Quine, W. V., "Kurt Godel," TT, 143–147.
76. Quine, W. V., "What's It All About?" in *American Scholar* 50:43–54 (1980).
77. Ibid., 53–54.
78. Quine, W. V., "Sellars on Behaviorism, Language and Meaning," *Pacific Philosophical Quarterly* 61:26–30 (1980).
79. Ibid., 27.
80. Ibid., 30.
81. Quine, W. V., "What Price Bivalence?" TT, 31–37.
82. Ibid., 32.
83. Quine, W. V., "The Pragmatists Place in Empiricism," in *Pragmatism*, ed. Robert Mulvany and Philip Zeltner (Columbia: University of South Carolina Press, 1981).
84. Ibid., 23.
85. Ibid., 23.
86. Quine, W. V., "Replies to Stroud and Chihara," *Midwest Studies in Philosophy* 6:453f, 473–475 (1981).
87. Ibid., 475.
88. Follesdal, Dagfinn, to Quine, MSAm2587(363), 11/5/81.
89. Quine, W. V., "Ontology and Ideology Revisited," *Journal of Philosophy* 80:499–502 (1983). F&Q/c 315–318.
90. Ibid., 317.
91. Ibid., 318.
92. Quine, W. V., "Relativism and Absolutism," *The Monist* 67:293–296 (1984). F&Q/c 319–322.
93. Ibid., 321.
94. Ibid., 322.
95. Ibid., 322.
96. Quine, W. V., "Sticks and Stones; or, The Ins and Outs of Existence," *On Nature: Boston Studies in Philosophy and Religion* 6:13–26 (1984). F&Q/d 312–324.
97. Ibid., 16.
98. Ibid., 17–18.
99. Ibid., 23.
100. Ibid., 24.
101. Ibid., 24.
102. Ibid., 25–26.
103. Transcripts of Stanford Meeting, MSAm2587(2749), Folder 1.
104. Ibid., Folder 2.
105. Ibid., Folders 1, 2.
106. Ibid., Folder 2.
107. Ibid., Folder 2.
108. Ibid., Folder 3.
109. Ibid., Folder 4.
110. Ibid., Folder 4.
111. Ibid., Folder 5.
112. Ibid., Folder 5.
113. Ibid., Folder 5.
114. Quine, W. V., "States of Mind," *Journal of Philosophy* 72:5–8 (1985). F&Q/c 323–326.
115. Ibid., 323.
116. Quine, W. V. "Events and Reification," in *Actions and Events: Perspectives on the Philosophy of Donald Davidson*, ed. E. Lapore and B. McLaughlin (New York: Blackwell, 1985), 162–171. F&Q/d 129–141.
117. Ibid., 134.
118. Ibid., 137.

119. Ibid., 139.
120. Ibid., 141.
121. Quine to Gibson, MSAm2587(407), 4/18/87.
122. Gibson, Roger, Letter to Davidson MSAm2587(287), 7/11/66.
123. Quine, W. V., *Pursuit of Truth* (Cambridge: Harvard University Press, 1990).
124. Ibid., vii.
125. Ibid., Sections 2–5.
126. Ibid., Sections 6–7.
127. Ibid., Sections 8–12.
128. Ibid., Section 13.
129. Ibid., Sections 14–19.
130. Ibid., Sections 20–23.
131. Ibid., Sections 24–31.
132. Ibid., Sections 32–37.82.
133. Ibid., Sections 38–43.
134. Quine, W. V., "Three Indeterminacies," F&Q/c 368–386.
135. Ibid., 370–375.
136. Ibid., 378.
137. Ibid., 381.
138. Ibid., 383.
139. Ibid., 385.
140. Quine, W. V., "Two Dogmas in Retrospect," *Canadian Journal of Philosophy* 21:265–274 (1991).
141. F&Q/c 393.
142. Ibid., 390–399.
143. Quine, W. V., "Structure and Nature," *Journal of Philosophy* 89:6–9 (1992). F&Q/c 401–406.
144. F&Q/c 403.
145. Donald Davidson to Gibson, MSAm2587(287), 1/13/93.
146. Quine, W. V., "In Praise of Observation Sentences," *Journal of Philosophy* 90:107–116 (1993).
147. Ibid., 411.
148. Ibid., 413.
149. Ibid., 414–415.
150. Ibid., 415.
151. Ibid., 419.
152. Quine, W. V., "Promoting Extensionality," *Synthese* 98:143–145 (1994).
153. Ibid., 443.
154. Ibid., 445.
155. Quine, W. V., "Naturalism; or, Living Within One's Means," *Dialectica* 49:251–261 (1994). F&Q/c 461–472.
156. Ibid., 466.
157. Ibid., 468.
158. Follesdal, Dagfinn, to Quine, MSAm2587(363), 12/15/95.
159. Quine, W. V., "Progress On Two Fronts," *Journal of Philosophy* 93:159–163 (1996). F&Q/c 473–477.
160. Ibid., 475.
161. Quine, W. V., "The Innate Foundations of Endowments," F&Q/c 176–181.
162. Ibid., 179.
163. Quine, W. V., *From Stimulus to Science* (Cambridge: Harvard University Press, 1995).
164. Ibid., 16.
165. Ibid., 19.
166. Ibid., 26.
167. Ibid., 29–42.

168. Ibid., 48.
169. Ibid., 54.
170. Ibid., 57.
171. Ibid., 67.
172. Ibid., 72.
173. Ibid., 76.
174. Ibid., 82.
175. Ibid., 86.
176. Ibid., 99.
177. Quine, W. V., "The Flowering of Thought in Language," in *Thought and Language*, ed. John Preston (Cambridge: Cambridge University Press, 1997). F&Q/c 478–484.
178. Ibid., 478.
179. Quine, W. V., "Confessions of a Confirmed Extensionalist," in *Futures Past*, ed. Juliet Floyd and Sanford Shieh (New York: Oxford Press, 2000), 215–221. F&Q/c 498–506.

Chapter 5

1. MSAm 2587(1001); Quine to Skinner 3/8/85.
2. Murphey, *Lewis* 238.
3. Cohen, Morris R., and Ernest Nagel, *An Introduction to Logic and Scientific Method* (New York: Harcourt Brace and Co., 1934), 219–221.
4. Quine, W. V., *Methods of Logic* (London: Routledge and Kegan Paul, 1952), xi.
5. Barrett, Marilyn, "Early Semantic Representations and Early Word Usage," in *The Development of Word Meaning*, ed. Stan Kuczaj and Martyn Barrett (New York: Springer Verlag, 1986), 39–67; George Butterworth and Leslie Grover, "The Origins of Referential Communication in Human Infancy," in *Thought Without Language*, ed. Lawrence Weiskrantz (Oxford: Clarendon, 1988), 5–24.
6. MSAm 2587(2749), Folder 5.
7. Op. cit., note 5.
8. Nelson, Katherine, "Constraints on Word Learning?" *Cognitive Development* 3:221–246 (1988); Elizabeth Spelke, "Perception of Unity, Persistence and Identity: Thoughts on Infants' Conceptions of Objects," in *Neonate Cognition*, ed. J. Mehler and R. Lox (Hillsdale: Erlbaum, 1985), 89–113; Rene Baillargeon, "Representing the Existence and Location of Hidden Objects: Object Permanence in 6- and 8-month Old Infants," *Cognition* 23:21–41 (1986); Jeffrey Fagen and Harlene Haynes, *Progress in Infancy Research* (Mahwah: Erlbaum, 2002), Vol. II, 95–112.
9. Bower, T. G. R., "The Object in the World of the Infant," *Scientific American* 225:30–38 (1971).
10. Gibson, Roger, *The Philosophy of W. V. Quine* (Tampa: University of South Florida, 1981), 81.
11. Ryle, Gilbert, *The Concept of Mind* (New York: Barnes and Noble, 1940).
12. JSL 12:97–122 (1947), 105.
13. Rosch, Eleanor, "On the Internal Structure of Perceptual and Semantic Categories," in *Cognitive Development and the Acquisition of Language*, ed. Timothy Moore (New York: Academic, 1973), 111–144; Eleanor Rosch and Carolyn Mervis, "Family Resemblances: Studies in the Internal Structure of Categories," *Cognitive Psychology* 7:573–605 (1975); Daniel Osherton and Edward Smith, "On the Adequacy of Prototype Theory as a Theory of Concepts," *Cognition* 9:35–58 (1981).
14. Cf. references 5 and 8 above.
15. Nunberg, Geoffrey, "The Pragmatics of Reference" (PhD diss., City University of New York, Ann Arbor: University Microfilms, 1978), 52–64.

16. Hershberg, James, *James B. Conant: Harvard to Hiroshima and the Making of the Nuclear Age* (New York: Knopf, 1993), 16–163.
17. Quine, *Pursuit of Truth,* 20.
18. Cf. note 2 above.
19. Armstrong, Sharon Lee, Lila Gleitman, and Henry Gleitman, "What Some Concepts Might Not Be," *Cognition* 13:263–308 (1983).
20. Gibson, Roger, *The Philosophy of W. V. Quine* (Tampa: University of South Florida Press, 1982), xx.

Name Index

Subject Index